SECURITY FOR WIRELESS AD HOC NETWORKS

THE WILEY BICENTENNIAL—KNOWLEDGE FOR GENERATIONS

Each generation has its unique needs and aspirations. When Charles Wiley first opened his small printing shop in lower Manhattan in 1807, it was a generation of boundless potential searching for an identity. And we were there, helping to define a new American literary tradition. Over half a century later, in the midst of the Second Industrial Revolution, it was a generation focused on building the future. Once again, we were there, supplying the critical scientific, technical, and engineering knowledge that helped frame the world. Throughout the 20th Century, and into the new millennium, nations began to reach out beyond their own borders and a new international community was born. Wiley was there, expanding its operations around the world to enable a global exchange of ideas, opinions, and know-how.

For 200 years, Wiley has been an integral part of each generation's journey, enabling the flow of information and understanding necessary to meet their needs and fulfill their aspirations. Today, bold new technologies are changing the way we live and learn. Wiley will be there, providing you the must-have knowledge you need to imagine new worlds, new possibilities, and new opportunities.

Generations come and go, but you can always count on Wiley to provide you the knowledge you need, when and where you need it!

WILLIAM J. PESCE
PRESIDENT AND CHIEF EXECUTIVE OFFICER

PETER BOOTH WILEY
CHAIRMAN OF THE BOARD

SECURITY FOR WIRELESS AD HOC NETWORKS

Farooq Anjum and Petros Mouchtaris

WILEY-INTERSCIENCE
A JOHN WILEY & SONS, INC., PUBLICATION

Wiley Bicentennial Logo: Richard J. Pacifico

Published by John Wiley & Sons, Inc., Hoboken, New Jersey
Published simultaneously in Canada.

For general information on our other products and services or for technical support, please contact our Customer Care Department within the United States at (800) 762-2974, outside the United States at (317) 572-3993 or fax (317) 572-4002.

Wiley also publishes its books in a variety of electronic formats. Some content that appears in print may not be available in electronic formats. For more information about Wiley products, visit our web site at www.wiley.com.

Library of Congress Cataloging-in-Publication Data:

Anjum, Farooq.
Security for wireless ad hoc networks / by Farooq Anjum & Petros Mouchtaris.
p. cm.
Includes bibliographical references and index.
ISBN: 978-0-471-75688-0

1. Wireless LANs - - Security measures. I. Mouchtaris, Petros. II. Title
TK5105. 59. A54 2007
005.8- -dc22 2006029342

Printed in the United States of America

10 9 8 7 6 5 4 3 2 1

Dedicated to:

My parents
(F. A.)

My late father
(P. M.)

CONTENTS

PREFACE

Wireless networks, whether cellular networks or wireless local area networks (LANs), have rapidly become an indispensable part of our life. Evidence of this is the widespread usage of such networks in several areas such as office, home, universities, hot-spots such as airports and hotels etc. In addition, the widespread availability of miniature wireless devices such as PDAs, cellular phones, Pocket PCs, and small fixtures on buildings, sensors are one step towards making possible the vision of wireless 'nirvana' a reality. Wireless 'nirvana' is the state of seamless wireless operation where any wireless device would be able to connect to any other wireless device or network at any time, in any place while satisfying the requirements of the user of the device. But as is obvious, we are still a long way off from the goal of wireless nirvana.

Technology under development for wireless ad hoc networks is enabling our march toward this end goal; however the security concerns in wireless networking remains a serious impediment to widespread adoption. The underlying radio communication medium for wireless networks is a big vulnerability that can be exploited to launch several attacks against wireless networks. In addition, wireless ad hoc networks usually cannot depend on traditional infrastructure found in enterprise environments such as dependable power sources, high bandwidth, continuous connectivity, common network services, well-known membership, static configuration, system administration, and physical security. Without adequate security, enterprises will shy away from the use of wireless ad hoc networks, governmental agencies will ban the use of wireless ad hoc networks, defense organizations might be unable to guarantee the safety of their personnel in battlefield scenarios and users will be liable for actions that they never committed.

Therefore, security of such wireless ad hoc networks is an important area that needs to be addressed if such networks are to be widely used. There are two ways of doing this. One way is for the researchers in this field to identify open problems and provide solutions to the identified open problems. Each such effort makes these wireless networks a little bit more secure. There have been several research efforts in the last couple of years exploring ways of making such networks more secure although much more work still needs to be done. We ourselves have also been engaged in this activity.

The second way to address the security issues of such networks is to disseminate widely the known results to the beginners in this field. This will allow more people to comprehend the problems and contribute towards expanding the knowledge in this area. Unfortunately there has not been any work done along these lines. Our effort in this book is focused on this approach of dissemination of known knowledge in the area of security in wireless ad hoc networks.

To our knowledge, this book is the first book that focuses exclusively on the topic of security for wireless ad hoc networks. The topic of security in wireless ad hoc networks itself is very vast. This topic spans areas such as securing networking protocols, operating systems on mobile devices, and applications etc. In this book we focus on the topic of securing network protocols in wireless ad hoc networks. Note that networking in ad hoc networks is concerned with enabling two devices with wireless interfaces to communicate with each other.

The objective of this book is to make the readers aware of the fundamentals of the area of security of wireless networks as well as the open problems. This will hopefully spur much more activity in this area in the upcoming years. This book provides a broad and comprehensive overview of the research that has been done to date on the security of wireless ad hoc networks and discusses the advantages and disadvantages of the various schemes that have been proposed in the literature.

Given the objective of this book, it is necessary to write it in a style that does not assume a detailed knowledge of many concepts. Therefore, in writing this book, the only requirement that we assumed from the reader is a basic understanding of networking concepts. Given this, we explain the concepts of wireless ad hoc networks at a fairly basic level. We also require limited knowledge of security concepts from the reader. We provide a chapter that introduces the basic security concepts that are required for the rest of the book.

This book will be of interest to a wide variety of people. A beginner in the field will benefit from a simple description of the various problems and solutions. Such a person will also gain by having a ready compendium of important results in this area thereby saving such a person from the problem of information overload. Thus, this book can be used as a textbook in the first class focusing on security in ad hoc networks.

Researchers focusing on wireless networks that would like to consider the security implications of the protocols they are designing would benefit from a description of known problems and solutions to these known problems. Additionally, researchers focusing on novel security schemes for wireless ad hoc networks that would like to become aware of existing research should also profit from the description of various schemes in this book. This will let them know about what is out there and what is needed. Finally, this should also be a valuable book for researchers focusing on applications of wireless ad hoc networks in a commercial or military environment. All these groups comprise the intended audience of the book.

Of course, we do not expect this effort to be perfect. Errors might have crept in; some other topics that you, the reader, feel are important might have been left out. In some cases, our comments on the problems and their solutions would have been biased due to our backgrounds. There will be other ways also in which the book could be improved. We would like to hear from you, the reader, on each of these aspects. Therefore, feel free to write to us on these or any other topic that you feel is relevant to the book. And if you enjoyed reading the book, we would like to hear about that also. We also have a blog at http://techraw.typepad.com/security/ where such errata or our responses to your comments will be provided. This title can be accessed at the following FTP site: ftp://ftp.wiley.com/public/sci_tech_med/security_wireless/

In the meantime, happy reading.

Farooq Anjum
fanjum@telcordia.com

Petros Mouchtaris
pmouchta@telcordia.com

FOREWORD

Rapid and automatic establishment of wireless networks and services in the absence of a fixed infrastructure is one of the big challenges of communication. The complexity of the problem is greatly compounded when the nodes of the network have to accommodate rapid and unpredictable motion, dynamically altering the connectivity of the network itself. The attractiveness and value of such 'ad hoc networks' rests on their ability to meet performance parameters hard to achieve otherwise and to do so while optimizing the use of resources such as spectrum, energy, and operations support functions at scale. A final dimension of this multidisciplinary problem is the achievement of a solution which in some sense minimizes the probability of disruption from natural and malicious threats and at the same time maximizes availability assuring authorized users access to critical services. This book captures the current state of the art in wireless ad hoc networks with an emphasis on security and assurance.

In the last decade researchers have explored many potential applications of wireless ad hoc networks. The research has ranged from basic theoretical investigations to prototypes and demonstrations. The largest body of work has been in the government arena. The Department of Defense has invested seriously in exploiting wireless ad hoc networks in its transformational programs. Telcordia has been at the forefront of both creating new network technologies and exploring newer approaches for securing such networks. This has involved leveraging ideas from basic science to propose engineering principles for designing and deploying such networks. It has also involved construction of proof-of-principle testbeds, prototypes, demonstrations, and the steps necessary to transition the technology to general use. While there are many problems still to be solved, it has been gratifying to see the technology move from a concept to reality. Over this span of time there has been stalwart support for these efforts from agencies such as the Army Research Lab (ARL), the Army Communications Electronics Research, Development and Engineering Center (CERDEC), and DARPA. On the commercial and public sector front the technology has been developing more slowly but is finding its way into many applications. These include transportation networks, emergency response, law enforcement, and sensor systems. Perhaps the greatest use of this technology will be to fill the gap in fixed infrastructure which will allow public wireless systems to really achieve the goal of delivering applications to any place through hybrid networks of cellular and ad hoc components.

In performing the research which is codified in this book, one of the ingredients that the authors bring to the table is the knowledge and intuition of 'real' communications systems and applications. In a commercial setting with clients who are used to hardened products that affect a large customer base the non-functional attributes of solutions are just as important as the functional aspects. The delivery of a service over a network can sometimes be demonstrated easily. When the requirement is to build it out at scale with high reliability and availability and with a high degree of security what may seem as an easy problem suddenly becomes hard. One of the values of this book is to expose the reader to such issues in an expository and complete way indicating the parts of the problem that have been solved and the parts that still require further investigation.

In closing I would like to commend both Petros Mouchtaris and Farooq Anjum for the professionalism and dedication they have shown in writing this book. They both have highly demanding jobs, so this task took a lot of extra effort. I hope that they get positive feedback and feel the satisfaction they deserve for the excellent job they have done in collecting, codifying, and explaining the advanced material that comprises this important book.

Adam Drobot
President,
Applied Research
Telcordia Technologies

ACKNOWLEDGMENTS

We would like to start by thanking the management of Telcordia who made it possible for us to not only entertain the idea of writing this book, but also executing the idea. Showstopper issues such as copyright were addressed very efficiently which again was due to the 'Telcordia' culture. We would like to thank in particular the president of Applied Research, Adam Drobot, who encouraged us to pursue writing this book.

The book itself would not have been possible if not for the efforts of the various people working in this field. These people not only identified the problems in this area but also provided solutions to such problems. In this book, we tried to identify various interesting research problems in the field and discuss approaches that have been pursued by various researchers. Names of such people are scattered all over the book in the form of references to their works that we explain in each of the chapters.

We also benefited much from comments on the various chapters of the book provided by various people working in this area. We are very grateful to Srdjan Capkun, Peng Ning, Adrian Perrig, Santosh Pandey, Rajesh Talpade, Ritu Chadha and Mike Little for their comments which resulted in a better manuscript.

Writing a book is never easy especially on the families of the authors. The hours of sacrifice needed on the part of the spouses and also their efforts to ensure that the authors are in the right frame of mind is something that we are sure every author appreciates and cannot put a cost to. On this front, the first author is very thankful to Ambareen while the second author is very thankful to Donna.

We are also very thankful to the team at John Wiley who made the entire process as painless as possible to us. We would be failing if we do not mention Whitney Lesch who was in constant touch with us to address any questions we had about the book.

1 Introduction

1.1 DEFINITION OF WIRELESS AD HOC NETWORKS

In the last few years we have seen the proliferation of wireless communications technologies. Wireless technologies are being widely used today across the globe to support the communications needs of very large numbers of end users. There are over 1 billion wireless subscribers of cellular services today utilizing wireless devices for voice communications (e.g. phone calls) and data services. Data services include activities such as sending e-mail and instant messages, and accessing the Web. In fact, in some areas of the world wireless technologies are more prevalent than traditional wireline communications technologies.

There are several reasons for the current popularity of wireless technologies. The cost of wireless equipment has dropped significantly, allowing service providers to significantly reduce the price of wireless services and making them much more affordable to end users. The cost of installing wireless networks in emerging markets has dropped well below the cost of installing wireline networks. The wireless technologies themselves have improved tremendously, making it possible to offer both voice and data services over such networks. The resulting allure of anytime, anywhere services makes such services very attractive for the end users.

In wireless networks, nodes transmit information through electromagnetic propagation over the air. The signal transmitted by a node can only be received by nodes that are located within a specific distance from the transmitting node. This distance is typically called the transmission range. The transmission range depends not only on the power level used for the transmission, but also on the terrain, obstacles, and the specific scheme used for transmitting the information. Typically, for simplicity, the transmission range of nodes is assumed to be a circle around the transmitting node, as shown in Figure 1.1.

Typically multiple nodes exist within an area and these nodes might need to make use of the wireless medium for communication. If many such transmissions happen at the same time within the transmission range of a node, then this will result in the transmissions colliding with each other. Such collisions make it impossible for receivers to interpret the data being transmitted by individual nodes. The effect here is similar to many people talking simultaneously to a person, in which case the person involved will not be able to understand any of them. Therefore, it is vital to prevent or minimize such collisions. This can be done by controlling access to the wireless medium. This is the approach typically followed by the collision avoidance or minimization schemes.

Many collision avoidance or minimization schemes have been developed for sharing the available wireless spectrum among wireless nodes transmitting concurrently.

Security for Wireless Ad Hoc Networks, by Farooq Anjum and Petros Mouchtaris

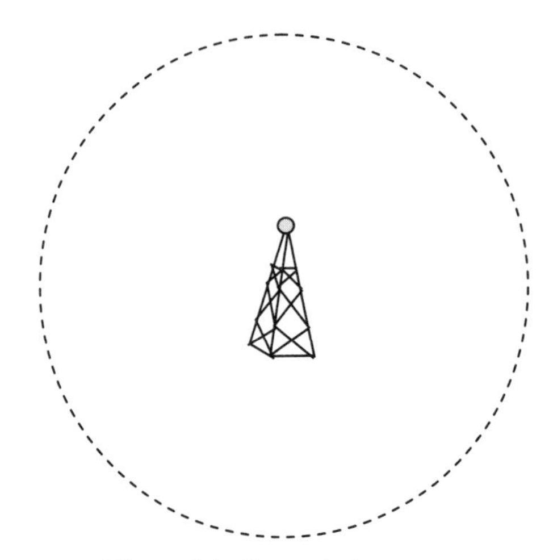

Figure 1.1. Transmission range.

Typical schemes include: (1) time division multiple access (TDMA), which divides time into small time slots and requires nodes to take turns transmitting data during separate time slots; (2) frequency division multiple access (FDMA), which provides for different frequencies such that each node transmits on a different frequency; (3) carrier sense multiple access (CSMA), which requires for every node to listen for transmissions on the wireless channel (on a given frequency) and transmit its own data when the node perceives the channel to be free of any other wireless transmissions; and (4) code division multiple access (CDMA), which allows nodes to transmit at the same time but requires them to use different spreading codes so that the signals from different nodes can be distinguished by the receivers.

Nodes might need to communicate with other nodes that are outside their transmission range. This is typically accomplished by having other nodes that are within the transmission range of the transmitting node receive and then retransmit the signal. As a result of this retransmission, nodes within transmission range of the node repeating the original signal receive the data. Depending on the location of the destination, multiple nodes may need to retransmit/repeat the data, as shown in Figure 1.2.

Various network architectures have been introduced based on the high-level concepts discussed so far. Such architectures allow wireless services and provide for end-to-end communication among users often located far away from each other. Figure 1.3 shows

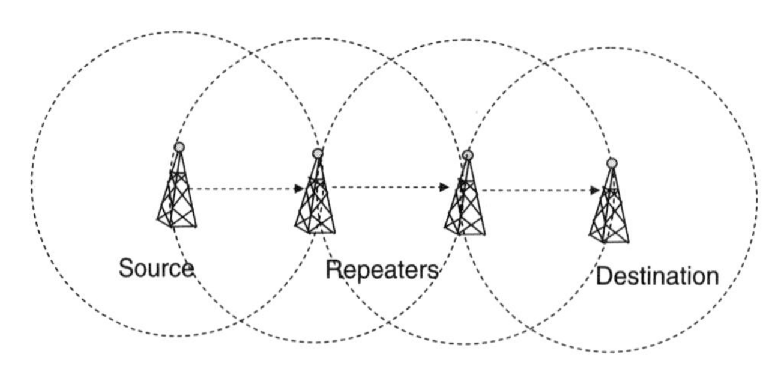

Figure 1.2. End-to-end transmissions.

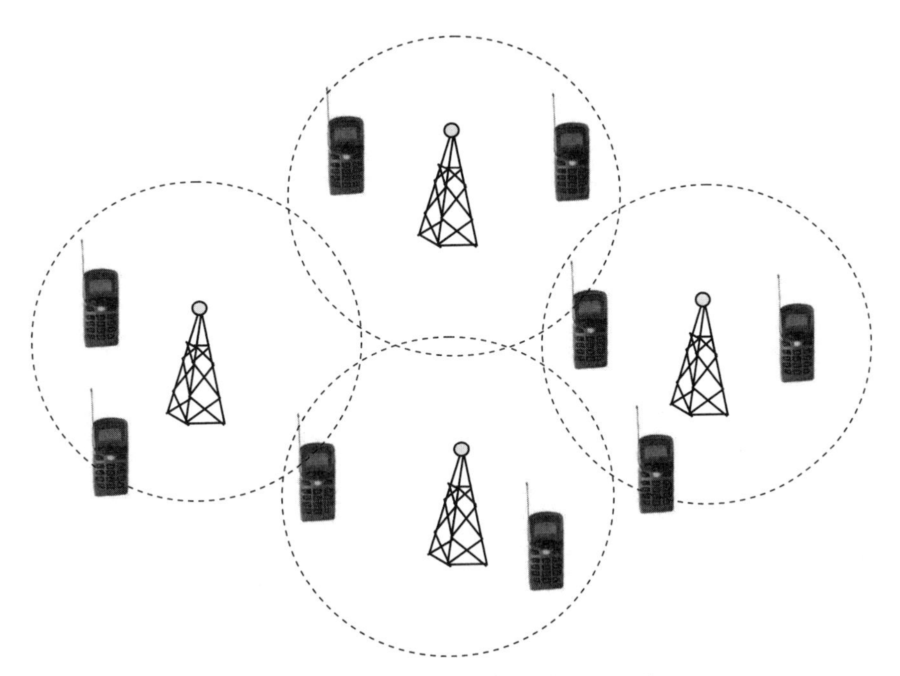

Figure 1.3. Architecture of cellular networks.

a typical architecture that is used for cellular networks. In a typical cellular architecture, radio transmission towers are placed across the area that the service provider desires to offer cellular service in. These towers are often built on top of buildings, on big towers, on high ground, and so on, and are hence stationary. These radio transmission towers are responsible for receiving the data transmitted by other nodes and then retransmitting the data as needed in order to reach the destination. The devices used by end users for accessing the service are typically small and mobile (e.g. mobile phones). End devices typically only communicate directly with the radio transmission tower that is closest to them. The radio transmission tower is then responsible for transmitting that information towards the node that needs to receive that information. The radio transmission tower might also enlist the help of other radio transmission towers in order to do this.

In a cellular network, towers are typically interconnected through a static wireline network (e.g. SONET network) with each other. An end device transmits information to the local tower. If the destination end device is unreachable from the local tower, then the local tower locates the tower closest to the destination. Following this, the local tower transmits the information to the tower closest to the destination end device through the wireline network. The tower closest to the destination is then responsible for transmitting the information to the destination end device.

Cellular technology is not the only wireless technology in existence. Another widely used wireless technology is IEEE 802.11-based wireless local area network (WLAN), also popularly referred to as Wi-Fi. Wi-Fi has mostly been used for providing wireless data connectivity inside buildings for personal computers and laptops. This technology allows such devices to communicate potentially at very high speeds (but over relatively smaller distances) as compared with cellular networks. In fact, these networks are called WLAN networks since they typically provide the equivalent of LAN connectivity

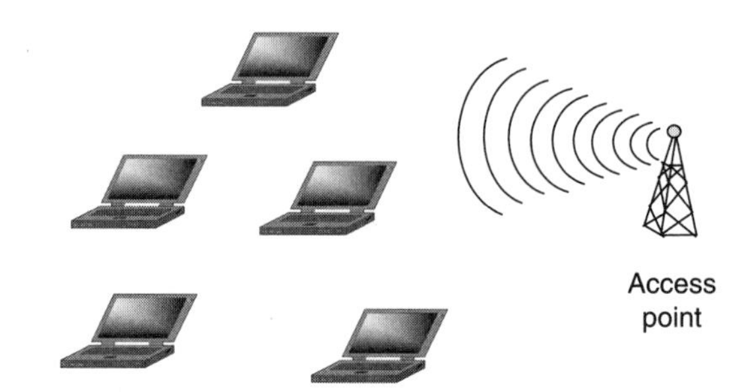

Figure 1.4. Typical enterprise architecture using 802.11 technology.

inside buildings. Figure 1.4 shows the typical network architecture used today for 802.11. This architecture utilizes fixed access points (APs) that play a similar role to that played by radio towers in the cellular environment. APs are responsible for receiving the signal from end devices and then retransmitting them to the destination. The APs also have the responsibility for interconnecting the wireless LAN to external networks such as the internet or other WLANs (through other access points to which they could be connected over wireline links).

The wireless networks that we have discussed so far are dependent on fixed nodes (the radio towers and access points) for connecting the mobile nodes. In addition, these networks require some fixed infrastructure to interconnect the fixed nodes with each other. This type of architecture has been very successful and widely deployed throughout the world for offering a variety of voice and data services, despite being inflexible (by requiring fixed nodes). This is because the architecture has been sufficient for services typically offered by service providers.

Having a communications network that relies on a fixed infrastructure, however, is not always acceptable for some applications (see Section 1.2). For example, when emergency responders move into an area (say to deal with a disaster), it is possible that the fixed infrastructure may have been destroyed or may be unavailable (e.g. in some remote areas). Emergency responders might not have enough time to establish a fixed infrastructure in such cases. A similar situation might also arise in a battlefield environment.

In the past few years, a new wireless architecture has been introduced that does not rely on any fixed infrastructure. In this architecture, all nodes may be mobile and no nodes play any special role. One example of this architecture is the "ad hoc" mode architecture of 802.11, as shown in Figure 1.5. In this architecture, 802.11 nodes do not rely on access points to communicate with each other. In fact, nodes reach other nodes they need to

Figure 1.5. Ad hoc mode architecture using 802.11 technology.

communicate with using their neighbors. Nodes that are close to each other discover their neighbors. When a node needs to communicate with another node, it sends the traffic to its neighbors and these neighbors pass it along towards their neighbors and so on. This repeats until the destination of the traffic is reached. Such an architecture requires that every node in the network play the role of a router by being able to determine the paths that packets need to take in order to reach their destinations.

Networks that support the ad hoc architecture are typically called wireless ad hoc networks or mobile ad hoc networks (MANET). We will use these two terms interchangeably throughout the book. Such networks are typically assumed to be self-forming and self-healing. This is because the typical applications of such networks require nodes to form networks quickly without any human intervention. Given the wireless links and mobility of nodes, it is possible that nodes may lose connectivity to some other nodes. This can happen if the nodes move out of each other's transmission range. As a result, it is possible for portions of the network to split from other portions of the network. In some applications it is also possible that some nodes may get completely disconnected from the other nodes, run out of battery, or be destroyed. For these reasons, nodes in a MANET cannot be configured to play any special role either in the way nodes communicate or in the way of providing communication services (e.g. naming services). This leads to a symmetric architecture where each node shares all the responsibilities. The network needs to be able to reconfigure itself quickly to deal with the disappearance (or reappearance) of any node and continue operating efficiently without any human intervention.

Routing in such networks is particularly challenging because typical routing protocols do not operate efficiently in the presence of frequent movements, intermittent connectivity, network splits and joins. In typical routing protocols such events generate a large amount of overhead and require a significant amount of time to reach stability after some of those events. The Internet Engineering Task Force (IETF), which is the main standardization body for the internet, has recognized that existing routing protocols cannot meet the unique requirements of MANET and has played a key role in the creation of novel MANET routing protocols. This is done through the IETF MANET Working Group, which has been a focal point for a lot of the related research. This group was established in 1997 and since then has created some of the most widely cited MANET routing protocols such as the ad hoc on demand distance vector (AODV) and optimized link state routing (OLSR) routing protocols (see www.ietf.org/html.charters/manet-charter.html). Its efforts are continuing with a focus on additional routing protocols and multicast.

1.2 APPLICATIONS OF WIRELESS AD HOC NETWORKS

So far we have discussed the unique concept of MANET. We next discuss the applications that have motivated much of the research on MANETs and are well suited for their use. Perhaps the most widely considered application of a MANET is battlefield communications. The Department of Defense (DoD) future transformation is based on a key initiative called Network Centric Warfare (NCW). It is expected that there will typically be a large number of nodes in the battlefield environment that need to be interconnected, including radios carried by soldiers, and radios mounted on vehicles, missiles, unattended air vehicles (UAV), and sensors. In such an environment the network plays a critical role in the success of the military mission. The vast majority of these nodes move around at varying speeds and nodes may lose connectivity to other nodes as they move around in the battlefield because of the terrain (e.g. obstacles may prevent line of sight), distance among the nodes, and so on. Because of the rapid pace and the large degree of unpredictability it is not possible to assume a fixed infrastructure in the battlefield environment. Network administrators have little time to react and reconfigure the networks. Existing networking technologies cannot support such an environment efficiently. MANETs are viewed as a potential solution for providing a much more flexible network in support of NCW. The DoD has been funding a large number of research efforts exploring the use of MANETs for battlefield communication. As a result, a large number of research papers are motivated by such applications.

The other widely considered application for MANETs is interconnection of sensors in an industrial, commercial, or military setting. Sensors are typically small devices measuring environmental inputs (such as temperature, motion, light, etc.) and often alerting users and/or taking specific reactions (e.g. starting an air-conditioner) when those inputs reach specific ranges. Sensors have been used extensively in industrial applications and even for applications inside the home (such as in security systems, heating systems, etc.). Most recently, advanced sensors are being considered for the detection of harmful agents (such as anthrax) or nuclear material. The availability of very inexpensive network interfaces has made it possible to provide network connectivity to sensors. Certain uses of sensors seem to be well suited for MANETs. For example, the military has considered scenarios where large numbers of sensors are dropped in an area of interest and those sensors then establish connectivity to each other and to the soldiers for providing advanced reconnaissance. In some cases, applications are considered where a very large number of sensors (hundreds or even thousands) is dropped in areas that need to be monitored closely. Sensors in such areas then establish a network. For example, "Smart Dust" which is a project at the University of California, Berkeley, (see http://robotics.eecs.berkeley.edu/~pister/SmartDust/) has focused on the development of small devices that have both sensor and communication capabilities and are smaller than 1 cubic millimeter. Typically in such applications it is not possible to have a fixed infrastructure and therefore these applications seem to be well suited for MANETS.

Another relevant application is that of emergency response. During major emergencies and disasters such as hurricanes or large explosions, the communications infrastructure in the immediate area of the disaster or emergency may be unusable, unavailable, or completely destroyed. When emergency responders first arrive in the disaster-struck area, it is critical for them to be able to communicate with each other. The communications make it possible for the team to coordinate the relief operations with each other. Since the communication infrastructure is often unavailable, first responders need to be able to establish connectivity immediately. MANETs are well suited for such an application because of their ability to create connectivity rapidly with limited human effort.

Several other applications of MANETs are also being considered. For example, municipalities are considering deployment of wireless ad hoc networks (in the form of so called mesh networks) for offering broadband access to end users including employees of the municipality, first responders, and even residents of the municipality. Such networks have already been deployed in a small (but increasing) number of municipalities. More recently researchers have considered the use of MANET in the vehicular environment. Making MANET networking capabilities available in such environments can enable a variety of new applications such as sharing of up-to-date traffic information between vehicles.

1.3 THREATS, ATTACKS, AND VULNERABILITIES

Having discussed the basic concept of wireless ad hoc networks, we next look at the threat, attacks, and vulnerabilities in such networks. Any system that has to be protected might have weaknesses or vulnerabilities, some or all of which may be targeted by an attacker. Hence, one approach to designing security mechanisms for systems is to look at the threats that the system faces and the attacks possible given the vulnerabilities. The designed security mechanisms should then ensure that the system is secure in the light of these threats, attacks, and vulnerabilities. While we look at the security mechanisms designed to achieve various objectives in ad hoc networks in several chapters throughout the book, we look at the threats, attacks, and vulnerabilities in this section. We start by providing definitions of the terms, threat, vulnerability, and attack.

- *Threat* is the means through which the ability or intent of an agent to adversely affect an automated system, facility or operation can be manifested. All methods or things used to exploit a weakness in a system, operation, or facility constitute threat agents. Examples of threats include hackers, disgruntled employees, industrial espionage, national intelligence services, and criminal organizations.
- *Vulnerability* is any hardware, firmware, or software flaw that leaves an information system open for potential exploitation. The exploitation can be of various types, such as gaining unauthorized access to information or disrupting critical processing.
- An *attack* is an attempt to bypass the security controls on a computer. The attack may alter, release, or deny data. The success of an attack depends on the vulnerability of the system and the effectiveness of existing countermeasures. Examples of attacks include actions such as stealing data from storage media and devices, obtaining illegitimate privileges, inserting data falsely, modifying information, analyzing network traffic, obtaining illegitimate access to systems through social engineering, or disrupting network operation using malicious software. Attacks can be divided into two main categories:

  Passive attacks—in these types of attack an attacker passively listens to the packet or frame exchanges in the wireless medium by sniffing the airwaves. Since an attacker only listens to the packets that are passing by without modifying or tampering with the packets, these attacks mainly target the confidentiality attribute of the system. However, this process of gathering information might lead to active attacks later on. Typically this attack is easier to launch than the next type of attacks.

  Active attacks—active attacks are those attacks where the attacker takes malicious action in addition to passively listening to on-going traffic. For example an attacker might choose to modify packets, inject packets, or even disrupt network services.

Security in wireless networks differs markedly from security for their wireline counterparts due to the very nature of the physical medium. While communicating over a wireless medium, the transmitted and received signals travel over the air. Hence, any node that resides in the transmission range of the sender and knows the operating frequency and other physical layer attributes (modulation, coding, etc.) can potentially decode the signal without the sender or the intended receiver knowing about such an interception. In contrast, in wireline networks, such an interception is possible only when one obtains access to the physical transmission medium (cable, fiber, etc.), which would typically involve tapping into such mediums.

Another problem with defending wireless ad hoc networks is that existing security technologies are more geared towards wireline networks, which are fairly static. Existing technologies often rely on the availability of traffic chokepoints (which most traffic goes through). Security devices placed at such chokepoints can inspect traffic for suspicious behavior and implement security policies and respond as needed. This is not true in ad hoc networks where the network entities often move around. This results in frequent changes in the structure of the network. Traditional security solutions also depend on a few centrally located devices for managing the security of the network. Such solutions are not applicable for wireless ad hoc networks on account of the features of these networks. The increased vulnerabilities of ad hoc networks and the limitations of existing security solutions designed for wireline networks will become clearer throughout the book.

Ad hoc networks that make extensive use of wireless links are vulnerable to several types of attack due to the inherent nature of the network. We would like to remark here that mechanisms such as encryption and authentication can greatly mask the vulnerabilities on the air-link, but these are not the only vulnerabilities in ad hoc networks. Since wireless ad hoc networks cannot depend upon infrastructure-based resources, such as stable power source, high bandwidth, continuous connectivity, or fixed routing, it is very easy to launch attacks on them. In the following subsections, we will briefly describe some vulnerabilities and attacks that are very common in the ad hoc network environment. Note that while the lists of vulnerabilities and attacks considered in here are by no means exhaustive, an attempt has been made to make the lists representative. Defenses against these vulnerabilities and attacks will be described in the remaining chapters of this book.

1.3.1 Threats

A pragmatic approach to building a secure system is to consider the threats that the system might face after deployment. We consider three main categories of threats:

- amateur adversary;
- professional adversary;
- well-funded adversary.

Some examples of amateur adversaries are script kiddies or hobbyist hackers. Crime syndicates or terrorist organizations can be considered as professional adversaries. Foreign intelligence services can be considered as an example of a well-funded adversary. The above categorization implicitly governs the types of attacks that can be launched by each type of adversary. Amateur adversaries can launch unsophisticated attacks such as wireless sniffing or denial of service. A professional adversary can launch more sophisticated attacks such as layer 2 hijacking, man-in-the-middle attack, or Sybil attack (explained in Chapter 4). A well-funded adversary does not have any constraints on money. Such an adversary can launch very sophisticated attacks such as rushing attacks, wormhole attacks (explained in Chapter 4), as well as capture devices that are part of the network.

1.3.2 Vulnerabilities in Ad Hoc Networks

Mobile computing has introduced new types of computational and communication activities that seldom appear in fixed or wired environments. For example, mobile users tend to be stingy about communication due to slower links, limited bandwidth, higher cost, and battery power constraints. Mechanisms like disconnected operations and location-dependent operations only appear in the mobile wireless environment. Application and services in a mobile wireless network can be a weak link as well. In these networks, there are often proxies and software agents running in intermediate nodes to achieve performance gains through caching, content transcoding, or traffic shaping. Potential attacks may target these proxies or agents to gain sensitive information or to mount denial of service (DoS) attacks, such as flushing the cache with bogus references, or having the content transcoder do useless and expensive computation. In this environment it is also difficult to obtain enough audit data. Mobile networks do not communicate as frequently as their wired counterparts. This can be a problem for intrusion detection systems attempting to define normality for anomaly detection.

Among the intrinsic vulnerabilities of ad hoc networks, some reside in their routing, others in their use of wireless links and still some others in their auto-configuration mechanisms. These key functionalities of ad hoc networks are based on complete trust between all the participating hosts. In the case of routing, the correct transport of the packets in the network relies on the veracity of the information given by the other nodes. The emission of false routing information by a host could thus create bogus entries in routing tables throughout the network, making communication difficult. Furthermore, the delivery of a packet to a destination is based on hop-by-hop routing, and thus needs total cooperation from the intermediate nodes. A malicious host could, by refusing to cooperate, quite simply block, modify, or drop the traffic traversing through it. By fooling the routing algorithm or even by choosing a strategic geographic positioning, a host can control the traffic to and from entire parts of the network.

Use of wireless links makes these networks very vulnerable to attacks ranging from passive eavesdropping to active interfering. An attacker just needs to be within radio range of a node in order to intercept network traffic. The current design of wireless networks places a lot of emphasis on cooperation. A very good example of this is the design of medium access control protocols used in these networks. Since these protocols follow predefined procedures to access the wireless channel, a misbehaving node can easily change the MAC protocol behavior, which may lead to a DoS attack.

The autoconfiguration mechanism also brings up new vulnerabilities. This functionality, whether it uses ICMP router advertisements, neighbor solicitation messages or simple DHCP autoconfiguration messages, is vulnerable to false replies. These processes use information given by the nodes on the network to either calculate an IP address or verify that a particular address is not already used. For example, in the case of duplicate address detection (DAD), a danger exists that a malicious node may pretend to be using any of the addresses chosen by an incoming host, thus denying the incoming host the right to join the network.

Constraints existing in ad hoc networks also add to the vulnerabilities. For example, such networks have limited computational ability, as evidenced by low processor frequencies and smaller memory sizes. The limitations on power usage are another major constraint. This implies that it might be very easy for an adversary to launch DoS attacks in such networks by trying to exhaust the battery of a legitimate node. The nodes in such networks are also vulnerable to being physically captured, which may result in the cryptographic keys being exposed. Another problem with protecting wireless ad hoc networks is on account of the fact that there is much more uncertainty in such networks. This makes it more difficult to discriminate between malicious behavior and acceptable behavior. For example, significant levels of packet dropping may be the result of the physical characteristics of the wireless links. These packet drops might not necessarily imply an attack. Nodes may appear and disappear from the network not because they are being attacked but because of mobility and power constraints.

In addition, ad hoc networks also suffer from the vulnerabilities present in their wired counterparts such as passive eavesdropping, spoofing, replay, or denial of service. Some of these vulnerabilities are accentuated in a wireless context. The topology of an ad hoc network is defined by the geographical position and by the wireless emission ranges of its hosts. A consequence of this is that these networks do not have a clearly defined physical boundary and thus no clearly identified entry point into the network (since typically adversaries try to launch their attacks from outside the network). Access-control to the network, as it is traditionally achieved by a LAN's firewall, thus becomes more difficult to deal with. Attention should thus be placed on the problems of IP masquerading and passive eavesdropping, and a protection against these attacks should be implemented.

To summarize, a mobile ad hoc wireless network is vulnerable due to its features of open medium, dynamic changing network topology, cooperative algorithms, lack of centralized monitoring and management point, and a lack of a clear line of defense.

1.3.3 Attacks

In this book we focus on the problem of securing wireless ad hoc networks and describe techniques and mechanisms that can make such networks less vulnerable against malicious attacks. Attacks against the network may come from malicious nodes that are not part of the network and are trying to join the network without authorization. Such nodes are typically called outsiders. Networks are typically protected from malicious outsiders through the use of cryptographic techniques. Such techniques allow nodes to securely verify the identity of other nodes and can therefore try to prevent any harm being caused by the malicious outsiders. We also consider attacks from nodes that are authorized to be part of the network and are typically called insiders. Insider nodes may launch attacks because they have been compromised by an unauthorized user (e.g. hacker) through some form of remote penetration, or have been physically captured by a malicious user.

We next discuss some possible attacks against wireless ad hoc networks. The list of attacks provided here is by no means a comprehensive list of possible attacks but provides a broad view of the attacks that need to be addressed which will motivate the subsequent chapters discussing approaches to defending against such attacks. Some example attacks that are possible in an ad hoc setting are:

1 *Routing Attacks*—in this case the adversary mounts attacks on the routing protocols or on the routing tables. For example, the adversary could disseminate false routing information. There are several attacks that fall into this category. We look at some of these in more detail in Section 1.3.3.1. We also provide ways of defending against these attacks in Chapter 4.

2 *Sleep Deprivation*—usually this attack [1] is practical only in wireless networks where battery life is a critical parameter. Battery-powered devices try to conserve

energy by transmitting only when absolutely necessary. In this attack a malicious user interacts with a node with the intention of draining the battery of the node. For example, an attacker can attempt to consume battery power by requesting routes from that node, or by forwarding unnecessary packets to that node, or by disrupting routing to route an excessive amount of traffic to that node.

3 *Location Disclosure*—a location disclosure attack can reveal information about the locations of nodes or the structure of the network. The information gained might reveal which other nodes are adjacent to the target, or the physical location of a node. The attack can be as simple as using the equivalent of the trace route command on Unix systems. As a result, the attacker knows which nodes are situated on the route to the target node. If the locations of some of the intermediary nodes are known, information can be obtained about the location of the target as well.

4 *Eavesdropping*—eavesdropping is a very easy passive attack in the radio environment. When one sends a message over the wireless medium, everyone equipped with a suitable transceiver in the range of the transmission can potentially decode the message and obtain sensitive information. The sender or the intended receiver has no means of detecting if the transmission has been eavesdropped. However, this attack can be prevented by using an encryption scheme at the link level to protect the transmitted data. Of course, this requires efficient key distribution strategies so that keys for encrypting the transmitted traffic can be transmitted to all nodes. We will look at such key distribution strategies in more detail in Chapter 3.

5 *Traffic Analysis*—the objective of an adversary launching this attack is to extract information about the characteristics of transmission. This could include information about the amount of data transmitted, identity of communicating nodes, or their locations. Prevention of this attack is not easy. One approach is to make use of routing protocols that make it difficult to get this information. Some examples of such routing protocols are given in Chapter 4.

6 *Denial of Service*—denial of service attacks are also easily applied to wireless networks, where legitimate traffic cannot reach clients or the access point because illegitimate traffic overwhelms the frequencies. DoS attacks are possible at various layers, namely, physical layer, MAC layer, and network layer, and also on the applications executing in such networks. For example, jamming of radio frequencies could be done at the physical layer similarly, violation of medium access control rules could lead to denial of service at the link layer.

7 *Sybil Attack*—in this attack [2] a single node attempts to adopt multiple identities. This attack will be discussed in detail in Chapter 4.

1.3.3.1 General Description of Routing Attacks Routing is a very important function in MANETS, as described earlier. It can also be easily misused, leading to several types of attack. We next describe some of the attacks on routing in MANETS.

Routing protocols in general are prone to attacks from malicious nodes. These protocols are usually not designed with security in mind and often are very vulnerable to node misbehavior. This is particularly true for MANET routing protocols because they are designed for minimizing the level of overhead and for allowing every node to participate in the routing process. Making routing protocols efficient often increases the security risk of the protocol and allows a single node to significantly impact the operation of the protocol because of the lack of protocol redundancy.

Below are some examples of attacks that can be launched against MANET routing protocols. The reader is referred to the literature [3–6] for a discussion of the various types of attacks against routing protocols and ways of categorizing those attacks. We would also like to remark here that we discuss several routing protocols that address one or more of these attacks in Chapter 4.

- *Black Hole Attack*—in this attack, a malicious node uses the routing protocol to advertise itself as having the shortest path to the node whose packets it wants to intercept. The attacker will then receive the traffic destined for other nodes and can then choose to drop the packets to perform a denial-of-service attack, or alternatively use its place on the route as the first step in a man-in-the-middle attack by redirecting the packets to nodes pretending to be the destination.
- *Spoofing*—a node may attempt to take over the identity of another node. It then attempts to receive all the packets destined for the legitimate node, may advertise fake routes, and so on. This attack can be prevented simply by requiring each node to sign each routing message (assuming there is a key management infrastructure). Signing each message may increase the bandwidth overhead and the CPU utilization on each node.
- *Modifying Routing Packets in Transit*—a node may modify a routing message sent by another node. Such modifications can be done with the intention of misleading other nodes. For example, sequence numbers in routing protocols such as AODV are used for indicating the freshness of routes. Nodes can launch attacks by modifying the sequence numbers so that recent route advertisements are ignored. Typically it is particularly difficult to detect the node which modified the routing message in transit. Requiring each node to sign each routing message can prevent these types of attacks. In such a case, if a node modifies routing packets, then it might escape undetected, but it will not be able to mislead other nodes because the routing messages will not have the appropriate signature. Other nodes can detect illegal modifications in the packet via the cryptographic protection mechanisms.
- *Packet Dropping*—a node may advertise routes through itself to many other nodes and may start dropping the received packets rather than forwarding them to the next hop based on the routes advertised. Another variation of this attack is when a node drops packets containing routing messages. These types of attacks are a specific case of the more general packet dropping attacks.
- *Selfish Nodes*—routing in MANET depends on the willingness of every node to participate in the routing process. In certain situations nodes may decide not to participate in the routing process. For example, nodes may do that in order to conserve battery power. If several nodes decide to do that then the MANET will break down and the network will become inoperable. Certain protocols have been proposed for encouraging nodes to participate in the routing process.
- *Wormhole Attack*—in this attack adversaries can collude to transport routing and other packets out of band (using different channels). This will interfere with the operation of the routing protocols. We will discuss this attack in more detail in Chapter 4.
- *Rushing Attack*—in this case, an adversary can rush some routing packets towards the destination, leading to problems with routing. We explain this attack and protection mechanisms against this attack in Chapter 4.

1.4 OVERVIEW OF THE BOOK

As discussed earlier, wireless ad hoc networks have attracted much interest in the research community due to their potential applications. The key characteristic of such networks is their openness, which makes it possible for nodes to come together and form a network with no human intervention and with no existing pre-established infrastructure. Unfortunately this characteristic that makes such networks so important also makes them vulnerable to a wide variety of attacks. In this book, we focus on the problem of securing wireless ad hoc networks and discuss potential solutions for protecting such networks. We focus on solutions that are unique to the wireless ad hoc networking environment. We attempt to explain a large number of solutions and techniques that have been discussed for securing wireless ad hoc networks. We discuss the advantages of these approaches and often their limitations. Securing such networks is a very challenging task, as discussed earlier. Often no perfect solution exists. In such cases we attempt to identify the limitations of the most promising approaches and discuss additional areas that require further research.

Typically, protection of networks is achieved using multiple overlapping approaches (multiple layers of defense) that make it difficult for an attacker to penetrate the network. The approaches used to secure wireless ad hoc networks can be considered to belong to three broad categories: (1) prevention approaches that try to prevent an attacker from penetrating the network and causing harm; (2) detection approaches that detect an attacker after the attacker has already penetrated the preventive barriers; and (3) response and recovery approaches that attempt to respond to an attacker once he/she has been detected to have penetrated the preventive barriers. We will discuss solutions and mechanisms that address all of these approaches. The book is structured as follows:

- Chapter 1 provides a general description of wireless ad hoc networks. Several applications that motivate the importance of this technology are also considered. The chapter also discusses the unique challenges associated with securing such networks.
- The next three chapters focus on protection mechanisms. A key protection mechanism is cryptography, which makes it difficult for malicious nodes to eavesdrop on traffic from other nodes, modify such traffic, or pretend to be somebody else. Chapter 2 discusses some of the fundamental concepts of cryptography that we leverage in later chapters. Several mechanisms for securing wireless ad hoc networks rely on cryptography and this chapter provides the foundation needed to understand such mechanisms.
- Communication is a basic function needed by the entities in the network. Further, communication between entities has to be done securely in order to protect against various attacks that can be launched by the adversaries. However, this is dependent on the sharing of cryptographic keys among the network entities. Chapter 3 discusses several schemes for sharing keys among nodes in wireless ad hoc networks.
- Routing is one of the fundamental requirements in a MANET. Most of the routing protocols have been built with the goal of establishing quick and efficient communication among nodes. Often those goals are orthogonal to the goal of providing secure connectivity. In Chapter 4 we consider some of the widely used MANET routing protocols and describe ways of securing such protocols.
- In spite of all the preventive mechanisms, adversaries might still be able to penetrate the network and launch attacks. Several mechanisms have been proposed in order to

detect such occurrences. Chapter 5 discusses intrusion detection techniques that can be used for detecting malicious behavior in MANETs.

- Once an attack is detected, it is important for appropriate response and recovery steps to be taken. Such actions need to be quick and preferably with limited or no human intervention. This will make it possible to have the network operational quickly so as to continue supporting the application of interest. Chapter 6 discusses the concept of policy management, which has been proposed as a way to automate management of networks. Such automation includes responding to specific events, including faults and attacks. Policy management allows network administrators to define the response of the network to the various events.
- The following chapter focuses on an interesting application, namely secure localization. Nodes in wireless networks are typically mobile. Identifying the location of a node is important for a variety of applications. Various approaches have been proposed for estimating the current location of a node. Several of those approaches are open to attacks from malicious users. Chapter 7 discusses some of the localization schemes and approaches for securing such schemes.
- Chapter 8 introduces another topic of special interest, namely the application of wireless ad hoc networks in vehicular networks. This is an area that we believe will attract a lot of interest in the future. This chapter also presents the conclusion of this book.

2 Basic Security Concepts

2.1 INTRODUCTION

Cryptography is the study of mathematical techniques concerned with keeping information/data protected from adversaries. For example, cryptographic mechanisms have been designed for keeping data confidential. Cryptographic schemes have been developed so that data that is transmitted over the air (e.g. via wireless systems) is encrypted (e.g. scrambled) and cannot be interpreted by adversaries. This is in spite of the fact that adversaries may obtain the encrypted data by eavesdropping on the data transmitted over the air. Cryptography can also be used for ensuring that the data was really created by the person claiming to have created the data. This property is also called data authentication. Cryptography can also be used for supporting other security services including data integrity and non-repudiation, as discussed later in this section. Cryptography encompasses several areas of computer science and mathematics such as number theory, complexity theory, algorithms, probability, computational theory, and information theory. A cryptographer focuses on designing and analyzing cryptographic algorithms and protocols. The analysis could give rise to discovery of ways to break existing cryptographic protocols.

In this book though, our focus is not on cryptographic algorithms and protocols. Rather our focus is on building secure systems by making use of cryptographic tools. More precisely, the systems that we focus on are the systems of ad hoc networks. Building secure systems does not consist of taking a good cryptographic algorithm and combining it with the system. Rather, a good cryptographic algorithm will have to be integrated intelligently into the system, keeping in mind the constraints and features of the system. The resulting system must then be analyzed to ensure that it does not have any undesirable vulnerabilities or weaknesses. In fact, a term has been coined for this—security engineering. As opposed to other branches of engineering where the focus is on designing systems and verifying that the system achieves the desired objective, security engineering is focused on designing systems and then analyzing these systems to ensure that there are no ways of circumventing the security defenses. In fact here the odds favor the attacker, as he/she has to find just one flaw in the system while the defenders have to verify that the system does not have any flaws, which might be difficult to achieve. Building a secure system is not easy and it typically involves several tradeoffs.

A very important factor to consider is related to costs. Even in cases where technology exists to achieve the security objective, the costs involved might make the solution impractical. For example, a defender may decide that the cost of implementing certain security technologies is too high relative to the potential risk and therefore not worth pursuing.

Security for Wireless Ad Hoc Networks, by Farooq Anjum and Petros Mouchtaris

Thus, the system might not be 100% secure. In addition, the system might also be vulnerable to attacks not thought of during the design phase. Other factors that govern the design include the risks and returns to the attacker as well as the degree of risk that is acceptable to the defender.

Even though cryptography is one of the main means that security engineers use for protecting information systems, it is not the only tool that can be used for such a task. Other technical measures such as biometrics or steganography can also be used. In addition, legal measures such as liability regulations or insurance might also be necessary to protect systems. Organizational measures such as proper security policies and correct information classification also play an important part in ensuring the security of the system. People-related measures such as screening, motivation, and education also cannot be ignored. In this book, though, our focus will mainly be on using cryptographic tools to achieve the desired security objectives for protecting information systems. Hence it is vital to understand the various cryptographic concepts, and this is the objective of this chapter.

We start off by looking at the basic concepts of cryptography in Section 2.2. We look at the various objectives that are attainable using various cryptographic mechanisms. We also discuss the fundamental cryptographic mechanisms. Readers familiar with the basic concepts of cryptography can skip this section. Note that our objective is to give a brief overview of the various cryptographic concepts. Readers who need more details should refer to other material [7]. These concepts are used in several other places in the book.

2.2 BASIC CONCEPTS

Cryptographic mechanisms are designed in order to achieve certain objectives. These objectives are typically referred to as the attributes associated with the cryptographic mechanism. We start this section by explaining some of the attributes in Section 2.2.1. We then explain the various mechanisms that cryptography provides in order to achieve these objectives (attributes) in Section 2.2.2.

2.2.1 Attributes

The security techniques that we explain later have been designed with one or more attributes in mind. There are several attributes but the following form the basic set. This is because the other objectives can be derived from these basic ones:

- confidentiality;
- integrity;
- authentication;
- non-repudiation;
- availability.

Confidentiality ensures that information content is never revealed to entities that are not authorized to receive it. Thus, a mechanism that has this attribute ensures that transmitted information can only be interpreted by the intended receivers. Secrecy and privacy are other terms that are considered synonymous with confidentiality. In the case of ad hoc networks, confidentiality is very important for protecting the transmission of sensitive information. This is especially vital considering the fact that the wireless links are easily susceptible to eavesdropping. Leakage of information related to data traffic or control traffic such as routing could have catastrophic consequences in certain situations, such as the battlefield environment.

Integrity ensures that data is not altered in an unauthorized manner during transmission. This alteration could be due either to accidental factors such as the vagaries of the wireless links or to malicious factors such as the presence of an adversary. An adversary could manipulate data by insertion, deletion, or substitution of data. Various protocols such as TCP and IP include mechanisms such as checksums which are designed to make such protocols robust to benign failures. These mechanisms, however, are not sufficient to protect against malicious adversaries that attempt to purposefully alter the content of data transmissions.

Authentication guarantees a node of the identity of the other party or parties that it is communicating with. There are two possible types of authentication, namely entity authentication and data authentication. Entity authentication is concerned with verifying the identity of the other communicating party. In systems that lack entity authentication mechanisms, an adversary can masquerade as an insider, thereby possibly gaining unauthorized access to network resources. In addition, this could also lead to interference with the operation of other nodes. Entity authentication involves corroboration of a claimant's identity through actual communication and typically involves no meaningful message other than the claim. On the other hand, data authentication is focused on providing guarantees as to the origin of data. Note that data authentication implicitly provides data integrity.

The fourth attribute, non-repudiation, ensures that a party cannot falsely deny its actions nor entities falsely claim commitments from other entities. For example, considering the transmission of data as an action, the originator of a message cannot deny having sent the message if it has indeed done so. Further, the receiver of a message cannot claim to have received a message falsely from an entity that has never sent the message.

The fifth attribute is related to availability. Availability ensures that the network services are available when required by the various entities in the network. This attribute is mainly geared towards attacks such as the denial of service attacks that attempt to prevent authorized users from accessing important services.

There are several other cryptographic attributes in addition to the fundamental ones given earlier. These include anonymity, which is the ability to conceal the identity of an entity involved in some process, authorization, which is the ability to convey to another entity the official sanction to allow something, timestamping, which is the ability to record the time, access control, which is the ability to restrict access to resources to privileged entities, revocation, which is the ability to retract authorization, and so on. We will not consider these attributes here, however, and will explain these in detail as needed in the various chapters.

Several cryptographic primitives of functions have been designed in order to achieve the above objectives. These primitives can be divided into three families:

1 symmetric key cryptography—involves the use of a single key;
2 asymmetric key cryptography—involves the use of two keys;
3 message digests—does not involve the use of any keys.

We next look at the three families of primitives in more detail.

2.2.2 Cryptographic Primitives

Cryptographic systems based on the use of keys can be thought of as analogous to combination locks. They involve both an algorithm and a secret value. The secret value is the key (analogous to the number that opens the combination lock) and this must be protected from adversaries. The cryptographic algorithm can be considered to be analogous to the working of the combination lock. The design of the algorithm is expected to be known widely. In fact, making the design of the algorithm public can lead to analysis of its weaknesses by the community. The security of the system should only depend on the secrecy of the key, and an adversary who desires to break the cryptographic system has to determine the key used to secure communications.

A basic approach that can be taken by the adversary is to try every key and see which breaks the system based on some information about the system. For example, if packets between two nodes are encrypted, then the adversary might have an idea as to the protocol being used, such as TCP or UDP. By trying every possible key to decrypt the packets, the adversary will come up with random strings. Only a few such strings will fit the format associated with a packet belonging to the protocol. More information about the data, such as the identity of the parties communicating or the port numbers being used for the communication, could be used to increase the probability of determining the right key being used. Such an approach is easy to pursue in wireless systems given the susceptibility of wireless links to eavesdropping, making all traffic going over the network available to the adversaries.

It is obvious that the effort required for such an attack to be successful is proportional to the length of the key. An 8 bit key length implies that there are 2^8 different keys possible while an 80 bit key length implies that there are 2^{80} different keys possible. This is similar to the case of combination locks. Even here, longer combinations of keys take longer to break as opposed to smaller combinations, assuming that all possible keys can be tried. However, we cannot conclude that longer key length always translates into better security for the system. In some cases, there can be other vulnerabilities in the system, including the encryption algorithm, that the adversary could take advantage of. For example, it has been shown that the security of WEP (the typical protocol used for securing 802.11b) can be broken irrespective of the size of the key used [8].

Hence, while key size is important, the evaluation of cryptographic primitives cannot be done solely based on the size of the keys. Other factors which are used to evaluate the various cryptographic primitives include:

- level of security;
- functionality;
- methods of operation;
- ease of implementation;
- performance.

The level of security afforded could be related to cost. A higher level of security could result in the system being secure even against well-funded adversaries. Note, though, that the level of security itself is a qualitative measure. The functionality would be related to the attributes that can be achieved using the proposed primitive. For example, nonrepudiation would need the use of concepts from asymmetric key cryptography.

The method of operation could also be different in order to achieve different functionality. For example, asymmetric key algorithms would require different modes when used to ensure confidentiality and nonrepudiation. Ease of implementation is another important factor. Performance also needs to be considered, especially in ad hoc networks given the resource constraints in such networks. It is well known that message digests, which we explain later, need the least amount of resources in terms of memory and computation power, while asymmetric key operations are the most expensive.

2.2.2.1 Attacks in Cryptographic Systems An adversary in any system protected by cryptographic primitives can have several objectives. For example, the adversary might aim for the total break whereby he/she obtains the secret keys being used to secure communication. A weaker objective of the adversary could be to obtain the ability to decrypt a ciphertext for which he/she has not seen the plaintext.[1] A still weaker objective could be to be able to distinguish between the encryptions of two plaintexts that the adversary has not seen before. A cryptosystem that does not permit distinguishability of ciphertexts is said to be semantically secure.

Given these goals of the adversary, there can be several types of attacks. Note that we consider generic attacks here instead of specific attacks (such as wormhole or sybil). An assumption here is that the adversary has access to all data transmitted over the ciphertext channel. In addition, we assume that the adversary knows all details of the encryption function except the secret keys. With these standard assumptions we can have the following types of generic attacks:

- ciphertext only;
- known plaintext;
- chosen plaintext;
- chosen ciphertext.

In case of the ciphertext only attack, the adversary possesses one or more strings of ciphertext. No additional information is available to the attacker. This is an easy attack to implement based on the standard assumptions for a wireless network. In such a case, the adversary has to search through all the possible keys to decrypt the ciphertext. Therefore, it is essential that the adversary be able to recognize when he/she has succeeded. This could be based on information about the plaintext that the adversary has. For example, if the adversary is aware of the protocol whose packets are being encrypted, he/she could look for decryptions that fit the structure of the expected packets. This, however, could lead to more than one candidate key. To further narrow down the list, the intruder might need to make use of more ciphertext values.

The goal of the adversary when trying to attack such systems will be to devise mechanisms where the plaintext can be recovered from the ciphertext without any knowledge of the keys being used. As explained earlier, one possible way to achieve this objective is to try all possible keys by exhaustive search. If the key space is large enough then this approach might become impractical. In fact, the designer of the system should aim towards making this brute-force approach the best approach to break the system. If such a brute force approach is impractical, it will guarantee the security of the system.

[1]Ciphertext is the encrypted version of the plaintext that a node is trying to transmit.

In a known plaintext attack, the adversary is assumed to have knowledge of one or more pairs of plaintext and ciphertext. These pairs could then be used to determine the key or determine the plaintext values corresponding to other ciphertext values. In case of a chosen plaintext attack, ciphertext corresponding to the plaintext chosen by the adversary is assumed available. A slight variation of this is the adaptive chosen plaintext attack where the adversary can choose the plaintext messages depending on the previous pairs of plaintext and ciphertext messages. Under the chosen ciphertext attack, adversaries are allowed access to plaintext–ciphertext pairs for some number of ciphertext messages of their choice. An adversary could then use this information to recover the key or the plaintext corresponding to some new ciphertext. Note that the ability of the attacker keeps increasing as the number of pairs available increases.

Given the several types of generic attacks on cryptographic systems, it is necessary to be able to determine the strength of these attacks. This can be done by looking at three aspects, namely data complexity, storage complexity, and processing complexity. Data complexity is given by the expected number of input data units required by the adversary to achieve his objective. These input data units could just be ciphertext or could be pairs of ciphertext and plaintext messages, depending on the type of attack. Storage complexity corresponds to the expected number of storage units required in order for the adversary to meet his objective. Finally, processing complexity is the expected number of operations required to process input data so as to meet the adversary's objective. The complexity of the attack then depends on the weakest of these three components.

Having discussed the generic attacks on cryptographic systems, we next focus on the cryptographic primitives used.

2.2.2.2 Symmetric Cryptography The working of a symmetric key cryptographic operation is shown in Figure 2.1. The plaintext messages are encrypted at the sender using the encryption key E. The resulting ciphertext can then be transmitted over the channel (wireless or wireline or a combination). Intruders are assumed to have access to the ciphertext on the channel as shown in the figure. Passive intruders can just record the ciphertext while active intruders can attempt to modify it. The ciphertext then reaches the receiver where it is decrypted using the decryption key D as shown. The decryption results in recovery of the original plaintext assuming that the ciphertext has not been modified in transmit by an active intruder. Typically both the E and the D keys are the same and this common key is typically called as the shared key. Such symmetric key schemes can be used to achieve confidentiality, integrity, and authentication.

A basic requirement for the symmetric key scheme is that the parties involved in the communication share a common key. This implies that the shared key must be distributed

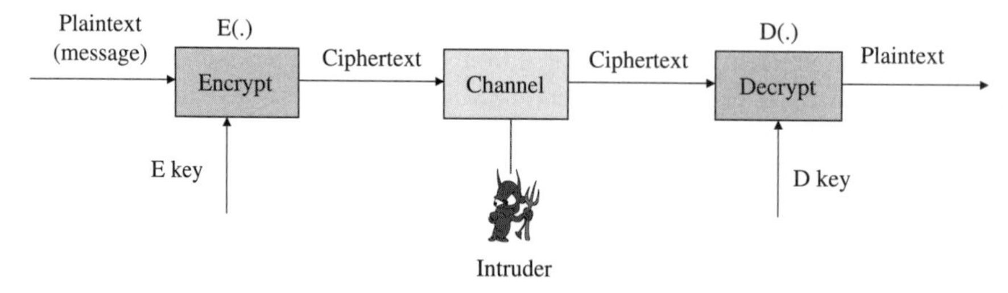

Figure 2.1. Basic operation of symmetric key cryptography.

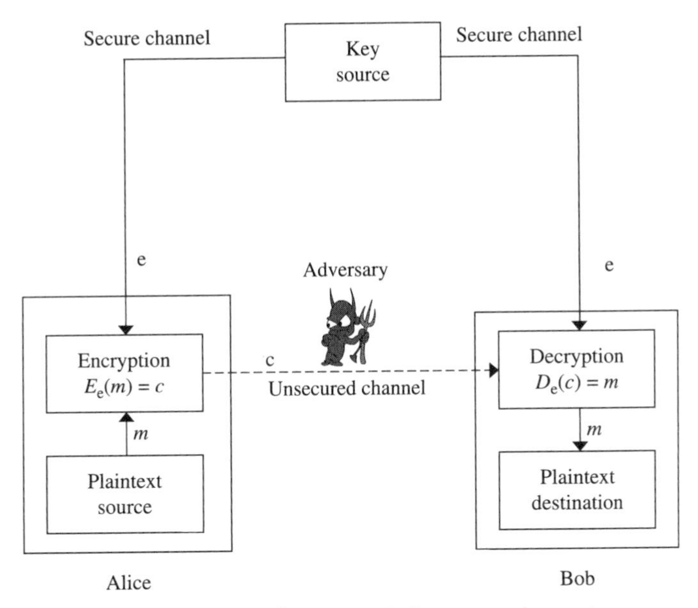

Figure 2.2. Model of a symmetric key encryption system.

over a secure communication channel as shown in Figure 2.2. This is related to the problem of key distribution and is a major problem, especially in wireless ad hoc networks. We will address the problem of key management in more detail in Chapter 3. A question might arise, however: Why not use the same secure channel used to transmit the secret keys to also transmit the data. This may not be possible, either because of bandwidth limitations on such channels or because such channels may not be available when data has to be transmitted.

Symmetric key algorithms are of two types, namely block ciphers and stream ciphers. Block ciphers work on blocks of data at a time. All of the bits constituting a block have to be available before the block can be processed. Thus, block ciphers can be viewed as functions that map an n-bit plaintext to an n-bit ciphertext. The function must be a one-to-one function in order to allow for unique decryption. The right size of block lengths is important from a security, performance, and complexity point of view. Short block lengths can possibly help the adversary construct the decryption table containing the plaintext–ciphertext pairs easily. Long block lengths can be inconvenient due to the complexity of performing the computations for encrypting and decrypting the data and also because of the performance penalties associated with such computations. The normally used block lengths are 64 or 128 bits. This is the most widely used type of cryptographic algorithm.

Stream ciphers work on a bit or byte of the message at a time. Thus, the data is processed as a "stream." Given the small size of the block (a bit or byte), stream ciphers might be inefficient. These are faster than block ciphers in hardware and require less complex circuitry. Many stream ciphers are proprietary and confidential.

We next look at block ciphers in more detail. There are two types of classical (traditional) block ciphers: (1) substitution cipher; and (2) transposition cipher. A substitution cipher makes the relationship between the key and the ciphertext as complex as possible. This is the property of confusion. Here the core idea is to replace symbols with other symbols or groups of symbols. A simple example of a substitution cipher is the Caesar cipher. In this case, every letter of the plaintext might be substituted with a different letter three away from the original letter to obtain the ciphertext. Other schemes, such as monoalphabetic substitution, polyalphabetic substitution, homophonic substitution, exist. Interested readers could look at [7] for details.

A transposition cipher (also called permutation cipher) spreads redundancy in plaintext over the ciphertext. This is the property of diffusion. More precisely, diffusion dissipates the statistical structure of the plaintext over the bulk of the ciphertext. As a result, a change in a single bit of plaintext can result in changes to multiple ciphertext bits. In other words, the effect of a change in the plaintext is spread to many parts of the ciphertext. Classical transposition ciphers obtain the ciphertext from the plaintext by rearranging the order of letters. The actual letters used in the ciphertext are not altered from the plaintext.

A problem with both substitution and transposition ciphers is that they are not strong enough by themselves. Therefore, in order to create harder ciphering algorithms, researchers did consider the use of several ciphers in succession. However, two substitutions make a more complex substitution. Similarly, two transpositions make a more complex transposition. On the other hand it was observed that combining both these type of ciphers (substitution followed by a transposition or vice versa) could lead to stronger symmetric key algorithms. Such hybrid ciphers are called product ciphers.

Product ciphers represent advancement over classical symmetric key cryptography and are considered to be the bridge from classical to modern symmetric ciphers. The basic idea behind product ciphers is to build a complex cipher by composing several simple operations. Each operation by itself offers insufficient protection while the combined operations offer complementary protection. Several modern symmetric ciphers such as data encryption standard (DES) and advanced encryption standard (AES) are product ciphers.

The current state of the art block cipher is AES. The other algorithm that was widely used but whose use is now deprecated is DES. DES was adopted in 1977 by what is now the National Institute of Standards and Technology (NIST). Originally, this was developed as the Lucifer cipher by a team from IBM led by Feistel. Lucifer used 64 bit data blocks with 128 bit keys. This algorithm was then redeveloped as a commercial cipher with input from the National Security Agency (NSA) and others. The revised Lucifer cipher was eventually accepted as DES. DES encrypts data in block sizes of 64 bits while using a 56 bit key. The 56 bit key is expanded to 64 bits using 8 parity bits. DES is efficient to implement in hardware and is relatively slow if implemented in software. DES has seen considerable controversy over its security. There was speculation that the reason for the short key was to have weak security. This could allow NSA to break the key. Additionally, the design criteria were also classified, strengthening the speculation. Subsequent events and public analysis show in fact that the design was appropriate. DES had come to be widely used, especially in financial applications.

Encryption in DES is done over 16 stages (rounds) with each round being functionally equivalent. Each round has eight fixed carefully selected substitution mappings and also a permutation function. Each round also makes use of a 48 bit subkey. These subkeys are generated from the initial 64 bit key. The structure of DES corresponds to a Feistel cipher [7]. DES has many desirable properties. Each bit of the ciphertext depends on all bits of the key and all bits of the plaintext. Further, there is no statistical relationship evident between plaintext and ciphertext. Altering any single plaintext or key bit alters each ciphertext bit with a 50 percent probability. Altering a ciphertext bit results in an unpredictable change to the recovered plaintext. This makes attempts by the adversary to "home-in" by guessing keys impossible. However, brute force approaches to break

DES are now practical. During 1997, it would take a few months to break DES using the brute force approach, while in 1999, an attempt to do so succeeded in 22 hours. In addition several theoretical attacks that can break DES have also been demonstrated. Thus, a replacement for DES was needed.

This replacement is the AES algorithm, which was selected in October 2000. AES has been designed to be resistant to known attacks and exhibits simplicity of design. Originally called as Rijndael, this was issued as a Federal Information Processing Standards (FIPS) standard in November 2001. AES uses three different key sizes, namely 128, 192, and 256 bits. The block size is 128 bits. It treats data in four groups of four bytes, each called the state. AES has been designed to have one of 9, 11, or 13 rounds. In each round the state undergoes four operations, namely byte substitution, shifting of rows, mixing of columns and XORing with the subkey of the round. All operations can be combined into XOR and table lookups. Hence these are very fast and efficient.

DES and AES are not the only block ciphers widely considered. In fact, there are several others, such as Triple-DES, Blowfish, IDEA etc. No block cipher is ideally suited for all applications. There are several factors that go into the selection of a cryptographic algorithm for a given scenario. These include the desirable key size, block size, complexity of cryptographic mapping, tolerance to error, and the estimated security level. As explained earlier, longer key sizes are desirable but this imposes additional costs associated with generation, transmission, and storage. An interesting technique used by block ciphers such as Khufu and stream ciphers such as SEAL is to make the task of changing cipher keys computationally expensive while allowing encryption to remain relatively efficient. Larger block size is desirable but might be more costly to implement. It might also affect performance, say by requiring padding. Complexity of cryptographic mapping affects implementation and development costs as well as the real-time performance. In fact some ciphers are more efficient when implemented in hardware and other ciphers are more efficient when implemented in software. The effects of corrupt ciphertext during decryption indicate the error propagation characteristics of the cipher. This is also an important factor to consider given the lossy wireless links in an ad hoc environment. The estimated security level of a cipher can be assumed to be proportional to the amount of expert cryptanalysis withstood.

2.2.2.3 Asymmetric Encryption Asymmetric encryption represents a very significant advance in the three thousand year history of cryptography. We have seen earlier that conventional symmetric key cryptography uses one key that is shared by both the sender and the receiver. Disclosure of this key will result in compromise of communications. In addition, this approach affords no protection if a receiver forges a message and claims such a message is from the sender. This is possible because two parties share the key and either of them could create a message using the shared key. Thus, symmetric key schemes do not lend themselves easily to verifying the authenticity of the sender. Another problem associated with such schemes is related to the secure distribution of keys.

These problems can be addressed using the approach of asymmetric key cryptography. The asymmetric approach uses two keys, a public key and a private key. The public key may be known to anybody while the private key is supposed to be known only to the entity creating a message. Every entity in the network wishing to send messages has these two keys, namely a private and a public key. Both the keys are different but related mathematically. Additionally, it is computationally infeasible to derive the private key knowing only the public key and the cryptographic algorithm. Derivation of the private key would require additional information. Note, however, that it should be computationally easy to encrypt or decrypt messages when the relevant key is known.

The public key defines the encryption transformation while the private key defines the decryption transformation. More precisely, either of the two related keys can be used for encryption while the other key is used for decryption, although we ignore this for ease of explanation. This approach is also called asymmetric because the encryption of messages uses a different key than that required for the decryption of messages. Thus, other entities can encrypt messages destined to a node using the public key of a node. Only the node can then decrypt such a message using its private key. In those cases, those who encrypt messages cannot decrypt the messages.

We have seen earlier that symmetric encryption schemes need a secure channel to transfer the key used for encryption. Public key encryption on the other hand does not require such a secure channel but has a weaker requirement on the channel. It only requires an authenticated channel. This is to ensure the genuineness of the public key of the other party. Information sent over this authenticated channel does not need to be confidential. The asymmetric key schemes can achieve attributes such as nonrepudiation, confidentiality, integrity, and authentication. However asymmetric encryption is substantially slower than symmetric key encryption given the same amount of computational resources. Due to this, generally public key encryption schemes are only used to encrypt small amounts of data such as the keys used for symmetric encryption. Thus, we can see that asymmetric key cryptography complements rather than replaces symmetric key cryptography.

Asymmetric key cryptography provides the ability to ensure nonrepudiation in addition to authentication. This is done using the concept called as digital signature which is intended to provide the digital counterpart to a handwritten signature. In this case, an entity can transform a message using its private key. This serves as a signature of the entity on the message. Any other entity can then verify the transformation on the signed message using the public key of the signer. This is the concept of digital signature. It is so called since only the entity is expected to have access to its private key. Thus, in this case the private key is used to create signatures while the public key is used to verify signatures. Note that a digital signature must have several features, such as it must depend on the message signed, it must use information unique to sender in order to prevent both forgery and denial, it must be relatively easy to produce, recognize and verify, it must be computationally infeasible to forge (either with a new message for an existing digital signature or with a fraudulent digital signature for a given message), and it must be practical to store.

We show the operation of an asymmetric key system in Figure 2.3, where we illustrate these concepts. In this case, we assume that Alice desires to transfer a message authentically and confidentially to Bob. The authenticity of the message gives a guarantee to Bob that the message has indeed originated at Alice. In addition, the confidential communication ensures that an adversary will not be able to eavesdrop on the communication between Alice and Bob. To achieve this, both Alice and Bob will have to have a public–private key pair each. In addition, they will also have to obtain the genuine public key of the other. Note that the key sizes used in the asymmetric key algorithms are typically very large (>512 bits).

We assume that Alice is the source of the message. The plaintext message is first transformed using the private key KR_a of Alice. By doing this Alice is digitally signing the message, thereby verifying that she is the source of the message. Next, in order to

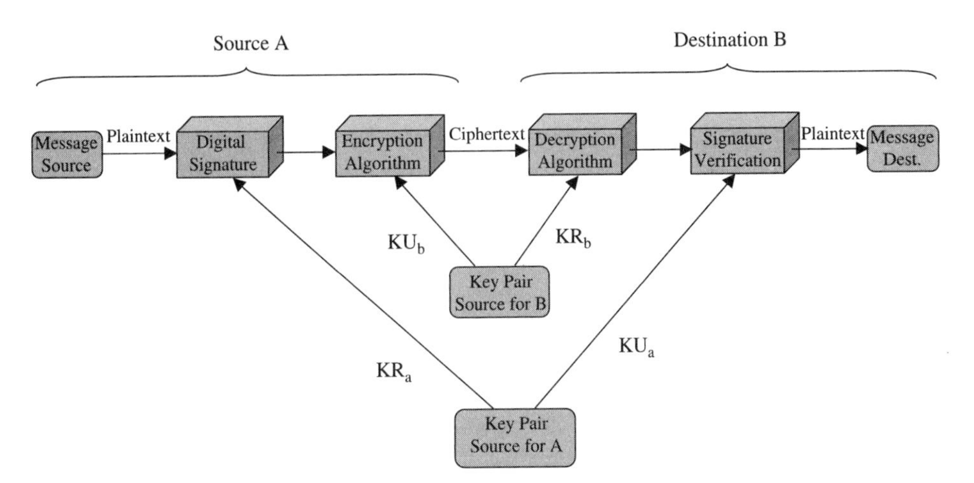

Figure 2.3. Operation of an asymmetric key system providing both confidentiality and authentication.

provide confidentiality, Alice encrypts the result from the previous step using the public key of Bob, KU_b, as shown in the figure. The resulting ciphertext is then transmitted over the channel. When Bob receives the message, he first decrypts it using his own private key. Following this, Bob checks the authenticity of the message by checking the signature on the message. This is done by transforming the message using the public key of Alice. If the authenticity of the message is verified then Bob can accept the message. Note that it is not necessary to have both confidentiality and authenticity simultaneously for every message as shown in this figure. Either of these attributes can also be achieved by itself. In such a case, the corresponding operations in the figure are omitted.

The security of asymmetric key schemes depends on a large enough difference in difficulty between a hard and an easy problem. The hard problem is what an adversary is expected to be faced with while the easy problem is the one used by the node which has access to both the private and the public key. Asymmetric ciphers can also be subjected to brute force exhaustive search just like the symmetric ciphers.

Numerous asymmetric key algorithms have been proposed, although only a few are both secure and practical. Further, some of these algorithms are more suitable for encryption while others are more efficient when considering digital signatures. There are three algorithms that work well for both encryptions as well as for digital signatures. These are the RSA algorithm, ElGamal algorithm, and the Rabin algorithm. We next describe briefly the most popular amongst these, namely the RSA algorithm.

RSA is the best known and widely used public key encryption algorithm. It was devised by Rivest, Shamir and Adleman of MIT in 1977. It can be used for both encryption and digital signature. It is based on exponentiation in a finite (Galois) field over integers modulo a prime. The security of this scheme is based on the intractability of the integer factorization problem. Note that the integer factorization problem consists of determining all the prime factors of a given number. It is obvious that this problem is trivial when dealing with small numbers but becomes intractable when dealing with large numbers. Hence, the numbers used in the RSA algorithm need to be very large (1024 bits and beyond). This is the reason for the large keys associated with these algorithms.

As with any public key scheme, a user has to generate a public–private key pair. In order to do this, the user selects two large distinct prime numbers p and q. The system modulus denoted N is the product of these two large primes. Following this, Euler's totient function, $ø(N)$, is calculated as the product of $(p - 1)$ and $(q - 1)$. The next step involves determining the public key component e such that $1 < e < ø(N)$ and $\gcd[e,ø(N)] = 1$ where gcd denotes the greatest common divisor. The public key of this user is then given by the pair of numbers (e, N). The private key is given by d where d satisfies the equation $e \cdot d = 1 \bmod ø(N)$ and $0 \leq d \leq N$. An adversary can determine the private key if he/she can factorize the system modulus. Hence, it is necessary to keep the factors of N, namely p and q, secret. This key setup is done once (initially) when a user establishes (or replaces) their public key. Note that the public key component e is usually fairly small.

Once the key pair is decided, the node publicizes the public key. Any other node that wishes to send a message M confidentially to this node obtains the public key of the recipient node, which is (e, N). The message M is then represented as an integer M where $0 \leq M < N$. The ciphertext C is then computed as $M^{\mathrm{e}} \bmod N$. The recipient on receiving the ciphertext uses the private key to decrypt the ciphertext as $C^{\mathrm{d}} \bmod N$, which results in the plaintext M.

As mentioned earlier, given the performance overhead associated with the public key algorithms, these algorithms are typically not used for encrypting data. Rather, they are typically used to transport symmetric keys between the parties that need to exchange data securely. In case of key transport, one of the parties involved decides the key. This party then encrypts the key using the public key of the other party and sends the message to the other party, who can then obtain the key. A problem with this scheme is that the key is decided by a single party. In some cases, it might be preferable to let both the parties contribute equally to the resulting key. This is referred to as key agreement. We next explain briefly one such widely used key agreement protocol, the Diffie–Hellman (DH) key exchange protocol.

The objective of DH protocol is to establish a common key between the participants of the protocol. The common key is expected to be used to exchange messages securely using symmetric key ciphers. Further, the common key is known only to the participants of the key exchange protocol. The Diffie–Hellman protocol itself does not ensure entity authentication. This implies that the participants in the protocol do not have guarantees of the identities of each other. Therefore, these participants need to use other techniques in order to verify the authenticity of each other. The DH key exchange protocol is based on exponentiation in a finite (Galois) field modulo a prime number. This requires the ability to find large primes as well as the ability to find primitive roots [7] for large primes. It also requires the ability to carry out efficient modular arithmetic. The security of the protocol depends on a hard problem. This hard problem is the difficulty of computing discrete logarithms.

We now discuss how the DH protocol works. We explain this in the context of two parties. The parties first have to select and share the public information, which is the prime number q and a primitive root α of this prime. In the next stage each of the two parties, Alice and Bob calculate their public and private parameters. The private parameter of Alice is a number XA chosen at random such that $XA < q$, while the public parameter is YA given by $\alpha^{XA} \bmod q$. Similarly, Bob also chooses the private parameter to be $XB < q$ randomly and determines the public parameter YB to be $\alpha^{XB} \bmod q$. They then exchange their respective public parameters and calculate the shared secret as $YB^{XA} \equiv S \equiv YA^{XB}$

(mod q). Note that any entity other than Alice and Bob that seeks to compute the shared secret could have access to the public information exchanged over the wireless links as well as the public parameters of both parties. Such an adversary would have to perform operations involving a discrete logarithm, which is a hard problem. Hence the key exchange is secure. Note, however, that both Alice and Bob have no guarantee that the other party is who he or she claims to be. In fact, an adversary can contact Alice pretending to be Bob and Alice will not be any wiser and vice versa. This is a result of the fact that the DH protocol requires no authentication.

2.2.2.4 Message Digests Another class of algorithms that are extensively used in cryptographic protocols is the class of hashing functions. Hashing algorithms are of two different types, as shown in Figure 2.4. The unkeyed hash functions do not require any secret key, as the name indicates, while the keyed hash functions require a secret key. We will refer to the unkeyed hash functions as hash functions and to the keyed hash functions as message authentication codes (MAC) in the rest of the chapter. We first look at hash functions.

Hash functions are functions with two important properties, namely compression and ease of computation. Compression condenses an arbitrary message to a fixed size. Thus larger domains are mapped to smaller ranges. This implies that collisions exist whereby multiple inputs have the same hash value. The other property, ease of computation, ensures that the hash value is easy to compute.

A hash value could be looked at as a compact representative image of any arbitrarily sized input. This is the basic idea leveraged by cryptographic hash functions. Hash functions which are used in cryptographic mechanisms also need to possess the following three properties:

1 Pre-image resistance.
2 Weak collision resistance.
3 Strong collision resistance.

Pre-image resistance, also called the one-way property, implies that finding an input which hashes to a prespecified hash value is difficult. Thus, given the hash value y, it should be difficult to find any x such that the hash of x equals y. In fact good hashing algorithms will make this computationally infeasible. For example a hashing function given as x^2 mod n for any input value x has the pre-image resistance property.

Collision resistance implies that finding any two inputs having the same hash value is difficult. Weak collision resistance implies that it is computationally infeasible to find any

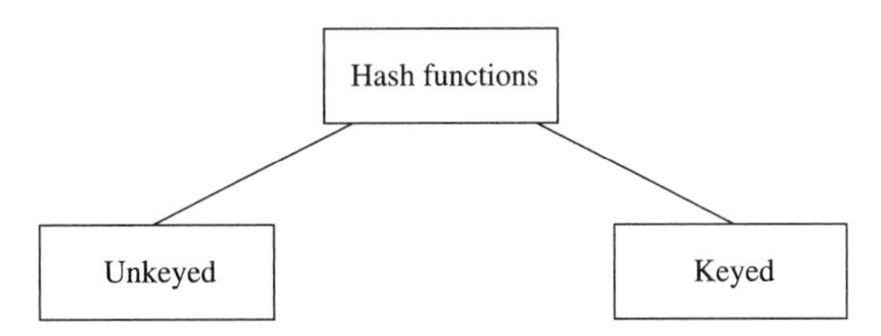

Figure 2.4. Hash function classification.

second input which has the same hash value as any given input. Thus given an input x, it should be infeasible to find any other input y such that both x and y hash to the same value. Note that the function $x^2 \bmod n$ does not have this property as both x and $-x$ have the same hash value.

Strong collision resistance implies that it is computationally infeasible to find any two distinct inputs which hash to the same value. Thus, in this case, it is infeasible to find any x and y values such that both x and y have the same hash value. Note that this is stronger than weak collision resistance since for the weak collision resistance property we fix one of the inputs while for the strong collision property we do not fix either of the inputs. Thus, strong collision resistance does imply weak collision resistance.

In addition there are other properties such as noncorrelation, near-collision resistance and partial pre-image resistance that are desirable in hash algorithms used as cryptographic primitives. Noncorrelation ensures that the input and output bits of the hashing function are not correlated. In fact, an avalanche property, whereby any change in one of the input bits causes at least half of the output bits to change randomly, is desirable. Near-collision resistance ensures that it is difficult to find any two inputs whose hash values differ in only a small number of bits. Partial pre-image resistance makes it as hard to recover any portion of the input as it is to recover the entire input given the hash value.

A hash value is very frequently used to create a digital signature. In this case, the hash value of a message is created and the signer signs the hash value rather than signing the message. This is more efficient from the point of view of performance since a hashing primitive is much faster than an asymmetric cryptographic primitive. In addition, a hash value can be used to detect changes to a message provided that the corresponding hashing algorithm has been designed such that collisions are computationally difficult to find. Thus, the hashing functions are used mainly to ensure data integrity and message authentication.

Secure hash algorithm (SHA) is a widely used hashing algorithm. SHA was designed by NIST and NSA in 1993 and revised in 1995 as SHA-1. This algorithm takes an arbitrarily sized input and produces 160 bit hash values. A message is processed in terms of 512 bit chunks over four rounds of 20 steps each. SHA-256, SHA-384, and SHA-512 have also been specified. These produce 256, 384 and 512 bit hash values.

Given that SHA-1 produces 160 bit hash values, the probability of finding a collision is negligibly small. More precisely, this probability is one in 2^{80} (follows from the birthday paradox). Thus, if we hashed 2^{80} messages, we would find one pair of messages that hashed to the same value. This is the brute force approach to finding collisions for hashing algorithms. It is clear that the difficulty of this approach depends on the size of the output hash value. It is desirable that the brute force approach be the best way to determine collisions for a given hash algorithm. To break a hash algorithm, an adversary will have to be able to find collisions while expending fewer cycles than needed by the brute force approach.

SHA-1 has been broken using this definition. The authors in [9] have shown the ability to find collisions in SHA-1 using about 2^{63} calculations. In order to understand the gravity of the problem, note that using today's technology it is expected that 2^{60} calculations might require a couple of days. On the positive side, however, note that these are collision attacks and not attacks on the pre-image property of SHA-1, and collision attacks are typically important in the case of digital signatures. As a result, many protocols will not be impacted by this result. In addition, leveraging these theoretical results to design software

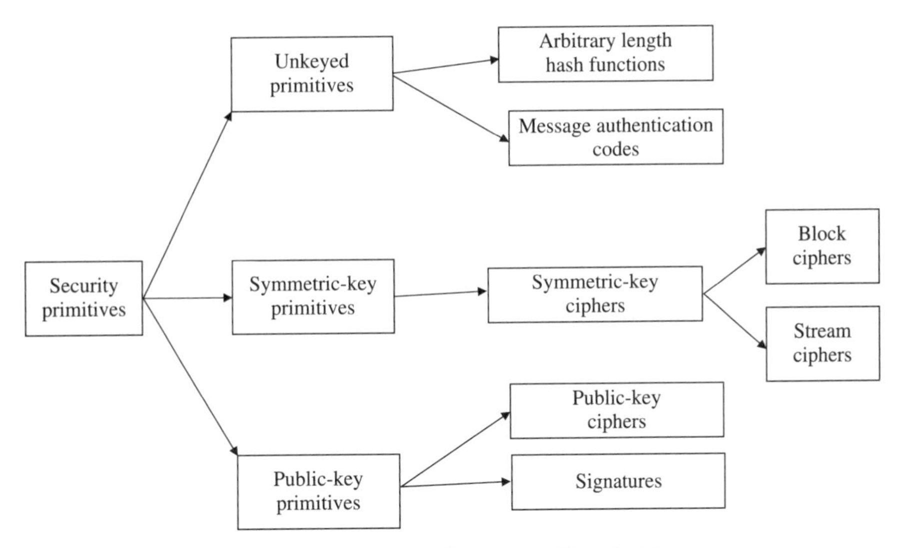

Figure 2.5. Summary of cryptographic primitives.

that will work with meaningful messages has also not been done yet. Thus, in summary, we should not use SHA-1 in any new protocols but rather use other possibilities such as SHA-256 or SHA-384.[2]

We next look at MAC functions. These functions take an arbitrarily sized message and also a secret key and produce a fixed size output. The function is designed such that it is infeasible to produce the same output without knowledge of the key. This allows efficient authentication of the source of the messages while also guaranteeing data integrity. MAC functions have similar properties as a hash function. In addition they also need to have the property whereby, given one or more pairs of text and the corresponding MAC value, it is computationally infeasible to compute a MAC value for any new input.

The widely used MAC algorithm is hashed message authentication code (HMAC). This is specified as an internet standard. HMAC has been designed such that it can use any of the several hash functions, such as SHA-1, MD5 etc. The security of HMAC therefore depends on the security of the underlying hashing algorithm used.

We would like to remark here that authentication as provided by use of hashing plus signature or MAC on the message can also be provided by either symmetric encryption or asymmetric encryption. This is because one can argue that message encryption by itself also provides a measure of authentication. For example, if symmetric encryption is used then the receiver would know that the sender must have created it, since only the sender and the receiver have access to the key used. Note that the receiver would have to depend on information such as the knowledge of message structure in order to detect if the content has been altered. This also assumes that the key has not been compromised. On the other hand, when using asymmetric algorithms, encryption provides no confidence in the identity of the sender since any entity that knows the public key of the entity of interest can send an encrypted message to it. However, if the sender signs the message using its private key, then authentication can be achieved. As opposed to these methods of achieving authentication, MAC algorithms can achieve authenticity while also ensuring

[2]Recommendation of a panel at the NIST First Cryptographic Hash Workshop, October 2005.

faster processing. In addition, MAC algorithms provide a separation of authentication and confidentiality that offers an extra degree of flexibility.

The cryptographic primitives described earlier help achieve the basic cryptographic attributes described earlier in Section 2.2.1. It is clear that both symmetric key and asymmetric key algorithms do indeed provide confidentiality. Integrity and authentication can be achieved by the use of message digests or cryptographic hash functions along with digital signatures on the hashed values. Nonrepudiation requires the use of public key algorithms to provide digital signatures. Availability is ensured by use of noncryptographic means such redundancy, physical protection, and the use of robust protocols. The various cryptographic primitives are shown in Figure 2.5.

2.3 MODES OF OPERATION

We have earlier seen examples of both symmetric ciphers as well as asymmetric ciphers. Symmetric ciphers like AES operate on fixed size input messages in blocks of 128 bits, while asymmetric ciphers like RSA also require the size of the input message to be smaller than the system modulus (the product of two large primes as discussed earlier). Other cryptographic algorithms have similar constraints. Given this, a question arises as to the mode of operation of these ciphers when messages which have arbitrary length (larger than the size of an input message) have to be encrypted. There are five modes of operation which are typically used in such cases and are applicable to any fixed-length encryption scheme. These are:

1 Electronic code book (ECB).
2 Cipher block chaining (CBC).
3 K-bit cipher feedback (CFB).
4 K-bit output feedback (OFB).
5 Counter mode (CTR).

We next explain these modes of operation starting with ECB. Note that some of these modes of operation also allow one to define stream ciphers from a block cipher.

The ECB mode is the simplest of these modes of operation and is fairly straightforward. A message is broken into independent blocks, with each block being encrypted with the secret key independently of the other blocks. The receiver gets encrypted blocks and decrypts each block in turn. As the name of this mode implies, each block is a value which is substituted like a codebook. We show the operation of the ECB mode in Figure 2.6.

This mode of operation ensures that bit errors in a ciphertext block affect the decryption of only that block. This reduces the complexity of the implementation of this scheme, but there are two serious shortcomings with this mode of operation. Repetitions in message being encrypted will be obvious in ciphertext if the repetitions are aligned with the message block. An adversary with access to the ciphertext can thereby gain some information about the content of the message by observing the repeated blocks. Thus, this mode does not offer semantic security. Another weakness is due to the fact that adversaries can rearrange the blocks or modify the blocks without being detected. This weakness follows on account of the encrypted message blocks being independent. On account of this, the ECB mode is not recommended for messages longer than one block.

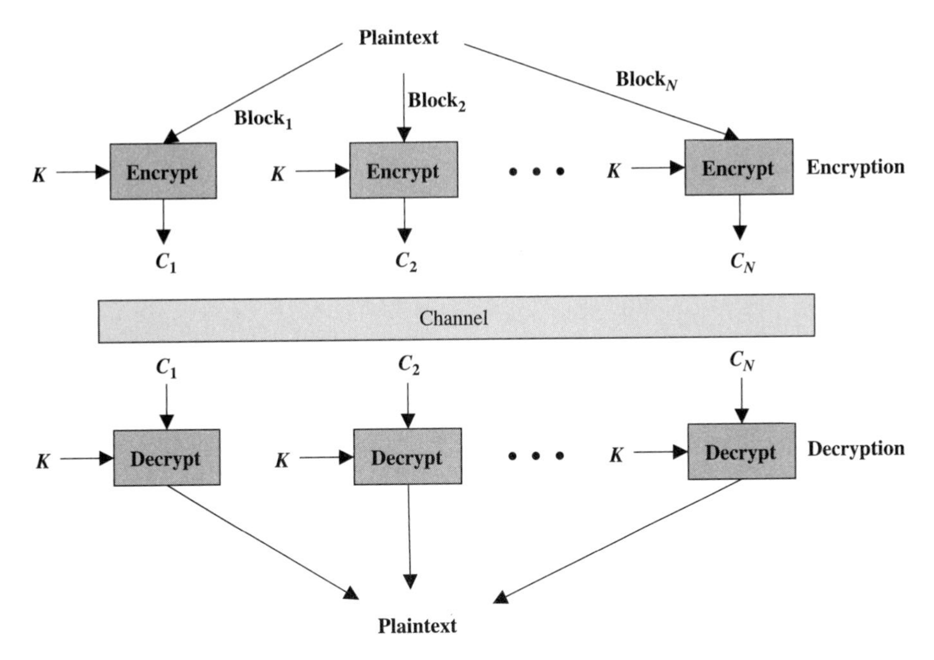

Figure 2.6. ECB mode of operation.

We next look at the CBC mode of operation. In this case, a message is broken into multiple blocks. The blocks are linked together during the encryption operation as shown in Figure 2.7. An initialization vector (IV) is used to start the process, as shown. The ciphertext resulting from every round is then chained (by an XOR operation) with the next block of plaintext. Following this the result is subjected to the encryption operation. The entire process is repeated in reverse at the receiver as shown in the figure. The efficiency of CBC mode is similar to that of the ECB mode since the cost of the XOR operation is negligible. This mode is typically used for bulk data encryption and for authentication.

The CBC mode does not suffer from the flaws that the ECB mode of operation suffers from as discussed earlier. Thus, repeated blocks in plaintext do not result in repeated ciphertext. Further, attackers also cannot move the ciphertext blocks around without being detected, as is the case for the ECB mode. Note also that the different values of IV prevent two identical plaintext messages from winding up with the same ciphertext. Of course, if the same IV value and the same key are used then identical plaintext messages will have the same ciphertext. In this mode of operation each ciphertext block depends on all the previous message blocks. Therefore, a change in the message will affect all ciphertext blocks after the change as well as the original block. Thus, this mode is not suitable for applications that need random read/write access to encrypted data since in this case both the encryption and decryption have to be done sequentially. Further, error in the jth ciphertext block affects decryption of both the jth and the $(j+1)$st blocks. The effect on the $(j+1)$st block is predictable while the error will garble the decryption of the jth block. The decryption of $(j+2)$nd and later blocks is not affected. In the CBC mode of operation it is important to protect the IV value since if IV is sent in the clear or if an attacker obtains the IV through some other means, the attacker can make predictable changes in the first plaintext block recovered. Therefore, either IV must be a fixed value or it must be sent encrypted in ECB mode before the rest of the message.

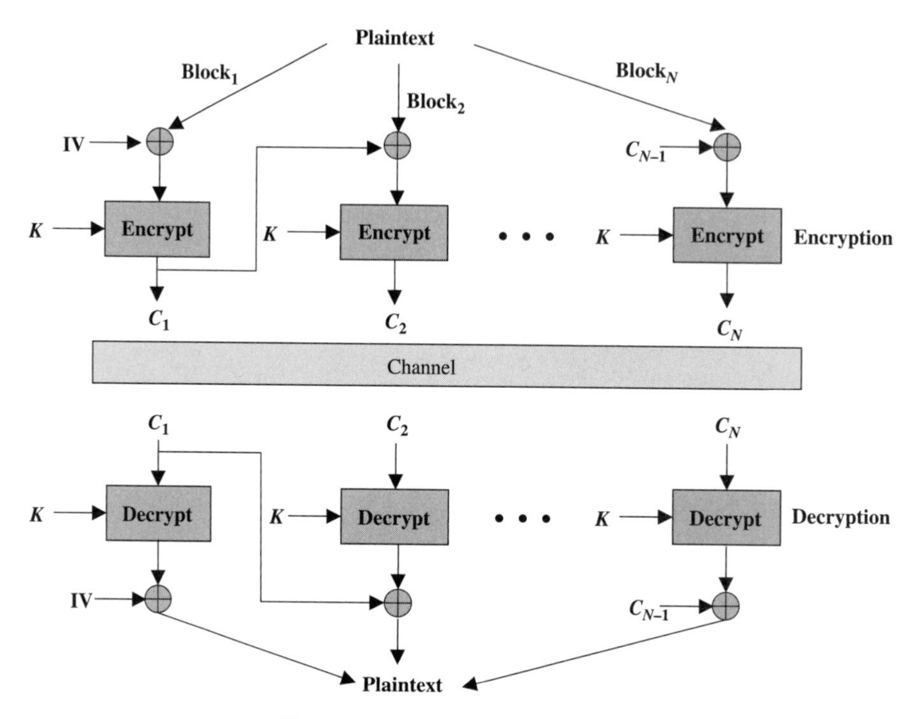

Figure 2.7. CBC mode of operation.

The K-bit CFB mode of operation is shown in Figure 2.8. The CFB mode treats the message as a stream of s bits. The plaintext message is chained (using XOR) to the output of the block cipher to obtain the ciphertext. The ciphertext is also used as a feedback for the next stage, as shown in the figure. The specification allows any number of bits (1, 8, or 64) to be used in the feedback. The CFB mode is typically used for stream data encryption and authentication.

CFB uses block cipher in either the encryption or the decryption mode at both ends. CFB mode also exhibits chaining dependencies. Therefore, reordering ciphertext blocks affects decryption. Proper decryption requires the preceding ceil(n/r) ciphertext blocks to be correct. This mode is also self-synchronizing but needs ceil(n/r) to do so. Errors in the jth ciphertext block affects the next ceil(n/r) blocks. The effect of such errors on the decryption of the jth ciphertext block is predictable. This property, however, could be used by an adversary to their advantage.

We next consider the K-bit OFB mode of operation. This mode can also operate as a stream cipher. In this case the keystream is generated independently of the plaintext and ciphertext messages, as shown in Figure 2.9. The output of the encryption operation is not only chained to the plaintext message (via an XOR operation), but also is used as a feedback for the next round. This mode is typically used to perform stream encryption.

Since the information which is used as the feedback is independent of the message, it can be computed in advance. Thus, the operations such as encryption or decryption can be done very fast if enough memory resources exist to store the keystream in advance. The OFB mode ensures that, if ciphertext is garbled, then only the corresponding plaintext is garbled. Further, the recovered plaintext will have predictable changes. This mode of operation, however, cannot self-synchronize after loss of ciphertext bits. In this case

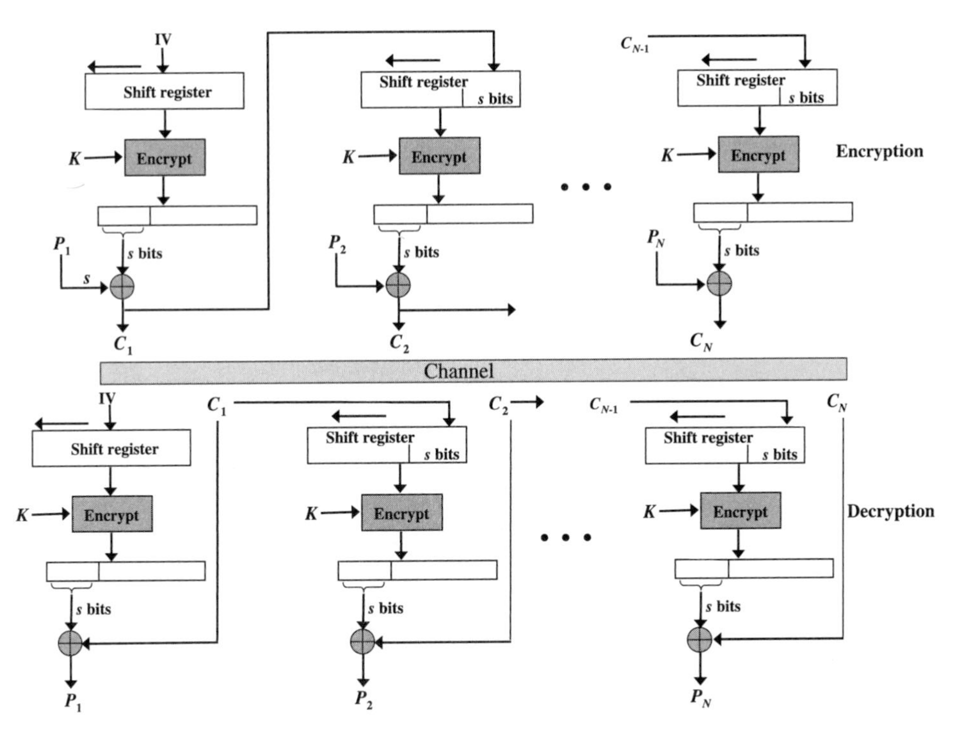

Figure 2.8. CFB mode of operation.

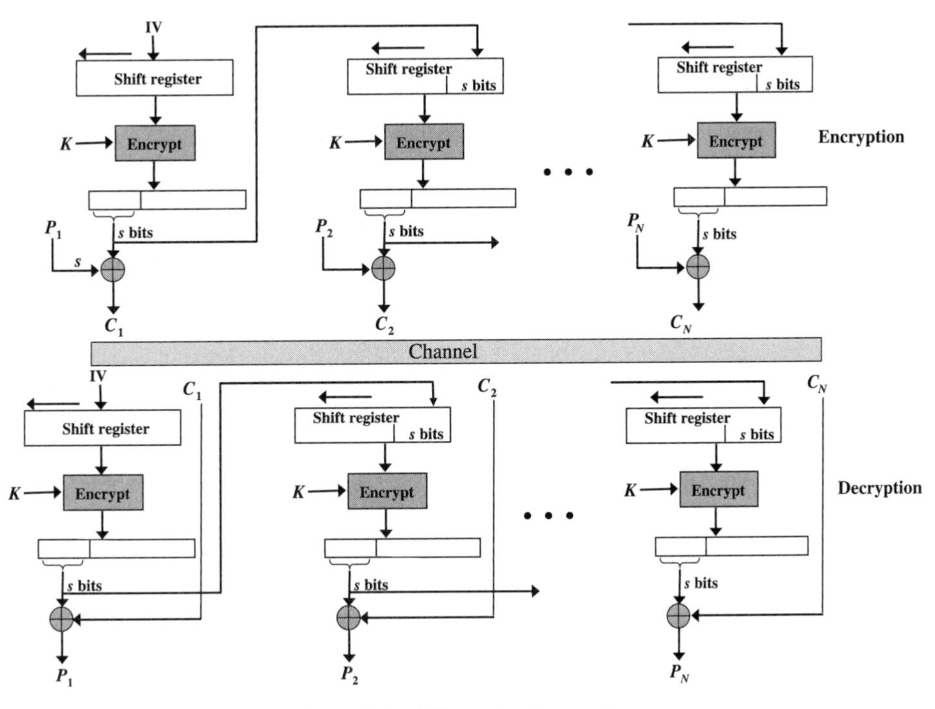

Figure 2.9. OFB mode of operation.

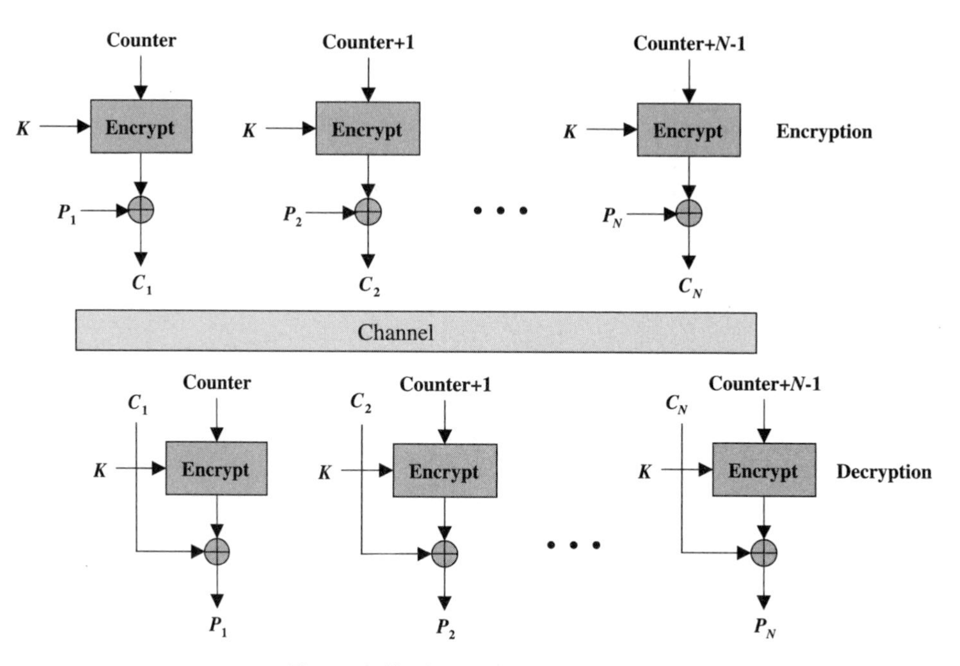

Figure 2.10. CTR mode of operation.

explicit self-synchronization is required. This implies that sender and receiver must be synchronized during transmission, which may increase the overhead associated with this scheme.

We next consider the CTR mode of operation, shown in Figure 2.10. This mode encrypts a counter value and the output of the encryption is then chained to the plaintext. Further, the counter value is changed for every plaintext block. Thus, the counter value is never reused. Typically, this mode is used to ensure high-speed network encryptions. This is because the key stream can be generated in advance and encryptions or decryptions done in parallel. In addition, this mode also allows random access to encrypted data blocks. Thus, decryption can start at any point rather than being forced to start at the beginning (assuming the value of the counter is known for the specific block of interest). In terms of error recovery, this mode has the property that, if a ciphertext block is garbled, then only the corresponding plaintext is garbled. When using this mode, it is important to ensure that the counter values are not reused assuming that the key does not change. Otherwise, the attacker can obtain extra information by using the XOR of two ciphertext blocks. By doing so, the attacker will obtain the value corresponding to the XOR of the matching plaintext blocks.

2.4 MISCELLANEOUS PROPERTIES

In addition to the fundamental properties discussed earlier, there are several other properties that are made use of in the various security mechanisms discussed throughout the book. In this section, we explain these properties. We start off by looking at hash chains. More specifically, we consider the pre-image resistance property of hashing algorithms and describe this again briefly. We do this because this property is used in several security mechanisms.

2.4.1 One-Way Property of Hash Chains

One application of the pre-image property of hash algorithms is in the context of one time passwords. Note, however, that this property is also used in other schemes such as TESLA, which we discuss later in this chapter. The one-time password problem consists of a party making use of passwords in order to prove his or her identity to the other party. Further, every password is used only once. We next explain this property.

Consider two parties, Alice and Bob. Alice is considered to be the claimant and is expected to verify her identity to the verifier Bob. We next explain how this can be done using the one-way hash function H. Initially Alice chooses a secret w as well as a constant t. The constant t indicates the number of times verification can be done. Then Alice hashes the secret wt times using the hash function H to get the value denoted as $H^t(w)$. Then Alice transfers $H^t(w)$ securely to Bob. This secure transfer can be accomplished through the transmission of a digital signature by Alice on the hash value transferred. This is the initial shared secret between Alice and Bob. Now during the operation of the protocol, when Alice has to identify herself to Bob for the ith time (let us say during the initiation of the ith session), Alice computes $H^{t-i}(w)$ and transmits the value to Bob. Bob checks that that the value is correct by calculating $H^i[H^{t-i}(w)]$ and verifying whether the outcome equals the initial value $H^t(w)$ transferred securely by Alice. If the verification holds, Bob accepts the claim of Alice and otherwise not. Note that this scheme requires that both Alice and Bob be synchronized in terms of the number of identification sessions completed. Further, this scheme also limits the number of identification sessions to t. Once this number of sessions expires, the whole process is repeated again. A scheme based on this idea, called Lamport's one-way function, based on one-time passwords, has been proposed [7]. In addition, this concept of hash chains also arises in several places when we look at the various functions in ad hoc networks.

2.4.2 TESLA

We next consider another primitive that is widely used, especially in the context of broadcast authentication. Broadcasting and multicasting are used widely in a variety of applications in wireless ad hoc networks. Example applications include IP multicast, situational awareness applications in the battlefield environment, and emergency response in ad hoc networks. Such broadcast/multicast applications also need security. However, security solutions designed for point-to-point communication are not often applicable to broadcast/multicast communication. This is particularly true regarding authentication.

To understand this, consider point-to-point communication, also referred to as unicast communication. Message authentication in such a case can be achieved through the use of symmetric, asymmetric schemes or message digests, as discussed earlier. A typical authentication approach using asymmetric cryptography is based on the use of digital signatures. As discussed earlier, though, the digital signature approach is typically expensive, particularly in terms of the computational overhead. The generation, verification, and communication costs associated with digital signatures are very high. This might make this approach impractical in resource-constrained networks such as ad hoc networks.

The other alternative in such a case is to make use of symmetric key concepts. The use of message digests can provide individual authentication in point-to-point communication requiring limited overhead. One such approach is based on the use of HMAC. In this case, the two parties share a secret which is then used in the HMAC to verify the authenticity of the message. Thus, the receiver can verify the originator of the data. This approach can be extended to the broadcast domain, where a single sender shares a MAC key with several receivers. The sender then calculates the HMAC of every message before sending it and attaches to the message. Each of the receivers of the message can check the MAC and verify the authenticity of the message.

Unfortunately there is a problem with applying this approach directly to the broadcast/multicast domain. Any receiver in the group can also use the shared key to provide an HMAC value on a message. The node can transmit this message claiming that the message was sent from another node. Therefore, a receiver does not have a guarantee that the source has indeed created the message. This implies that symmetric schemes cannot be used for providing individual authentication. This is because users share the key and therefore it is not possible to verify which one of the users sharing the secret actually created the message. Of course, an alternative here is to have the source share pairwise keys with every receiver and use these pairwise keys to authenticate the message. However, such an approach is highly inefficient.

Therefore, more efficient schemes to ensure broadcast authentication are needed. One such scheme is TESLA (timed efficient stream loss-tolerant authentication), [10], which makes use of only symmetric cryptographic primitives. TESLA requires loose time synchronization among the entities participating in the network. Further it achieves its objective by making use of delayed key disclosure as we explain next.

To understand TESLA, consider Figure 2.11. We consider a single sender and multiple receivers. The sender in the group determines a secret key K, which is then hashed to get the hashed value $H(K)$. The sender discloses the hashed value authentically (using a digital signature for example) to all the receivers. The sender then transmits all packets P such that the transmitted packet contains the HMAC of P in addition to P. The HMAC calculation is done using the key K. The receivers on receiving these packets cannot verify the HMAC since they do not have the key K, so receivers will have to buffer all the packets.

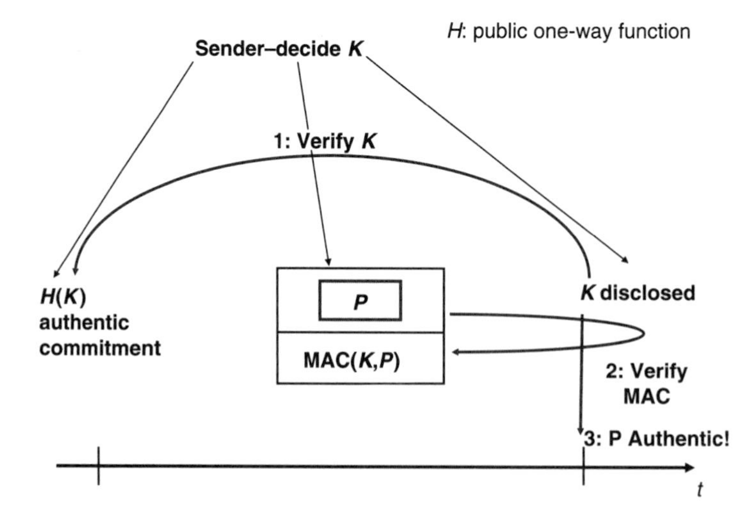

Figure 2.11. Basic operation of TESLA.

After a certain duration, the sender discloses the key K to all the receivers. At this point every receiver can check if the disclosed key K is indeed valid. This can be done by checking if the hash of K corresponds to the hashed value sent authentically by the sender earlier. If this is true, then every receiver can verify the authenticity of every packet. Note that the receiver must not accept any packet the HMAC of which has been calculated using key K after the disclosure of the key K. This is achieved by the sender disclosing the timeout schedule to the receivers and by having loose time synchronization between the receivers and the sender.

The approach given above can be easily extended (and is in fact more efficient) when considering multiple time intervals. In this case, each sender splits time into multiple intervals. It then chooses a secret key and generates a one-way key chain through repeated use of the one-way hash function property. The last value of the hash chain is then transferred to the receivers authentically. Each generated key that is one of the values of the hash chain will be used in one time interval. The keys are used in the reverse order of generation. Thus, the message authentication keys used for packets in the previous interval are derived (via hashing) from the message authentication keys used in the current interval. During each interval the sender calculates the HMAC of each packet using the key corresponding to that interval. The transmitted packet contains the original contents of the packet, the calculated HMAC value over the original packet and the most recent one-way chain value that can be disclosed. Thus, the sender discloses the keys used after the time interval of their use.

The receivers must know the key disclosure schedule, so when the receiver receives a packet, it checks that the key used to compute the HMAC on the received packet is still secret. It can do so using its knowledge of the key disclosure schedule since the receivers and the sender are assumed to be loosely synchronized in time. As long as this key is still secret, the receiver buffers the packet. If the key is not secret then the receiver has to drop the packets. When the key is disclosed, the receiver checks the correctness of the key (using the one way property of hash functions) and then authenticates the buffered packets. This operation of TESLA is illustrated in Figure 2.12.

Thus, TESLA achieves asymmetry needed for broadcast authentication through clock synchronization and a delayed key disclosure. Note also that TESLA is robust to packet loss. A drawback, though, is the need for receiver to buffer packets transmitted during an

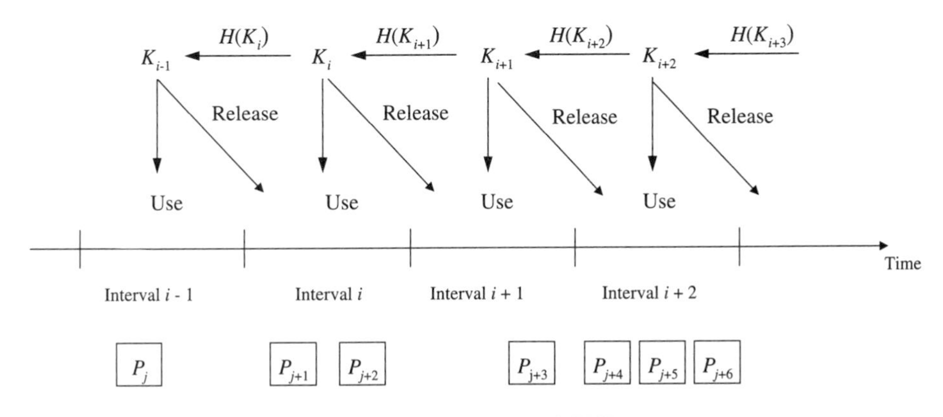

Figure 2.12. Operation of TESLA.

interval. In addition, the time synchronization (loose though it may be) might also be problematic. The sender also needs to compute the entire chain before communication starts.

In addition to the above shortcomings, TESLA has a few additional inadequacies when used in sensor networks [11]. Firstly, TESLA authenticates the initial packet with a digital signature, which is impractical in sensor nodes. Secondly TESLA discloses a key in each packet, which is inefficient given the severe bandwidth constraints in sensor networks. This leads to an overhead of approximately 24 bytes per packet. This is significant considering that sensor nodes send very small messages that are around 30 bytes long. Thirdly, it is expensive to store a one-way chain in a sensor node.

To address these limitations of TESLA, μTesla has been proposed in [11] for sensor network deployments. In this case, a sink or basestation is assumed to be present in the network. The base station is expected to execute many of the functions needed in TESLA. μTesla requires loose time synchronization just like TESLA, but the synchronization is only between the base station and sensor nodes. Packets are also broadcast by the base station. The initial key is transmitted authentically by the base station to the various sensor nodes. This could be done by having the base station share keys with every node in the network. The shared key between the base station and a sensor node is proposed to be used for authentic transmission of the initial key. Additionally, it is not necessary for the key to be included in every packet. Finally, if a sensor node needs to broadcast data, then it either sends the data to the base station, which broadcasts it after setting up the keys in the manner of TESLA or the sensor node gets the authentication keys from the base station. Thus, a sensor node does not have to either store the one-way chain or perform the hashing operations.

2.5 SUMMARY

In this chapter we have described the basic concepts of cryptography. These concepts are very important for providing security services in a wireless ad hoc networking environment and will be used throughout the book. Note, however, that we have skipped several other important concepts such as secret sharing, threshold cryptography, and identity-based keying, since we do not need these for the schemes described in this book.

Cryptographic techniques can be used for providing encryption, authentication, and integrity in such an environment. A key problem with utilizing cryptographic techniques in such an environment is the management of cryptographic keys, as remarked earlier. Network dynamics such as the degree of change in the network and the lack of a central infrastructure make this problem more difficult. We focus on key management techniques in the next chapter.

3 Key Management

3.1 INTRODUCTION

Nodes in a network need to be able to communicate securely with each other. The existence of secure communication channels is especially crucial in ad hoc networks on account of the use of wireless links and other characteristics of such networks. These channels are required for many operations such as exchanging data or exchanging control packets in the case of functions like routing. To make such secure communication possible, it is necessary for nodes to have access to the proper keying material. This is the objective of the key management process. This has lately been a very active area of research in ad hoc networks.

The importance of key management cannot be overemphasized for both traditional and ad hoc networks. When employing cryptographic schemes, such as encryption or digital signatures, to protect both control and data traffic, a key management service is always required. For secure communication between any two entities, both the entities should possess a secret value or key. The possible ways in which secure communication can be established are for the entities concerned to share a key (symmetric-key system) or for the entities concerned to possess different keys (asymmetric-key system). Key management is the process by which those keys are distributed to nodes on the network and how they are further updated if required, erased, and so on. There are several steps that key management has to be concerned with for both symmetric key systems as well as asymmetric key systems. These include [7]:

1 Initializing the system users.
2 Creating, distributing and installing the keying material.
3 Organizing the use of the keying material.
4 Updating, revoking, and destroying the keying material.
5 Archiving the keying material.

The purpose of the first step is to bootstrap the system. This could include various noncryptographic operations such as verifying the information of the users, providing identities to the users of the system, ensuring that they have the proper software needed to participate in the key management process, and so on. This is then followed by creation and distribution of the keying material. The creation and distribution of the keying material can be done in a centralized fashion or can be done in a decentralized fashion. The keying material which has been distributed is then installed on the various nodes. This is followed

Security for Wireless Ad Hoc Networks, by Farooq Anjum and Petros Mouchtaris

by the third stage where the keying material is used to protect the communication between the various nodes. This is done by using the keys to encrypt the data and control traffic exchanged between the various nodes in the network. The fourth step is crucial in addressing various threats that could result in compromise of keys. Key compromise could lead to lack of confidentiality or authenticity as well as the unauthorized use of keys. In such instances the key management processes have to ensure that the compromised keys are cancelled (revoked). In addition it might sometimes be necessary to replace the compromised keys. Of course, replacement of compromised keys is not necessary for the nodes under the control of the adversary. Finally, the fifth step might be needed in cases where the keying material has to be saved. This is more relevant in situations where such keying material might be needed for auditing purposes, such as in the case of legal proceedings.

As explained in Chapter 2, there are two main categories of cryptographic systems, namely symmetric cryptographic systems and asymmetric cryptographic systems. The key management process involves different techniques for these two types of cryptographic systems. Kerberos (see Section 3.2) is a widely used key management system in traditional networks for symmetric key systems. Similarly the purpose of the public key infrastructure (PKI) (see Section 3.2) is to ensure that different steps of the key management process are executed when considering asymmetric cryptographic systems.

Key management in ad hoc networks, however, is more difficult than in traditional networks. This is because of several factors, such as the vagaries of wireless links, lack of a central authority, constraints on resources such as power, memory, and bandwidth availability, and the inability to predetermine the neighbors of a node after deployment, which is further worsened on account of the mobility of nodes in such networks.

In this chapter we consider the key management problem in ad hoc networks with a focus on sensor networks and explain several proposed approaches to addressing this problem. We also point out the advantages and disadvantages of the different schemes. The plan of the chapter is as follows. We start by summarizing briefly the approaches to key management in traditional networks in Section 3.2. Following that, we consider the proposed approaches for solving this problem in ad hoc networks in Section 3.3.

3.2 TRADITIONAL SOLUTION

The typical approach to the key management problem in wireline networks is based on the use of trusted third parties. A trusted third party (TTP) is an entity trusted by all users of the system. Three different types of TTPs can be categorized depending on the nature of their involvement. These are the in-line TTP, on-line TTP, and off-line TTP, as shown in Figure 3.1.

An in-line TTP is an active participant in the communication path between the two users, as shown in the figure. Such an entity is not only involved in deciding the keying material, but is also involved during the data transfer. An on-line TTP on the other hand is also an active participant, but only for the purpose of deciding the keying material. Thus, once the keying material is decided, the on-line TTP no longer participates in the data communication between the users. The off-line TTP communicates with the users prior to the setting up of the communication link. During such a time, the keying material would be distributed to the various entities present in the system. As a result, during the actual communication, the off-line TTP need not be present in the system at all.

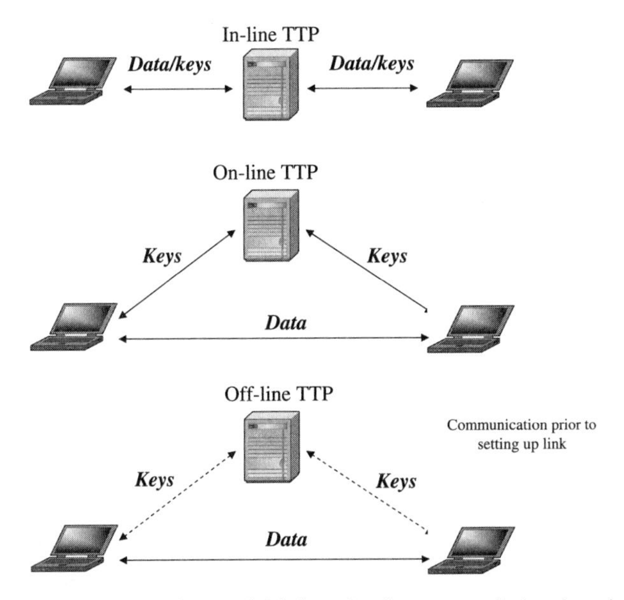

Figure 3.1. Categories of trusted third parties for symmetric key based systems.

Examples of TTPs include key distribution centers (KDC) and certificate authorities (CA). A KDC, which is more like an on-line TTP, is typically used in case of systems depending on symmetric key cryptography, while a CA, which is more like an off-line TTP, is typically associated with systems depending on asymmetric key cryptography.

A TTP simplifies the key management process. In the case of symmetric key-based systems, every user in the system is assumed to share a key with the TTP. In the case of asymmetric key-based systems, every user in the system might have access to the public key of the TTP. Users do not need to share secret keys with each other or know the public keys of each other. The keys of other users can be obtained when necessary with the help of the TTP. This results in a scalable solution since the number of symmetric keys in a system with n users is reduced from $n(n-1)/2$ (when each user has a shared secret key with every other user in the system) to n (when making use of the TTP). In addition, revocation and updating of keys is also straightforward when using a TTP, since the TTP is akin to a central authority that can easily carry out these actions. In addition, the smaller number of keys is another alleviating factor.

The detailed operation of an on-line TTP is shown in Figure 3.2 in the context of symmetric key-based systems. In this case a user denoted A in the figure desires to set up a secure communication channel with user B. Then in the first step (indicated as message 1 in the figure), the user requests the TTP for help in meeting this objective. The communication of user A with the TTP is assumed to be encrypted using the key shared between user A and the TTP. On receipt of such a request the TTP performs checks to verify the authenticity of the incoming request. If the checks are satisfied, then the TTP generates the session key, that is, the key to use for the session between user A and user B. The TTP then encrypts the session key using the key the TTP shares with user B, and includes this token in the message sent to user A (indicated as message 2 in the figure). In addition, this message also includes the session key outside the token. This entire message (message 2 in the figure) is encrypted using the key shared between user A

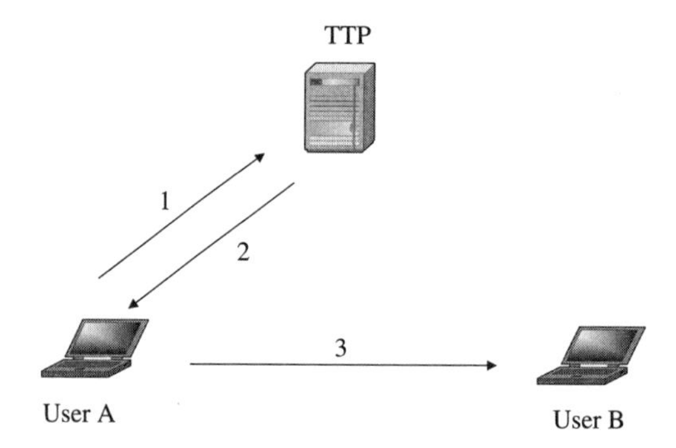

Figure 3.2. Principles of operation of an on-line TTP such as Kerberos.

and the TTP. Following this, user A decrypts the incoming message, performs checks to verify that it is communicating with the authentic TTP, and sends the token received to B in a separate message, shown in figure as message 3. On the receipt of this separate message, user B can decrypt it using the key that it shares with the TTP and thereby recover the session key. Following this, secure communication between users A and B is possible using the session key. We would like to remark here that, if the TTP has information that a particular user is compromised or if any of the checks fail then it can refuse to proceed with the key distribution. Similarly, either of the users A or B can also refuse to proceed with the protocol if the checks that they carry out are not satisfied. The approach described here is followed by Kerberos, the widely used symmetric key management system for wireline networks.

Key distribution when dealing with asymmetric keys can also follow a similar approach. The difference, though, is that in this case the public keys of nodes have to be distributed to other nodes while the private keys will have to be kept secret by each node. Since public keys do not need to be hidden, authenticity of the distribution channels is needed while confidentiality is not required at all. This is as opposed to the distribution of symmetric keys, where the distribution channel has to guarantee both authenticity and confidentiality.

The traditional and best known approach to solving the problem of authentic distribution of public keys has been in the form of public key certificates. A public key certificate is a statement issued by some trusted party also called the certification authority, which guarantees that the public key indeed belongs to the claimed user. The trusted party (i.e. the CA) then digitally signs this statement. In order to do this, the CA is assumed to have its own public–private key pair. Every node in the system is assumed to know the authentic public key of the CA. This could be done for example by using out-of-band techniques such as direct contact.

In order to sign certificates binding the public key of a node to the identity of the node, the CA might use a procedure that includes verifying the identity of the node and also verifying that the node has the private key corresponding to the public key. Any node that wishes to verify the public key of another node will then just have to request the latter node to provide the certificate signed by the CA. A new node entering the system can contact the CA to obtain such a certificate. A certificate can also be revoked by the CA. The revocation can be based either on timeouts present in the certificate or on explicit revocation lists. Thus we see from this approach that it is necessary that every node in the network should just be able to obtain the authentic public key of the CA, which makes the problem more manageable. Note that the CA can be considered as an off-line TTP.

Figure 3.3 shows the information in an X.509 certificate, which is a public key certificate scheme widely used today. It consists of several information items. The serial number is used to uniquely identify the certificate. The issuer name is the name of the trusted party that has issued the certificate. The validity period specifies the duration of time for which the certificate is valid. The subject is the entity to whom the certificate is issued and whose public key is being certified. The public key information field contains the public key while the key usage field provides details about the ways in which the key is to be used. This could be either for encryption or for signature. The extension field supports possible extensions. Finally, the last field contains the digital signature of the certificate authority along with any information needed to verify the signature, such as the algorithms used.

It is also clear from this description that a CA is required in order to determine the authenticity of the public key of any node, but a single CA cannot be expected to be able to verify the authenticity of the public keys of all the nodes. Hence, several proposals have been made in order to distribute the functions required of a CA. The steps that a CA needs to follow before certifying a public key are also specified in detail. All this together forms what is generally called as the public key infrastructure. We do not go into the details of this concept. A PKI for our purpose is assumed without loss of generality to consist of a group of CAs that can verify the authenticity of the public keys of the nodes in the network.

Several key management protocols based on the use of public keys have been proposed for wireline networks. Examples include IKE [12] and JFK [13]. However these solutions that have been proposed for wireline networks are not always applicable to ad hoc networks. This is due to the differences in the basic characteristics of the wireline and wireless ad hoc networks discussed earlier. Hence, it is necessary to study the solutions proposed for such networks. We look at this in the next section.

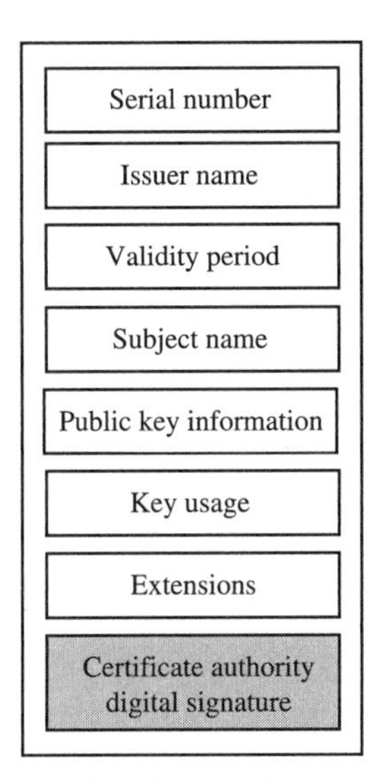

Figure 3.3. X.509 certificate format.

3.3 SOLUTIONS FOR AD HOC NETWORKS

Protocols that depend on the use of trusted third parties are not well suited to the ad hoc network scenario. The main reason is the lack of trusted infrastructure in ad hoc networks. However, some proposals based on modifications of this approach have been made. Such proposals have been made, especially for asymmetric key systems in ad hoc networks. We will explain this in detail later in Section 3.3.1.

Another approach for key management in ad hoc networks that has been proposed prominently for symmetric key systems is based on the use of prior context. In this case, it is assumed that nearly all the nodes share prior context at the same time before the network operation begins. This prior context is generally in the form of an offline secret key predistribution before network deployment. This approach has received a lot of attention, especially in the context of sensor networks [14–18]. Of course, the requirement that all nodes share a context prior to deployment might not always be practical.

A third approach which might be useful in such a situation is that based on self-organization. In this case the nodes deployed in the network do not depend on any prior shared context. There have been several proposals using this approach. A basic assumption common in many of these proposals is related to the existence of an out-of-band authenticated communication channel [1, 19].

Another approach is dependent on the use of identity-based public key systems [20–22]. In this case, the public key of a node in the system is derived from the identity of the node. We do not consider these solutions in this chapter.

We next explain the proposed solutions in detail, starting with solutions based on the use of asymmetric keys.

3.3.1 Asymmetric Key-Based Approach

The traditional approach towards developing an asymmetric key-based system is based on the use of a CA, as discussed earlier. However, such an approach is not practical in ad hoc networks for several reasons. Firstly, a CA will be a vulnerable point in the network, especially if it is not distributed. Compromise of a CA will allow an adversary to sign any certificate, thereby paving the way for impersonation of any node or for revocation of any certificate. More importantly, in order to carry out the key management operations, the CA will have to be accessible all the time. If the CA is unavailable, then the nodes in the system might be unable to update/change keys. New nodes will also not be able to obtain certificates. An approach to improving the availability would be to replicate the services of the CA, but a naive replication of the CA could lead to more problems. Compromise of any single replica could lead to collapse of the entire system. An approach to solving this problem is to distribute the trust reposed in a single CA over a set of nodes, thereby letting the nodes share the responsibility of key management. This is the approach that we explain in the next few subsections.

We start by looking at approaches proposed to partially distribute the functionality of a CA in Section 3.3.1.1. Approaches to fully distribute the functionality of the CA are explained in Section 3.3.1.2. We also discuss self-issued certificates in Section 3.3.1.3 and some other schemes in Section 3.3.1.4.

Before that, we would like to remark briefly about the applicability of public key systems for resource constrained networks. It is widely accepted that the limited resources associated with sensor nodes make it impossible to execute public-key cryptographic algorithms on them. The authors in [23], however, show that this need not be the case. In this paper the authors demonstrate the tractability of public key algorithms on sensor networks. The authors make use of elliptic curve cryptography and show that public keys operations can be executed within 34 s. While this is a tremendous improvement over earlier numbers, it is still a significant number. A potential solution is to utilize asymmetric keys for setting up symmetric keys for subsequent communications. Note though that, in a few years, the use of asymmetric techniques in sensor networks might become viable.

3.3.1.1 Partially Distributed Authority We start this section by explaining briefly the concept of threshold cryptography (TC). A TC scheme makes it possible for n parties to share the ability to perform a cryptographic operation. For example, consider the digital signature on a message. We have seen techniques whereby a single user creates the digital signature. A problem occurs, however, when this user is compromised or cannot be trusted. A better approach then is to distribute the trust placed on a single user among multiple users. This indeed is what threshold cryptography strives to achieve.

The objective of theshold cryptography is to protect information by distributing it among a set of n entities. In addition, there is a threshold t associated with the TC schemes such that any t of the n parties can execute the cryptographic operation. Such schemes are referred to as (n,t) TC schemes. In case of an (n,t) TC scheme, fewer than t parties will not be able to execute the cryptographic operation successfully. Thus, TC can be considered to be an approach for secure sharing of a secret. We see from here that, even when some number of entities (less than the threshold t) in the network is compromised, the system is not at risk. Nonavailability of certain number of nodes (at most $n - t$ nodes, to be precise) in the network will also not have an impact on the working of the system. Note that the TC schemes perform the cryptographic operation in a distributed manner.

In [24], the authors propose using a scheme based on the technique of threshold cryptography to distribute the private key of the certification authority. Knowledge of this key is distributed over a subset of the nodes in the network. The system, made up of the nodes in the network, is expected to have a public–private key pair. This key pair is created initially by a trusted authority before deployment of the nodes. Following that, the private key is divided into n shares using an $(n, t + 1)$ threshold cryptography scheme. These n shares are then allocated to n arbitrarily chosen nodes by the authority that created the public–private key pair. These chosen nodes are called servers. Following this distribution of the shares of the private key to the servers, the central authority is no longer needed. Thus, the central authority is only needed during the bootstrapping phase. Each server also has its own key pair and stores the public keys of all the nodes in the network. In particular, each server (chosen node) knows the public keys of other servers. As a result, the servers can establish secure links among themselves. We show the initial configuration of such a service in Figure 3.4. The service as a whole has a public–private key pair K–k. The public key K is known to all nodes while the private k is divided into shares $s_1, \ldots, s_n$, with each server having one share. Each server also has a public–private key pair K_i–k_i.

Whenever a certificate has to be signed using the private key of the system, the servers are contacted. Each server generates a partial signature for the certificate using the share of the private key that the server has. The partial signature is then submitted to a combiner

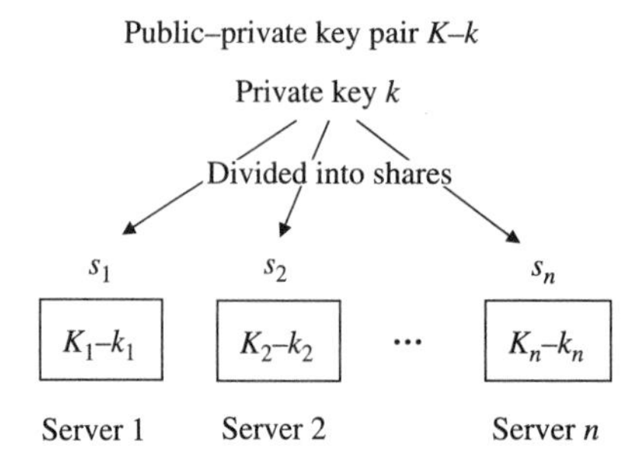

Figure 3.4. Configuration of partially distributed authority.

that computes the overall signature from the partial signatures. Note that the combiner will not be able to create the overall signature without the partial signatures.

Since this scheme is based on the concepts of threshold cryptography, the system can tolerate a certain number of compromised servers. Thus, t or fewer than t compromised servers will not be able to derive the private key of the system. This is because compromised servers (assuming there are at most t of them) cannot generate correctly signed certificates by themselves, since they can generate at most t partial signatures. Further, not all n servers are needed to generate the overall signature, but any $t+1$ servers will suffice. Figure 3.5 shows how servers generate a signature using a (3, 2) threshold signature scheme. Each server generates a partial signature PS(m,s_i) for message m using its share of the key. The combiner c can generate the signature $<m>_k$ even though server 2 does not provide a partial signature.

In addition to the above, a compromised server could generate an incorrect partial signature. Use of this partial signature would yield an invalid signature. Therefore, a combiner might have to verify the validity of a computed signature using the service public key. In case verification fails, the combiner tries another set of $t+1$ partial signatures. This process continues until either the combiner constructs the correct signature from $t+1$ correct partial signatures or the combiner reports failure. More efficient ways to achieve this can also be designed.

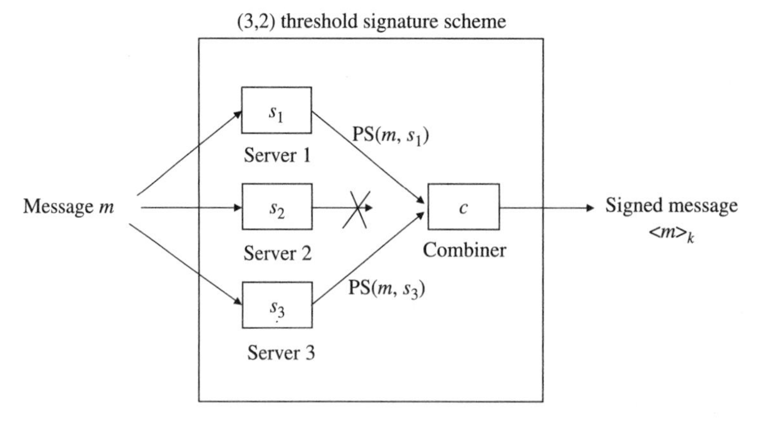

Figure 3.5. Example of signature generation with threshold cryptography.

The authors also propose to refresh the keys proactively in order to tolerate mobile adversaries, because the adversary will have to capture multiple shares during the same interval. Thus, if the old keys under the control of the adversary are refreshed, then they will be of no use with the new keys that the adversary might compromise. Additionally, the system can also be designed to adapt its configuration to the changes in the network. For example, a key management service might start with a (7, 3) configuration but then modify this later to a (4, 2) combination since some servers were detected to be compromised while other servers were unavailable.

3.3.1.2 Fully Distributed Authority The studies [25] and [26] describe an approach for distributing the functions of a certification authority. The difference compared to schemes in Section 3.3.1.1, however, is that in this case any node instead of a chosen few can contain a share of the private key of the service. A new node that does not have a certificate will have to contact at least $t + 1$ servers. These servers can issue a certificate to such a node after establishing the identity of the node. Any $t + 1$ nodes can also renew a certificate. In addition, a node that does not possess a share can obtain one by contacting any group of at least $t + 1$ nodes that already possess the share. The bootstrapping of the system, however, needs a trusted authority that provides the initial shares to the first $t + 1$ nodes.

3.3.1.3 Self-Issued Certificates We next describe an approach where the security of the system does not rely on any trusted authority. This is the self-organized approach that allows users to create, store, distribute, and revoke their own public keys without the help of any trusted authority. Such a self-organized public key management approach for ad hoc networks has been considered in [27, 28].

Each user decides their own public–private key pair. Other users issue certificates to each other based on other factors such as personal acquaintances. Each user then maintains in their personal repository the list of valid certificates. When a user A needs to authenticate a certificate belonging to user B, then user A checks if a certificate has been issued by itself to B earlier. If not, user A checks if the repository contains a certificate to any other user C who in turn has B's certificate. Thus, a chain of valid public key certificates is obtained by user A such that the first certificate of the chain is one issued by A. Further, each remaining certificate in the chain can be verified using the public key contained in the previous certificate of the chain. The last certificate of the chain should then contain the public key of user B. Thus, certificate chaining is used in order to authenticate public keys of users. If A cannot form a certificate chain to B, then A cannot authenticate B's public key. In addition to this, another problem is related to bootstrapping the certificate repository of a node without depending on a separate authentication channel.

3.3.1.4 Other Schemes Other approaches for solving the key management problem in ad hoc network have also been proposed. In [29] and [30] it is proposed to derive the IP address of the node from the public key of the node. This is done by hashing the public key and then using part of the hash value as the part of the IP address of the node. As a result, it is easy to verify if a public key corresponds to a node with a given IP address, thereby addressing the key authentication problem associated with public keys. A problem, however, arises when the private key corresponding to a public key is compromised, thereby making the public key unusable. As a result a new public–private key pair will have to be created, but this will also have to result in changing the IP address of the node. This can cause problems in some applications. In addition, this approach does not lend itself easily to those cases where node addresses are not numbers but some meaningful strings.

In [31] the authors propose to make use of geographical nearness of the parties involved to solve the key management problem partially. The authors assume that the various nodes execute the DH protocol (see Chapter 2). As discussed there, a problem with the DH protocol is the absence of authentication during the initial key agreement. As a result, an active adversary can launch a man-in-the-middle attack on a system based on the DH protocol. The authors propose to address this aspect of the DH protocol using physical presence.

The basic idea is simple. The authors propose that users set up keying material or more precisely execute the initial key agreement aspect of the DH protocol when they are in the same location, since users are assumed to recognize each other. The authors devise mechanisms to ensure the integrity of the parameters (derived from the random secret that each user selects) of the DH key agreement during this initial key agreement. The shared key resulting from a successful key agreement can then be used even when the users are not in close proximity to protect communications between the two users. Note that in this case a man-in-the-middle attack on the parameters of the DH protocol is assumed to be infeasible on account of the small geographical separation between the entities.

The authors proposed three mechanisms intended to ensure the integrity of the parameters of the DH key agreement. A basic assumption for all three mechanisms is that the users are in very close proximity. The first mechanism proposes that both the users check the validity of the exchanged DH public parameters by comparing them directly. For example, this comparison can be done by representing the DH public parameters as a string and then by looking at the screen of the party or reading aloud the values to be compared. The string can also be formed by hashing the public key parameters to ensure extra security. Thus, this solution belongs to the family of solutions requiring users to compare strings of words. [32–35] also propose a similar method for key verification in which key hashes are converted to readable words or to an appropriate graphical representation.

The second mechanism requires the users to estimate the distance between them. This could be done using a mechanism such as distance bounding (see Chapter 7). The measured distance appears on both displays, which is verified by the users. Finally, the users also visually check to see if there are other users/devices in the same area. If this is not the case, then the exchanged DH public parameters are accepted. The third mechanism makes use of integrity codes to exchange the DH public parameters. These codes ensure that it is not possible to block emitted signals without being detected. As a result any modifications to the DH public parameters are detected.

An advantage of this approach is that it can be used in situations where the users do not share any authentication information in advance. A drawback is that the system depends on the users verifing the keys or the keying material; devices in the same area might be visually undetectable due to their small size; people might be unreliable (for accidental or malicious reasons) when dealing with meaningless strings. Perrig and Song [36] address this using the concept of hash visualization to improve the security of such systems. In this case the users compare structured images instead of meaningless strings. Of course, comparing complex images might also not be easy. We would like

Pre-authentication over the location-limited channel

1. A→B: $addr_A$, $h(PK_A)$
2. B→A: $addr_B$, $h(PK_B)$

Authentication continues over the wireless channel even when the location limited channel is not available

1. A→B: PK_A...

Figure 3.6. Use of location-limited channels for authentication.

to remark here that such channels which depend on the close proximity of users are also referred to as location-limited channels.

The location-limited channel as a privileged side channel to perform out-of-band pre-authentication is also considered in [7] and [19]. In those schemes either physical contact or the infrared channel is used to establish a shared key. Stajano and Anderson [7] propose the use of the resurrecting duckling security policy model. In this case two devices that wish to establish a key do so based on physical contact. Cryptographic mechanisms are not used to protect the information exchanged. Thus, a drawback of this scheme is that it is vulnerable to passive (eavesdropping) attacks in the location-limited channels.

Baltanz *et al.* [19] also propose using a location-limited channel such as physical contact, infrared, or ultrasound to exchange pre-authentication data, but in such a manner as to address the drawbacks of [7]. Use of the pre-authentication data is proposed in such a manner that only authentication guarantees on this need to be provided. The pre-authentication data does not need any confidentiality guarantees. Authenticity of the exchanged data is guaranteed due to the use of the location-limited channel. Any standard key authentication protocol is then used over the standard radio channel along with the pre-authentication information to set up the keys. For example (see Figure 3.6), one set of pre-authentication data could consist of the identity of the node (i.e. address shown in the figure) along with a hash of the node's public key. The actual public key can then be exchanged later as part of the messages of any standard key authentication protocol. The receiving node can then easily verify the authenticity of the public key by using the hash value exchanged over the location limited channel. We show this in Figure 3.6. The authors also propose protocols to address the situation in which either or both of the devices are incapable of executing the public key operations. Group key exchange protocols are also considered.

A drawback though of all of these schemes that depend on the location-limited channel is the need for devices to be close to each other at some point in time, which might not always be possible.

3.3.2 Symmetric Key-Based Approach

In this section we look at symmetric key based key management schemes proposed for systems of ad hoc networks. Much of the work in this area has focused on sensor networks, as will much of our discussion in this section. As in the case of the asymmetric key based solutions, it is not possible to set up an infrastructure to manage keys used for encryption in the traditional style (Kerberos). This is due to constraints mentioned earlier which are especially severe in case of sensor networks.

A practical solution given these constraints is to load the keys on the nodes before the nodes deploy. Thus the nodes have some secret information on them; using this they set up secure communication infrastructure for use during the operation of the network.

Several solutions based on predeployed keying have been proposed, including approaches based on the use of a global key common to all nodes, approaches in which every node shares a unique key with one or more nodes in the network, and approaches based on each node being deployed with a random set of keys. These approaches can be loosely divided into two main categories: deterministic schemes and probabilistic schemes. Deterministic schemes have a deterministic relationship between the keys loaded on a node and the identity of the node. More precisely, the existence of a secure link between any two nodes in the network can be predicted exactly. Compromise of nodes in such networks can result in secure communication between noncompromised nodes also becoming vulnerable, but the non-compromised nodes that will be impacted as a result of such compromised nodes can generally be precisely determined. In case of probabilistic schemes, the keys loaded on a node are chosen randomly from a pool. Thus, a secure link between any two nodes in the network exists with a certain probability. Capture of nodes by an adversary in such networks will also result in the compromise of secure communication between non-compromised nodes, but the non-compromised nodes that will be impacted as a result cannot be precisely determined.

We would like to remark here that, if the set of neighbors of a node after deployment is known perfectly, then key predistribution becomes trivial. In this case, given a node, we need to generate a pairwise key that this node will share with each of its neighbors and load this on the node as well as on those nodes that will become the neighbors of this node. However, a problem is that such an assumption (of perfect knowledge of the neighbors of a node) is unrealistic since nodes in networks such as sensor networks are randomly deployed. In addition, mobility might be a factor especially in ad hoc networks.

We look at the deterministic and probabilistic key distribution schemes next.

3.3.2.1 Deterministic Schemes When considering deterministic schemes, we have several possible options. The easiest ones among these are

- use of a single key for the entire network;
- use of pairwise sharing whereby each node has a separate key for every other node in the network.

In [37] the authors adopt the first approach. They consider sensor networks consisting of tamper-resistant nodes (with the nodes being called pebbles). Hence, the keys on the nodes are not compromised even when the nodes fall into enemy hands. All nodes are initialized with a single symmetric key before deployment. This saves on storage while also minimizing the costs associated with key management. This single key is then used to derive the keys used to protect data traffic.

The key derivation happens in stages with the nodes being organized into clusters in the first stage, as shown in Figure 3.7. Leaders of every cluster, called clusterheads (shown as squares in the figure), are also selected during this stage. In the second stage the clusterheads organize into a backbone, as shown in Figure 3.8. A key manager is chosen probabilistically from these clusterheads. The key manager decides the key to be used to

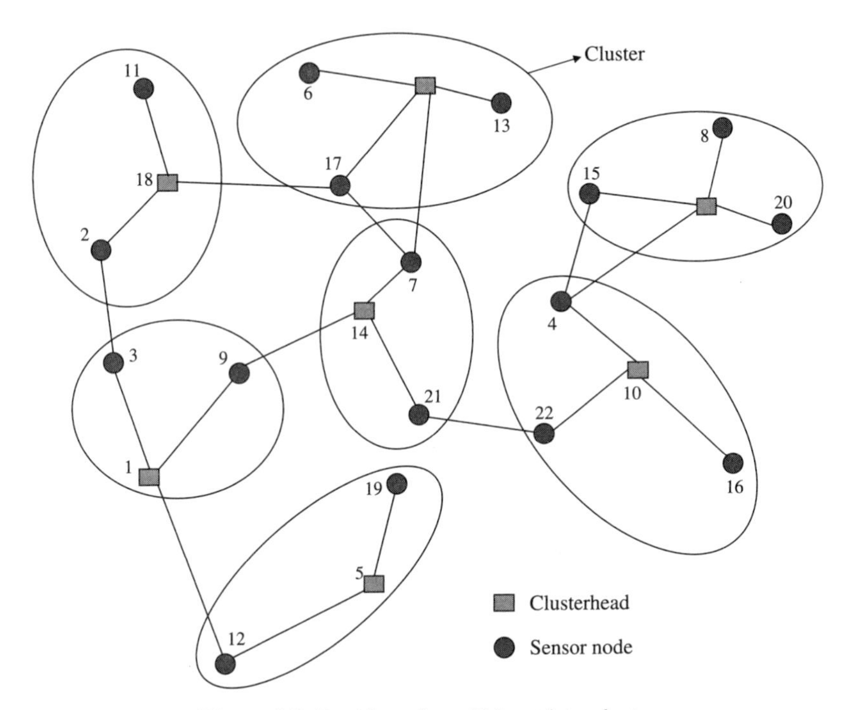

Figure 3.7. Partition of a pebblenet into clusters.

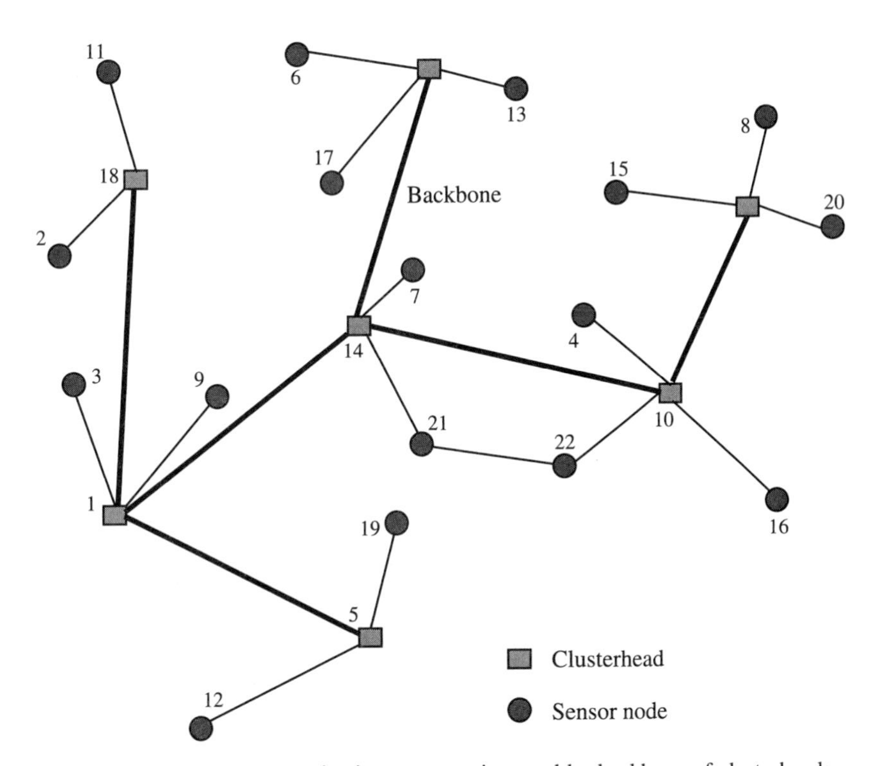

Figure 3.8. Secure communication structure imposed by backbone of clusterheads.

encrypt data in the next cycle and distributes this key in a secure fashion (based on a one-way chain derived from the single system wide key) to the nodes in the network via their clusterheads. The protocol considers factors such as the capabilities of the sensor nodes in deciding the clusterheads as well as the key manager. A key manager is selected each time the protocol is run. The fittest node becomes the key manager and the key manager changes every time increasing the security. Having a single key shared by all nodes is efficient in terms of storage requirements. Further, as no communication is required amongst nodes to establish additional keys, this also offers efficient resource usage.

This approach of using a single key in the entire network is problematic from a security standpoint since the compromise of a single sensor will break the security of the entire network communication. This will also make selective key revocation difficult. Ensuring security in such cases would involve complex revocation and rekeying mechanisms, but this is not an attractive option on account of the energy consumption associated with communication. This is in addition to the need for mechanisms authenticating the rekeying process. Absence of such mechanisms will allow an attacker to rekey or zero keys remotely. Note, though, that the compromise of a single sensor is not an issue in [37] because of the tamper-resistant nodes assumed. The tamper-resistant nodes, however, will increase the costs associated with this solution.

Pairwise secret sharing avoids this problem of compromise of communication in the entire network following the leakage of a single key. In fact, this scheme is very resilient since compromising any node does not impact the security of communication among any other noncompromised nodes. However, the scheme places great demands on the amount of storage needed on each sensor node. This makes it an impractical solution for large networks. For example, with networks of n nodes, each node will contain $(n - 1)$ keys for a total of $n(n - 1)/2$ keys in the entire network, which makes it impractical for networks with, say, 10,000 nodes. Note also that many of these keys would not be used since direct communication between nodes is possible only if the nodes are neighbors. All the nodes in the network cannot be neighbors of each other since the neighborhood sizes are limited due to the communication range and density of nodes. This gives rise to inefficient usage of memory storage while also requiring large memory sizes on the nodes. This solution also makes it difficult to add more nodes to a deployed system than intended initially since this involves rekeying all the nodes already deployed with the keys corresponding to the new nodes to be deployed. The procedure to load keys into each sensor node also adds to the costs associated with key management.

There have also been other deterministic schemes for key management in sensor networks. In [11] the authors propose an approach where every node in the network shares a pairwise key with a single special node (called the sink) in the network. The authors consider the scenario where the communication is typically between the nodes and the sink (also called the base station). The authors mainly provide two security building blocks. The first is a secure network encryption protocol (SNEP) in order to provide data confidentiality, data authentication, and data freshness guarantees in sensor networks. The second is μTesla, a protocol for providing authenticated streaming broadcast, which has been explained in Chapter 2. The secure communication facilitated by SNEP is possible by having each node share a separate secret key with the basestation. This secret key is loaded in each node at the time of deployment. All other keys needed during the operation of the network are derived from this key. Thus, this approach can be considered as a special case of the pairwise key approach where a node shares a pairwise key with only one another special node (which is the base station).

SNEP has several advantages such as low communication overhead, semantic security, data authentication, and replay protection. Compromise of sensors is also partially addressed since a compromised node can only break the secure link between the node and the base station. The authors also address other communication patterns such as between nodes or node broadcast. In this case the base station is expected to act as the trusted entity. Therefore, the base station distributes secret keys to the parties that need to communicate. This is possible since each of the parties share a secret key with the base station.

Thus, this approach relies on the existence of a resource rich key distribution center (the base station) to act as a trusted arbiter for key establishment. A problem, though, is the need for every node in the network to be able to communicate with this special node. This might result in an unbalanced load on the nodes closest to the special node thereby resulting in decreased network lifetime. This communication pattern also makes it easy for an adversary to perform traffic analysis. In addition, the special node is also a single point of failure. Compromising the base station can render the entire network insecure. Some of these drawbacks can be addressed partially by having multiple such special nodes in the entire network.

Another deterministic technique called LEAP (localized encryption and authentication protocol) has been described in [38]. This is a key management protocol for sensor networks that is designed to support in-network processing,[3] while at the same time providing security properties similar to those provided by pairwise key sharing schemes.

LEAP provides support for four different types of keys, namely an individual key, a pairwise key, a cluster key, and a group key. Each node maintains these four types of keys. The need for different types of keys is due to the observation made by the authors that the security requirements of different messages are different. Hence a single key would not be sufficient to meet these requirements. The individual key is used by a node to communicate with the base station, the pairwise key is used to communicate with a neighboring node, the cluster key is used to communicate with many neighboring nodes, and the group key is used for secure communication with all nodes in the network.

The individual key on every node is generated and preloaded on the node by the central controller prior to deployment. In addition, each node is also preloaded with a different key common to all the nodes. This key is used by the node to derive pairwise keys with its neighbors. After deployment each sensor node can use this common key to derive a different key with each of its neighbors. Following this, each sensor node is expected to delete this common key after a timeout period. An adversary is assumed to be unable to compromise any node before the expiration of this timeout period. Deletion of this common key prevents the adversary from being able to access it by compromising any sensor node after the initial secure time interval. Thus, each link is secured by a different key. Hence compromise of a node will only affect the links that the compromised node is involved in.

Following this every node creates a cluster key and distributes this key to its immediate neighbors using pairwise keys that it shares with each of the neighbors. The cluster key can

[3]Such processing would be needed when resorting to data aggregation in sensor networks. Passive participation by nodes, where a node does not report on an observation since a similar observation is reported by some other node, also requires network processing. Such in-network processing optimizes on the resources available in the network, which is a very important factor, especially in sensor networks.

be used to support in-network processing. The group key is also proposed to be preloaded on every node in the network. The rekeying of group keys makes use of cluster keys and rekeying of cluster keys is done using the pairwise shared keys.

LEAP also includes an efficient protocol for inter-node traffic authentication based on the use of one-way key chains. While LEAP offers deterministic security, the bootstrapping phase is quite expensive. Further the storage requirements on each node are also significant since every node must set up and store a number of pairwise and cluster keys. However, this number is proportional to the number of actual neighbors that the node has. Another problem with this approach is that it requires that all the nodes in the network be deployed at the same time. This approach cannot be used in scenarios where the sensor nodes are deployed at different points in time since the nodes that have been deployed earlier have destroyed the common key and hence will not be able to communicate with the nodes deployed later. Finally, note that this scheme is robust against outsider attacks but susceptible to insider attacks in which an adversary only needs to compromise a single node to inject false data.

3.3.2.2 ***Probabilistic*** We now discuss several probabilistic schemes. A probabilistic key sharing approach was first proposed in [17]. In that scheme each node is loaded with one or more keys before deployment. These keys are randomly chosen from a pool of keys. After deployment a secure link can be established between a pair of neighboring nodes provided a key happens to be common to both these nodes. We explain this scheme in detail next.

The proposed probabilistic scheme, also called the random key predistribution scheme, consists of three phases namely:

- key predistribution;
- shared key-discovery; and
- path-key establishment.

The *key predistribution phase* is executed by a TTP, also called a trusted authority (TA), before deploying the nodes. This phase consists of generating a large pool of P keys along with their key identifiers. Following this, for each node, the TA randomly draws k ($k \leq P$) distinct keys from this large pool of P keys and installs these k keys on each node along with the key identifiers. These k preloaded keys on each node are collectively referred to as the node's key ring. The size of the key pool is chosen such that any two random subsets of size k (each of this subset could represent a key ring on some node) will share at least one common key with some probability. In addition, the TA also saves information about the identity of the node on which the keys are installed as well as the keys themselves on a trusted controller node. At this point, the function of the TA is over and the key predistribution phase ends.

The next phase is the *shared-key discovery phase*, which happens after the nodes are deployed and when the network has to be initialized. In this case, every node discovers its neighbors and determines the keys that it shares with these neighbors. A pair of neighboring nodes can then use any common key to initiate communication between themselves. Note that a node might have some neighbors with whom it has no keys in common. In such a case, the node will not be able to set up secure links directly with such neighbors.

A simple way for any two nodes to discover whether they share a key is to let every node contain the identities of the other nodes with which it shares keys. Thus, after deployment a node only needs to know the identities of its neighbors to determine whether it shares any keys with those neighboring nodes. A problem with this simplistic method is that it is difficult to implement in situations where nodes can dynamically join the network. This is on account of the fact that the TA has to inform existing nodes about the addition of new nodes. In addition, an attacker by compromising a node might know the distribution of keys. Therefore the attacker can determine the nodes from which it can get new information and the nodes which will not provide any new information. This facilitates the adversary in terms of targeting the nodes to compromise. The memory needed to store the identities of other nodes, especially in large networks, is also another disadvantage.

Another method for nodes to discover whether they share a key is to have an identifier associated with every key. Every node would then have to broadcast the identifiers of the keys that it possesses. A disadvantage of this approach is that the adversary would know the identifiers of keys on a given node by eavesdropping. If this is of concern then an alternative method based on random numbers can be used. In this approach a random number is encrypted using each distinct key on the node. Following that the node transmits the random number as well as the multiple ciphertexts corresponding to the random number. Nodes on receiving this transmission search their own set of keys for common keys by decrypting the received encrypted data with their keys and comparing with the random number (sent in the clear). Verification of these claims can then be based on a challenge–response protocol between the two nodes. A disadvantage with these latter schemes is the additional communication as well as computational overhead.

On account of the random allocation of keys on the various nodes, it might be possible that a shared key might not exist between some pairs of neighboring nodes. In this case, such nodes make use of the third phase of *path-key establishment* to set up keys between themselves. The idea here is that, if a path of nodes sharing keys pairwise exists between the two neighboring nodes, then the two nodes can use this path to exchange a key in order to establish a direct link. This achieves the same effect as that provided by pairwise sharing between every pair of neighboring nodes in the system, but at lower cost. Of course, given that the models of connectivity are probabilistic, it is possible that the network may not be fully connected, thereby preventing some pairs of neighboring nodes from communicating with each other securely.

During the operation of the network, if a sensor node is detected to be compromised, then the entire key ring present on that node has to be removed. This is accomplished by having a trusted controller node (on which the TA has saved information about the keys on each node during the key predistribution phase) broadcast a revocation message that contains a list of the identifiers of the keys to be revoked. This message is encrypted using the keys shared between the controller and each of the nodes. A single controller node can be used for the network if the range of the controller node is large or if the controller node is allowed to be mobile. Each node on receiving this encrypted message from the controller removes the corresponding keys from its key rings after verifying the authenticity of the message. Note that, as a result, some encrypted links might disappear and hence those nodes might have to resort again to the shared key discovery and path-key establishment phases.

Using random graph analysis and simulations, the authors show the promise of this scheme. As an example, they show that, to establish almost certain shared-key connectivity for a 10,000 node network, a key ring of only 250 keys on each node drawn from a key pool of 1,000,000 keys is sufficient. The size of the key pool P is important for ensuring both better connectivity and greater resilience of the scheme. A smaller size of P will ensure that two nodes share a key with a higher probability, thereby leading to greater chances of ensuring connectivity. On the other hand, this could also increase the susceptibility of the network since, by compromising fewer nodes, the adversary would be able to eavesdrop successfully on a larger number of communications amongst the noncompromised nodes in the network. For example, consider one extreme when the size of the key pool is 1. In this case, this scheme reduces to the simple deterministic case of a single key being used in the entire network. While this scheme ensures high connectivity, it suffers from no resilience against node compromise since capture of a single node compromises the communication in the entire network. Having a very large key pool causes the opposite behavior in terms of connectivity and resilience. In such a case connectivity will be very low but resilience will be very high.

Several proposals have been developed based on this idea of random key predistribution. The various proposals in this family of probabilistic key sharing approach differ in terms of the structure of the key pool, the number of common keys required, their applicability to a dynamic deployment scenario and the method of determining the common keys. These schemes offer network resilience against capture of nodes since a node has very few keys deployed on it. Thus, there is no need to consider tamper resistant nodes (though note that nontamper-resistant nodes might not be acceptable in some situations such as military deployments). This is an advantage given the problems such as cost and complexity associated with tamper-resistant nodes. In addition this approach is more scalable for large-scale, dense deployments of sensor networks. The drawbacks of these schemes are that the key setup is based on probabilistic arguments. Additionally, setting up of a secure channel between nodes which do not share common keys might require communication over multiple hops. This increases the workload and the latency associated with setting up secure links. We next look at several solutions based on this idea of probabilistic key sharing.

In [39] the authors focus on very large, dense sensor networks. The authors propose three random key predistribution schemes for such networks. They are:

- the q-composite random key predistribution scheme—two nodes need to share more than one common key in order to set up a secure link;
- the multipath key reinforcement scheme—An attacker has to compromise many more nodes to achieve a high probability of compromising any communication.
- the random-pairwise key scheme—enables node-to-node mutual authentication whereby any node can determine the identity of the other node it is communicating with.

We first consider the q-composite scheme. This scheme is very similar to the one in [17]. The difference, though, is that, in order to form a secure link, this scheme requires at least q common keys, $q > 1$, between any two nodes, as opposed to the scheme in [17], where a single common key is needed for any two nodes to communicate securely. By requiring more common keys between any two nodes, the connectivity of the network is impacted. This is due to the fact that it is now more difficult for two neighboring nodes to set up a secure link between themselves. At the same time, the resilience of the network against node capture is increased. This is because increasing the common key threshold makes it harder for an attacker with a given set of keys to break a link between two noncompromised nodes.

Now in order to make it possible for two nodes to establish a secure link with some probability, it is necessary to reduce the size of the key pool for a given size of the key ring on each node, but this makes it possible for the adversary to compromise a larger percentage of keys in the key pool by controlling nodes beyond a certain threshold number. This implies that the network is more vulnerable when the number of nodes that have been compromised is larger than this threshold. Thus, these opposing factors result in better security for small number of compromised nodes and worse security for large number of compromised nodes as compared with the basic scheme in [17]. Thus q-composite scheme ensures that small-scale attacks are less productive as compared with large-scale attacks. This is advantageous given that it might be easier to detect large-scale attacks which are also more expensive to launch.

In the q-composite scheme, during the initialization phase, a certain number of keys are selected randomly from a key pool and installed on a node. During key-setup, a node must discover all the common keys it possesses with each of its neighboring nodes. This can be achieved by using options as described earlier with the basic scheme. Once a node determines a neighboring node with which it shares at least q keys, a secure link can then be formed between the two. The actual key used to secure the link can be a hash of all the shared keys. Secure links cannot be formed between nodes that share fewer than q keys. Other operation details are similar to the scheme in [17].

Determining the right size of the key pool for a given size of the key ring is critical here. This is because, if the key pool size is too large, then the probability of any two nodes sharing at least q keys might be less than desired. On the other hand, if the size of key pool is small, then it will result in easier compromise of the security of the entire network as the adversary has to determine a smaller number of keys.

The second scheme proposed by the authors is a *multipath reinforcement scheme*. This is a method to strengthen the security of a secure link by establishing a new key to protect this link. The new link key is set up through multiple disjoint paths. It is assumed that the initial key setup is completed. Thus, there exist secure links between the various nodes. Consider two nodes A and B that do not need to be neighbors. Assume that they have a secure link between themselves. This secure link between A and B would be compromised if nodes elsewhere that have a superset of the keys shared between A and B were captured by the adversary. One way to address this is to derive a new key between A and B once an initial secure link is formed. If the new key is derived by sending the key fragments over multiple independent paths, then the updated key will be secure unless the adversary is able to eavesdrop successfully on all the paths. The final key can then be derived from these fragments such as by XORing all the fragments. This is the basic idea behind the multipath reinforcement scheme. The more paths are used to update the key, the more secure it is. On the other hand, the longer the path is, the less secure it might be. This is due to the fact that, if any one link on the path is insecure, the entire path is insecure and longer paths provide more compromise opportunities to the adversary. Note that, if the key update is done only over a single path, then it might be possible for the adversary to record all traffic used to set up the keys and decrypt the key update message after it compromises the necessary keys. As a result, the new updated key would be vulnerable.

It is clear that the communication overheads associated with this scheme can be significant due to the need to transmit the key fragments over multiple disjoint paths. The authors show that this scheme has approximately 10 times more communication overhead over the basic scheme in [17], but then it provides an approximately 150 times improvement in the security of the scheme. Note that this scheme is similar to the q-composite key in that both schemes increase the difficulty of compromising a given link by requiring that the adversary possess multiple keys in order to eavesdrop on a given link. The tradeoff, though, is different in both cases. For the q-composite key scheme the tradeoff is in terms of the requirement for a smaller key pool to achieve certain connectivity in the network. For the multipath scheme, the tradeoff is in terms of the larger communication overhead.

Combining both the schemes (q-composite and multipath reinforcement), however, might not be very effective since the combination compounds the weaknesses in both the schemes. The smaller key pool size required for q-composite key scheme implies that more links are rendered insecure by compromising nodes greater than a threshold number. This could undermine the effectiveness of the multipath reinforcement scheme. This is due to the fact that, by compromising nodes beyond a threshold number, the adversary would have control over nearly all links in the network. If the adversary has captured the traffic on these links sent earlier, he/she can decrypt the traffic and obtain the shares used to set up the keys. This compromises the keys.

A problem with the two schemes given above is that they do not have mechanisms to allow one node to authenticate the identity of the other node. The third scheme, called the *random pairwise scheme*, addresses this. This scheme is a modification of the basic pairwise scheme. Here each node is provided with its own ID and initialized with a certain number of keys in its key ring. Each key on a node is shared with only one other node in the system. With each key contained in the key ring of a node is also associated the ID of the other node which shares the same key. After deployment a node broadcasts its node ID. This information is used by the node's neighbors to search in their key rings to verify if they share a common pairwise key with the broadcasting node. A cryptographic handshake can then be performed with nodes that have been identified to share common keys so that the nodes can mutually verify knowledge of the key. As a result the nodes are aware of the authentic identity of the corresponding party. This scheme is quite resilient against node capture since each pairwise key is unique. Thus, capture of any node results in compromising only those secure links that the compromised node is directly involved in. No other secure links in the networks are compromised. A drawback, though, is that this scheme does not scale to large network sizes. In addition, dynamic deployment of sensors also leads to inefficiency. In this case the setup server either has to reserve a space for nodes that may never be deployed or has to update the key ring installed on nodes already deployed. The former increases the memory overhead while the latter increases the communication overhead.

We next look at another scheme belonging to the probabilistic key deployment family. This is the hashed random preloaded subsets (HARPS) scheme proposed in [40]. The idea here is for the TA to choose P secrets (in the key pool) as earlier. These secrets are also called as root keys and represented as $[M_1, M_2, \ldots, M_P]$. The TA then derives L keys from each root key by repeated application of a one-way function on each root key. Following that, the TA needs to determine the subset of these $P \times L$ keys to be deployed on each node. This is done by the TA using a function of the node identity. We explain this using an example. Consider a node A with identity given by $\mathrm{ID_A}$. A public function $\mathrm{F}(\cdot)$ is applied to the node identity in order to yield a sequence of k-length ordered pairs. In this case assume $\mathrm{F}(\mathrm{ID_A})$ equals $[(\mathrm{A1,a1}), (\mathrm{A2,a2}), \ldots, (\mathrm{A}k,\mathrm{a}k)]$. Here $(\mathrm{A1}, \ldots, \mathrm{A}k)$ is a partial random permutation of integers between 1 and P (the number of root keys) while $(\mathrm{a1}, \ldots, \mathrm{a}k)$ is a sequence of randomly distributed numbers between

1 and L (the depth to hash each root key). The k keys preloaded on the node are such that the jth key is given by $h^{aj}(M_{Aj})$ where $h(\cdot)$ denotes a one-way hash function. Thus, the jth key corresponds to the Ajth root key hashed to a depth of aj. This is done for each node by the TA. Following this the nodes are deployed. We would like to remark here that determining the keys deployed on a node as a function of the node identity will also make life easier for the adversary. Using this information, the adversary can determine which nodes to compromise so as to increase his (the adversary's) knowledge of the keys from the key pool.

After deployment, every pair of neighboring nodes needs to figure out whether they have any common keys. If the two nodes have common keys then they need to set up the session or link keys. Both these steps are executed by the nodes based on information about their identities and then using the same public function $F(\cdot)$ from earlier. The identity is hence the only information exchanged by the nodes. We explain this again using an example. Consider a node B which has to decide if it has any common keys with node A given earlier. Let the identity of node B be ID_B. Both the nodes, A and B, exchange their identities. Following this each node can calculate the sequence of k-length ordered pairs corresponding to the other. Let this sequence for node B be given by $F(ID_B)$ which equals [(B1,b1), (B2,b2), ..., (Bk,bk)]. As in case of node A, (B1, ..., Bk) is a partial random permutation of integers between 1 and P (the number of root keys), while (b1, ..., bk) is a sequence of randomly distributed numbers between 1 and L (the depth to hash each root key). By comparing (A1 ... Ak) and (B1, ..., Bk), the two nodes can determine the indices of the shared root keys that are present on both. Let these indices of shared root keys be given as [s1, ..., sm] where $m \leq k$. Note that it is possible that $m = 0$, which implies that the nodes do not have any common keys. But otherwise if $m > 0$, then the two nodes will have to decide the link key or the session key used to encrypt communication between themselves. To do that the nodes determine di = max(ai,bi) where the ai and bi values correspond to the depth of the hash for node A and node B, respectively, for the ith root key of the m shared root keys. The session key between node A and node B is then obtained as

$$K = h[h^{d1}(M_{s1})|h^{d2}(M_{s2})| \cdots |h^{dm}(M_{sm})]$$

We illustrate the above using a figure. Consider Figure 3.9, where we assume that the TA has 10 root keys. Further we assume that $L = 5$ and there are three keys predeployed on every node. In this figure, nodes A and C share root key indexes 5 and 10. For the key indexed 5, node A has to hash forward two times to reach the same depth as node C. For the key indexed 10, node C has to hash forward one time. Note that the scheme considered earlier in [17] is a special case of HARPs obtained when $L = 0$.

The authors in [18, 41] propose another probabilistic key distribution strategy. The schemes here are based on the polynomial-based key predistribution protocol in [42]. While the protocol in [42] was intended for group keys, the strategy has been enhanced in [18, 41] to make it feasible for a two-party case in sensor networks. Consider the basic protocol in [42] applied to a two-party case. In this case there exists a TA which generates a t-degree polynomial in two variables x and y denoted $f(x,y)$. Further, this polynomial is symmetrical due to which $f(x,y) = f(y,x)$.[4] The TA then computes the

[4]We ignore some details for simplicity such as the fact that $f(x,y)$ is over a finite field corresponding to a large prime number.

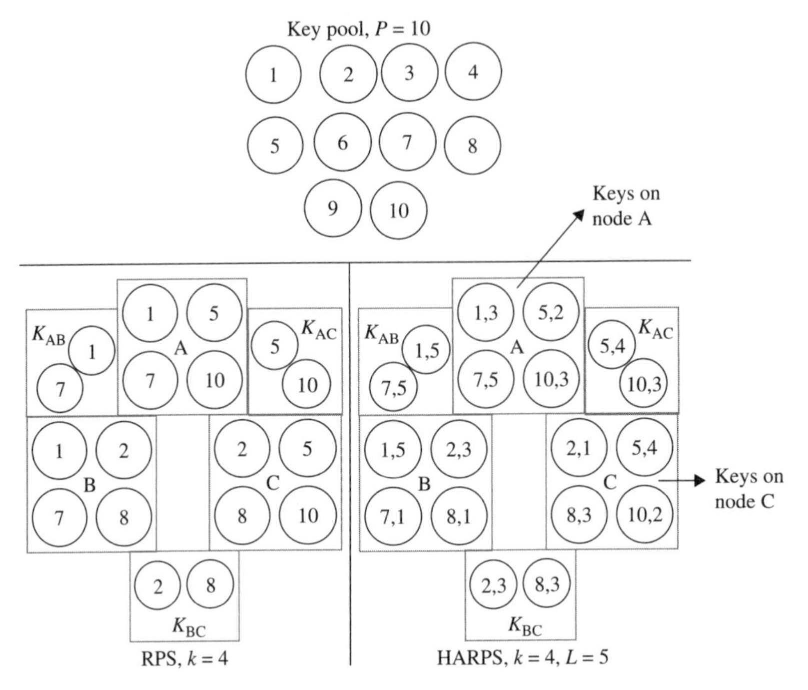

Figure 3.9. Illustrating the working of RPS and HARPs.

polynomial share of each node as f(I,y) where I is the unique ID corresponding to the node and stores these polynomial shares on the node I. Now to establish a pairwise key, any two nodes will have to evaluate the polynomial at the ID of the other node using the polynomial shares present on the node. As a result, any two nodes with IDs I and J compute the common key as $f(I, J) = f(J, I)$. Thus node I evaluates the stored polynomial at point J while node J evaluates the stored polynomial at point I. Thus, each node will need to store the coefficients of a t-degree polynomial. Blundo *et al.* [42] proves that such a scheme is unconditionally secure. Further a coalition of less than $t + 1$ compromised nodes will not be able to determine a pairwise key used by any two non-compromised nodes. A coalition of $t + 1$ or more compromised nodes though can pool their shares together and can hence determine the t degree polynomial. As a result, such a coalition of $t + 1$ or more nodes can determine the pairwise keys used by any pairs of nodes.

Thus, we see that the basic scheme in [42] can tolerate at most t compromised nodes. To strengthen this further, the authors in [18, 41] enhance this scheme using the concept of random key predistribution. In this case the TA generates a pool of multiple distinct bivariate symmetric polynomials. Each polynomial has a unique identity. As earlier, the pairwise key establishment is performed in three phases namely key predistribution, shared key discovery and path key establishment. The goal of each phase is the same as earlier except that we are now dealing with polynomials instead of numbers. Thus, we see that this general framework encompasses the schemes in [42] and [17] as special cases. Specifically, when the polynomial pool has only one polynomial we obtain the scheme in [42]. On the other hand when all the polynomials are of zero degree, we obtain the scheme of [17].

The authors propose two methods for allocating the polynomial shares to the nodes in the setup phase. In the first method, the TA selects a random subset of polynomials in the pool and assigns the shares corresponding to these polynomials to the node. After deployment, two neighboring nodes set up secure links between themselves if they have at least one common polynomial. Thus, this scheme is similar to the basic probabilistic scheme in [17], where instead of randomly selecting keys from a large key pool and assigning them to nodes, polynomials are chosen from a pool and the polynomial shares are assigned to nodes. However, the difference is that, while in [17] the same key will be shared by multiple nodes after the shared key discovery phase, in this scheme there will be a unique key between each pair of nodes after the shared key discovery phase. Further, if fewer than $t+1$ shares of the same polynomial are compromised, then keys constructed using that polynomial will not be disclosed.

The first method provides for more uncertainty in terms of key distribution as compared with the other probabilistic distribution schemes. Hence, it can be expected to perform better than the basic scheme [17] for a larger number of compromised nodes, but once the number of compromised nodes exceeds a threshold, this scheme will degenerate at a faster rate than the other schemes with respect to the security of the network. The results shown by the authors do indeed confirm this. In fact the authors show that, when the fraction of compromised direct links is less than 60 percent, given the same storage constraint, the proposed scheme provides a significantly higher probability of ensuring secure communication between noncompromised sensors than the earlier methods. Note also that unless the number of compromised nodes sharing a polynomial exceeds a threshold, compromise of nodes does not lead to the disclosure of keys established by noncompromised nodes.

The second method to distribute polynomial shares in the setup phase is the grid-based key predistribution. We explain this method with the aid of Figure 3.10. Let the maximum

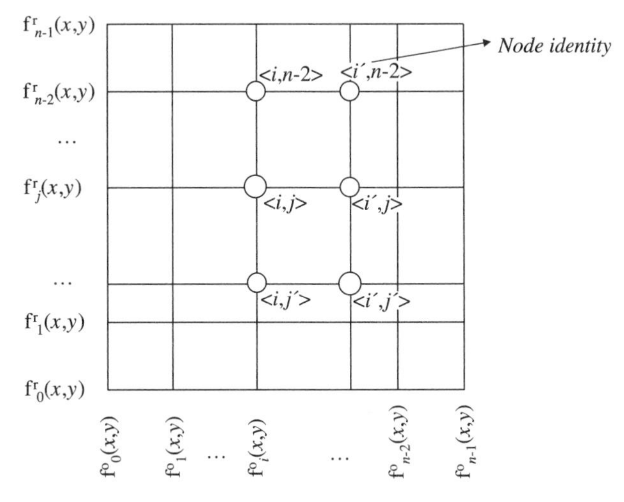

Figure 3.10. Grid-based key predistribution.

number of nodes in the network be n. Assume for simplicity that n is a perfect square. Then the TA constructs a $\sqrt{n} \times \sqrt{n}$ grid of a set of $2\sqrt{n}$ polynomials, as shown in the figure. Each row and each column in the grid is associated with a polynomial as shown in the figure. The TA then assigns a unique intersection of this grid to each node. For the node at a coordinate (i,j), the TA distributes the polynomial shares of the polynomial corresponding to the ith row and jth column. The nodes can then use this information to perform share discovery and path discovery very efficiently. This can be facilitated by ensuring that the node ID corresponds to the row and column associated with the node. In Figure 3.10 we show the identities of several nodes. We would also like to remark here that this method has also been extended to a hyper-cube-based scheme in [18].

This method of predistribution has a number of attractive properties. This scheme guarantees that any two sensors can establish a pairwise key by using intermediate nodes when there are no compromised nodes and assuming that the network is not partitioned. This scheme also allows the nodes to determine efficiently if they can establish pairwise keys with another node and if so which polynomial to use. Thus this results in minimal communication overhead during polynomial share discovery. Further, each node only needs to store the shares corresponding to two t-degree polynomials. The drawback, though, is that this scheme might provide extra information to the adversary. This information would allow the adversary to select the nodes to compromise in order to recover the polynomial. Thus, the adversary might try to target all nodes that lie on the same row or column so as to compromise at least t of them in order to obtain the polynomial.

A similar approach to the one in [18, 41] has been independently proposed in [15, 16]. The minor difference though is that the approach in [41] is based on the two-party scheme in [42], while the scheme in [15, 16] is based on the scheme in [43]. As mentioned earlier, [42] proposes several schemes to allow a group of n parties to compute a common key that is resistant to any coalition of t other parties where $t < n$. When $n = 2$, the scheme in [42] is equivalent to the scheme in [43]. Therefore, the schemes in [15, 16], as well as in [41], are equivalent.

In [15, 16], the authors consider the probabilistic analog of the scheme in [43]. The scheme in [43] is not based on a polynomial representation like the one in [42], but is based on a matrix representation in a finite field GF(q) where q is some large prime number. The proposed scheme is obtained by combining the key predistribution scheme of [43] with the random key predistribution methods discussed previously. Blom's scheme [43] requires every node to store $m + 1$ keys, where m is much less than the total number of possible nodes in the network. As a result, it is much more efficient in terms of memory storage as compared with the pairwise keys scheme. The drawback, though, is that, when an adversary captures more than m nodes, communication links between any nodes not under the control of the adversary are also compromised.

Blom's scheme is based on the use of a single key space. Du *et al.* [15, 16] propose extending this using multiple key spaces. Thus, during the key predistribution phase, a certain number of key spaces are created using Blom's scheme. Following this, each node is loaded with information from a certain subset of these key spaces. After deployment, in order for a node to compute a shared pairwise key with another node, it is necessary that they both carry information from the same key space. On account of the use of multiple key spaces, there is only a probabilistic guarantee that two nodes can generate a pairwise key. Note also that keys generated in this scheme are all pairwise keys and hence can also be used for authentication purposes as long as the key space is not compromised.

This scheme exhibits a threshold property similar to the earlier schemes such as the q-composite scheme [39] and the scheme in [41]. Thus, when the number of compromised nodes is less than the threshold, the probability that the secure links formed by noncompromised nodes are affected on account of some compromised nodes is close to zero. A difference as compared to the q-composite scheme or the basic scheme in [17] is that the thresholds at which this scheme breaks down is much larger. For example, the authors show that, in order to eavesdrop successfully on 10 percent of the secure links formed by noncompromised nodes, the adversary has to break five times as many nodes as he has to break under the q-composite scheme or the basic scheme in [17]. Thus, this scheme substantially lowers the initial payoff to the adversary from compromising a small number of nodes. Hence, the adversary needs to compromise a significant fraction of nodes in the network to be able to also eavesdrop successfully on communication between noncompromised nodes. This scheme also allows nodes to be deployed in increments and does not need all nodes to be deployed at the same time.

The random key schemes that we have seen so far ensure that communication costs per node are constant irrespective of the number of nodes in the network, but these schemes require that the average number of neighbors of a node be above a threshold in order to ensure that secure links can be established by each node in the network. Thus such schemes will not be suitable for networks where nodes are not densely distributed nor for networks where the node density is nonuniform. This is due to the probabilistic nature of key establishment in such networks. This could result in a disconnected network on account of the fact that some critical pairs of nodes could not successfully perform key establishment.

To address key establishment in such sparse networks, the authors in [44] present a key distribution scheme called Peer Intermediaries for Key Establishment (PIKE). PIKE establishes a key between any pairs of nodes in the network irrespective of the topology or the density of the network as long as the network is not partitioned. Under this scheme every node in the network shares a pairwise key uniquely with each of $O(\sqrt{n})$ other nodes in the network, assuming that the network has n nodes. Each key is unique and shared only between two nodes. These keys are set up during the predeployment phase. After the nodes are deployed, if two nodes A and B need to set up secure communication link between them, they have two options. If A and B share a pairwise key set up during predeployment they can use that. The second possibility is when A and B do not share any pairwise key. In such a case, it would be necessary to find some node C that shares a unique pairwise key with both A and B. Then C is used as a trusted intermediary to establish shared keys between nodes. To accomplish this, A encrypts the new key to be shared with B using the key that A shares with C and sends this message to C. C decrypts this message and re-encrypts the result using the key that C shares with B and sends the resulting message to B. As a result, B is able to get the new key shared with A securely provided that C has not been compromised by the adversary. Note that this is also a result of the fact that the pairwise keys are shared between exactly two nodes. Note that B can finally send a message to A encrypted using the new key to confirm receipt. Thus, the basic idea in PIKE is to use nodes as trusted intermediaries to establish shared keys between nodes.

00	01	02	03	04	···	09
10	11	12	13	14	···	19
20	21	22	23	24	···	29
30	31	32	33	34	···	39
·	·	·	·	·		·
·	·	·	·	·		·
·	·	·	·	·		·
90	91	92	93	94	···	99

Figure 3.11. PIKE scheme.

A problem at this point is to design a scheme to share keys with the $O(\sqrt{n})$ nodes in the network. The scheme has to ensure that the communication costs involved in setting up shared keys are minimal. We explain a scheme proposed in [44] with the aid of Figure 3.11. Here we illustrate how the scheme works in a network of 100 nodes. Let the maximum number of nodes in the network be n. Assume for simplicity that n is a perfect square. Then each node is associated with an ID of the form (x,y) where x and y vary between 0 and $\sqrt{n} - 1$ as shown in Figure 3.11. Each node (x,y) is then loaded with a secret key shared pairwise with all other nodes with IDs given by $(*,y)$ and $(x,*)$ where the asterisk indicates a wildcard character. Thus, in the figure a node shares a secret key pairwise with all other nodes that have IDs in the same row and same column. Thus, a node with ID 91 shares a different key with each of the nodes in the tenth row and second column. Thus each node stores $2(\sqrt{n} - 1)$ keys and the total number of unique keys generated is $n(\sqrt{n} - 1)$. Thus two nodes A and B with ID (xa,ya) and (xb,yb), respectively, will share pairwise keys with two nodes having IDs (xa,yb) and (xb,ya). We would also like to remark here that the authors have also proposed a method which reduces the memory storage overhead in every node by a factor of 2. This method requires that every node stores half of the pairwise keys while it generates the other half of the pairwise keys dynamically using a hashed combination of a single secret and the identity of the other node.

A square grid deployment such as shown in Figure 3.11 provides two possible intermediaries for every pair of nodes that desire to set up a secure communications channel. Note that intermediaries are not needed if both the nodes lie on the same column or row. A less redundant version can also be created by using node IDs from only one side of the diagonal of the grid. In this case, atmost one intermediary will exist between any two nodes. A more redundant version can also be formed by adding an additional axis of nodes which share keys with each other. Figure 3.12 shows an additional axis which runs diagonal to both of the original axes. As a result, there are three choices of axes to traverse first followed by another choice of two axes to traverse next, yielding six possible intermediary nodes (shown as circles), as shown in the figure. In general for q nonparallel axes on a two-dimensional plane, the number of potential intermediaries will be $q(q - 1)$. This redundancy also makes the scheme less vulnerable to failed nodes. Of course, this results in increased memory overhead since now

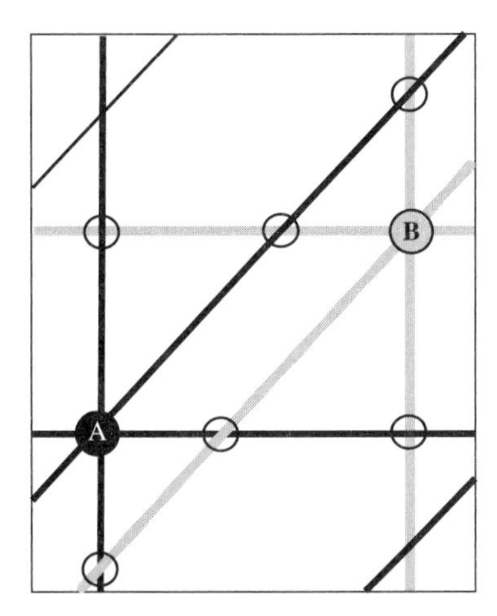

Figure 3.12. Increasing redundancy in PIKE.

each node will have to store more keys. In addition, compromise of a node also results in leakage of more keys, thereby resulting in a more vulnerable scheme.

3.3.2.3 Location-Based Key Management A problem with the schemes explained so far is that they do not take the location of sensor nodes after deployment into consideration. Even imperfect information about deployment might help to enhance significantly the performance of the various schemes. This is because, in this case, the scheme can ensure that nodes closer together have a larger number of common keys. Nodes which are far away need not have any common keys. This can be justified due to the fact that direct long-distance communication among nodes would be rare, unnecessary, and also impossible in many scenarios. This is the approach taken in [14]. The authors assume that information about deployment can be derived from the way the nodes are deployed. The authors consider nonuniform sensor deployment and show how it can help to improve the performance of key predistribution.

The scheme proposed in this paper consists of three phases as for the other random key predistribution schemes earlier, namely key predistribution, shared key discovery, and path-key establishment. The last two phases are the same as those considered in the earlier scheme in [15, 16]. It is only the first phase that is considerably different. Hence, we focus on explaining only the first phase here.

The authors consider that nodes are deployed in groups. Consider a possible deployment as shown in Figure 3.13. Here the nodes are assumed to be deployed in groups around each deployment point shown as a dot. Thus, 36 different groups are shown in this figure. Some deployment points are shown as points A, B, C, D, E, F, G, H, and I. Now, the nodes are grouped into several such groups. Further, the global key space pool is also divided into subgroups of key spaces. Each subgroup of key spaces is associated with a subgroup of nodes. Thus, a key-space subgroup is associated with each deployment point in the figure. What is now needed is a scheme that ensures that, when the deployment points of two groups are closer to each other, the amount of overlap

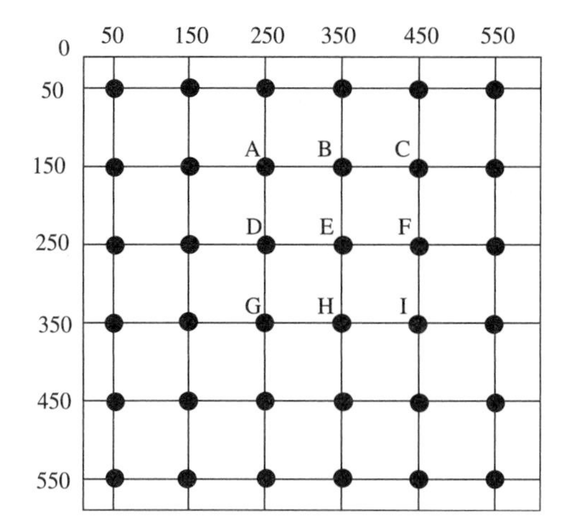

Figure 3.13. Deployment of nodes.

between the corresponding key-space subgroups is large. Further, when the deployment points of the two groups of nodes are farther away from each other, the amount of overlap between the corresponding key spaces is very small or even zero.

The authors then suggest one possible scheme that meets the above requirement. In this case, two horizontally or vertically neighboring key-space pools share exactly a fraction a of the key space. This fraction a can vary between 0 and 0.25. Similarly, two diagonally neighboring key space pools share a different fraction b of the key space. Even this fraction b varies between 0 and 0.25. Two nonneighboring key space pools do not share any key spaces. We show this in Figure 3.14. Here the nodes deployed in group E share a fraction a of the keys with the nodes deployed in groups B, D, F, and H. Further, they also share a fraction b of the keys with nodes in groups A, C, G, and I.

In the same figure we describe a possible method proposed in [14] to select key spaces for the subgroups from the global key space pool, assuming a deployment of nodes in 16 subgroups in a 4×4 arrangement. In this case, the key spaces for the

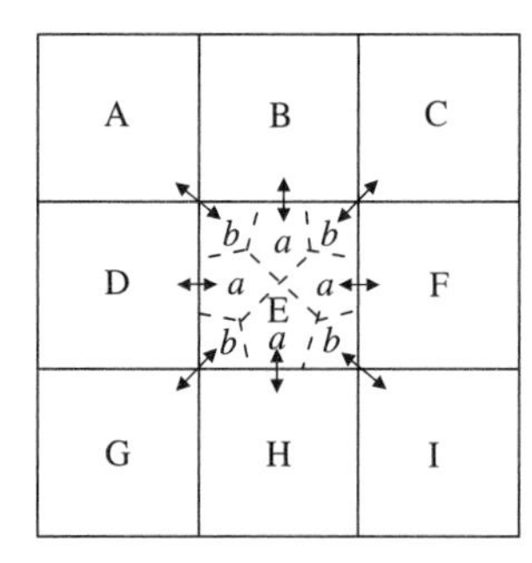

1	1-a	1-a	1-a
1-(a+b)	1-2(a+b)	1-2(a+b)	1-(2a+b)
1-(a+b)	1-2(a+b)	1-2(a+b)	1-(2a+b)
1-(a+b)	1-2(a+b)	1-2(a+b)	1-(2a+b)

Figure 3.14. Sharing of key spaces based on deployment knowledge; a and b denote different fractions.

first subgroup (upper left corner) are selected from the global key space without any restrictions save for the size of the key-space subgroup. The key spaces for the other groups in the first row are then selected from the global key space such that a fraction a of key spaces is shared with the neighbors on the left. Thus, for these sub-groups a fraction $1 - a$ of key spaces is chosen independently from the global key space, as shown in Figure 3.14. Key spaces corresponding to groups in the other rows are selected from their neighbors with a fraction of the remaining being selected from the global key space as shown in the figure. In each row the process is conducted from left to right. Given such a deployment of nodes and assignment of key spaces, the authors show that the performance in terms of the connectivity of the network and the resilience to node capture improves significantly. The memory resources needed for this scheme are also shown to be lower.

Liu and Ning [41] also use a similar approach for key management. The difference as earlier is that, instead of dealing with key spaces, the authors use polynomial shares as also random pairwise keys. Thus, sensor nodes that are expected to be neighbors are loaded with overlapping polynomials (or possess pairwise keys), while sensor nodes that are expected to be deployed far apart have no common polynomials (or pairwise keys). The authors in [45] have also provided an approach to making use of postdeployment knowledge for key management. The idea here is to associated keys (or polynomials) with locations. Excessive numbers of keys (or polynomials) are randomly loaded on a sensor node. After deployment, a sensor node determines its own location and then prioritizes every key (or polynomial) preloaded on it. The priority of a key is based on the distance between the sensor node's location and the location associated with the key. Keys with lower priority are then discarded while keys with higher priority are expected to be used to set up secure links as in case of other schemes. This therefore frees up the sensor memory for other tasks.

Liu *et al.* [46] propose an approach which does not require knowledge of the expected locations of sensors after deployment. This approach, however, requires knowledge of the groups of sensors that are going to be deployed together. The expectation is that sensors that are deployed as a group will be located in the same region. Hence, such sensor nodes will be deployed with common keys before deployment.

3.3.2.4 Mobility Typically, it is expected that mobility causes more problems in ad hoc networks, especially when it comes to key management. However, in [27], the authors leverage mobility to provide peer-to-peer security in mobile networks. Mobility could result in two nodes becoming neighbors of each other. In such a case the authors propose that the one-hop links be made use of either to set up keys that can be used later when the nodes are no longer neighbors or to cross-check on keying information provided by an off-line certification authority. In both cases the nodes that have become neighbors can make use of other secure side channels such as visual contact in order to authenticate each other. This secure side channel is typically of a short distance. It also might require line of sight to allow visual contact. Note that confidentiality on this side channel might not be needed if there is no possibility of eavesdropping on it. This could be because there is no adversary in the region. The presence of no adversary can be verified by visual confirmation. This approach can be used for both symmetric keys as well as asymmetric keys. In this sense mobility can be looked at as a positive factor for key management.

3.4 SUMMARY

Key management in ad hoc networks is a hot topic in current research. In this chapter, we have considered several different key management schemes designed for ad hoc and sensor networks. We have pointed out the advantages and disadvantages of each of these schemes. Ways in which some of these schemes can be improved have also been pointed out.

4 Secure Routing

4.1 INTRODUCTION

Routing is an important function in any network, be it wired or wireless. The protocols designed for routing in these two types of networks, however, have completely different characteristics. Routing protocols for wired networks typically do not need to handle mobility of nodes within the system. These protocols also do not have to be designed so as to minimize the communication overhead, since wired networks typically have high bandwidths. Very importantly, the routing protocols in wireline networks can be assumed to execute on trusted entities, namely the routers.

These characteristics change completely when considering ad hoc wireless networks. Mobility is a basic feature in such networks. Resource constraints also govern the design of routing protocols for such networks. Ad hoc networks also do not have trusted entities such as routers, since every node in the network is expected to participate in the routing function. Therefore, routing protocols need to be specifically designed for wireless ad hoc networks. This indeed has been an area of research for the last few years.

In the beginning of this chapter, we provide an overview of different types of routing in ad hoc networks. Following that, we discuss in detail several routing protocols proposed for such networks and explain the various proposals to enhance the security of each of these protocols. We conclude this chapter by looking at generic attacks on routing protocols in ad hoc networks.

4.1.1 Distance-Vector and Link-State Routing

There are two main categories of traditional routing protocols: distance-vector and link-state. Distance-vector routing is the type of routing protocol that was originally used on the internet. In distance-vector routing, each node maintains a table that contains the distance from that node to all other nodes in the network. Whenever a node receives a routing update from one of its neighbors, it examines that table to see whether it can reach additional destinations through that neighbor, or whether the path to some destinations through that neighbor is shorter than the existing route. If that is the case, the node updates its own routing table and then sends the updated table to all of its neighbors. Those in turn update their own tables and may send updates to their own neighbors. DSDV is a typical example of a MANET distance-vector routing protocol and will be discussed in Section 4.4.

The challenge with the distance-vector routing protocols is that they are usually slow to converge. A simple example of slow convergence [47] is the case when a node is

Security for Wireless Ad Hoc Networks, by Farooq Anjum and Petros Mouchtaris

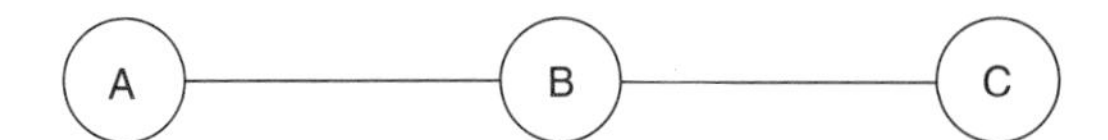

Figure 4.1. Distance vector example.

disconnected from the network. Consider the three nodes shown in Figure 4.1. Assuming the cost of each link to be 1, the distance of B from C is 1 and the distance of A from C is 2. If C is disconnected from the network (e.g. because the link connecting B and C is broken), B will realize that it is not directly connected to C any more. B, though, will receive an advertisement from A stating that A can reach C at a cost of 2. Therefore, B will advertise a distance of 3 from node C. Since A's path goes through B, when A hears that the distance of B from C is now 3, it will update its table and advertise a distance of 4 from C. This process will continue until the distance from B to C reaches infinity (typically represented by a large number in routing protocols) and both A and B will realize that C is unreachable. This process may take several steps.

The second category of routing protocols, namely link-state routing protocols, addresses the limitations of distance-vector routing protocols though they have their own disadvantages. Link-state routing protocols typically work as follows. Each node discovers its neighbors through broadcast advertisements (typically called Hello messages) that each node transmits and its neighbors can hear (if they are within the transmission range of that node). Once a node has discovered its neighbors, it transmits a message, typically called a link-state advertisement (LSA) to all other nodes in the network that lists its neighbors and the cost to get to those neighbors. Each node then can use those LSAs to calculate the topology of the complete network and routes to all other nodes. OLSR is a typical example of a MANET link state routing protocol (see Section 4.5). The link state routing protocols tend to converge more quickly than distance vector protocols. In the example described before (see Fig. 4.1), B will advertise that C cannot reach B. Once this information is flooded throughout the network, the other nodes (including A) will realize immediately that C is not reachable. This is a much faster process than that described earlier for distance-vector routing protocols. Link-state routing protocols, however, typically generate more overhead since routing tables are flooded throughout the network.

Both distance-vector and link-state routing have their own advantages and disadvantages, as described in [47]. The specific routing protocol that works best depends on the topology, the application requirement, and the node capabilities.

4.1.2 Proactive vs Reactive Routing

A different classification of routing protocols is based on the instance at which the routes are set up. Based on this there are two different categories namely: proactive and reactive. In proactive routing protocols nodes typically try to create routes proactively before there is a need to route traffic from a specific source to a specific destination. Nodes usually accomplish that by sending periodic routing updates. In addition, routing updates are also sent whenever the topology changes. Such updates ensure that nodes have up-to-date paths to other nodes. Optimized link-state routing protocol (OLSR) is a good example of such protocols (see Section 4.5).

Reactive protocols on the other hand set up routes between two nodes only when there is a need to send actual traffic between those two nodes. Nodes using reactive routing protocols usually accomplish that by flooding the network with route request messages requesting information on the route from the source to the destination. These route request messages originate at the source and are flooded throughout the network when the source needs to send data to the destination. Eventually the destination (or a node that has recently communicated with the destination) receives the route request message and responds to it with the necessary path information. Ad hoc on-demand distance vector (AODV) is a good example of such protocols (see Section 4.2).

Proactive protocols are mostly concerned with minimizing the initial delay that data traffic experiences in reaching the destination from the source. Since the routes are established before there is a need to send traffic between the source and destination, there is no need to discover the route when data traffic actually needs to be sent. Therefore the traffic doesn't get delayed initially while waiting for the route establishment. On the contrary, reactive routing protocols result in traffic being delayed because the route needs to be discovered before the data can be routed. This delay only happens the first time that traffic between the source and the destination needs to be routed. After that, the routes are established and do not need to be recreated until there are node movements or link failures resulting in route changes.

The problem with proactive protocols is that they generate a lot of routing overhead. This is especially so when there are frequent topology changes. This is highly inefficient when there are routing updates for routes that carry traffic rarely. A reactive protocol is much more appropriate for such situations since it generates significantly lower overhead in terms of the bandwidth used. In fact, reactive routing protocols reduce (or eliminate) routing overhead in periods or areas of the network where there is little data traffic (or none at all).

The problem with reactive routing protocols is that route updates as discussed earlier often delay traffic initially. If latency is not a concern for a specific application then reactive protocols are probably more suited for that application. Since routing requests are flooded throughout the network, when there are a lot of changes in the network topology and a lot of requests for sending traffic among multiple nodes then overhead is a significant concern for reactive protocols.

It can be inferred from the above descriptions that the routing function in ad hoc networks is a cooperative function where all nodes in the network cooperate with each other. As a result, it would be easy for an adversary to launch attacks on the routing protocols used in such networks. Most routing disruption attacks are caused by modification of the routing data. In order to prevent such attacks, it is necessary for a node that receives routing information to verify the origin and integrity of the routing data. The various proposals to secure routing in ad hoc networks mainly focus on providing mechanisms to perform this verification. Note that confidentiality might typically not be needed to protect the routing information.[5]

We next look at several secure routing protocols proposed for ad hoc networks. The secure routing protocols typically are extensions of well-known routing protocols. The extensions add in various security attributes to the basic protocol. Hence, in such cases, before explaining the secure routing protocol we briefly summarize the basic protocol. We start with AODV and the secure routing extensions proposed to AODV.

[5]We will see exceptions to this in Section 4.6.

4.2 AD HOC ON-DEMAND DISTANCE VECTOR

AODV [48] is a reactive routing protocol that has been developed specifically for MANET. In AODV, whenever a node needs to send data to another node that it does not have a route to, it tries to discover a new route. Route discovery is accomplished using a route request (RREQ) message. Thus, a node that needs to discover a route to another node sends out a RREQ message. The RREQ contains the IP address of the node originating the request and the IP address of the destination node, that is, the node the originator is trying to reach. The RREQ message is broadcast through the network until a node that has a fresh route to the destination or the destination itself responds to the request with a route reply (RREP) message. The RREP message is sent back towards the originator and as the message traverses towards the originator it helps establish the end-to-end path between the originating and destination node. RREP messages contain a lifetime field that defines how long the route will be valid. This ensures that nodes do not send packets through stale paths. To minimize overhead, the originator may limit how far the RREQ may be distributed via the time-to-live (TTL) value in the RREQ message. The RREQ message will be propagated in the network until the number of hops that the message has traversed is less than the TTL value.

We now describe the protocol in more detail. As mentioned earlier, whenever a node needs to discover a path to another node, it sends out a RREQ message. One of the fields contained in the RREQ message is the *hop-count* field, which indicates the number of hops traversed by the RREQ message. Whenever a node receives an RREQ message, it checks the hop-count field contained in the RREQ message indicating the number of hops the RREQ message had to travel to get to this node from the originator of the RREQ message. The node then creates a new route towards the originator of the RREQ (unless it has a shorter route to it already), which will be used for routing the RREP messages. The route to the originator from this node goes through the immediate node from which the RREQ message was received. The length of the route is equal to the hop-count field contained in the received RREQ message. The node then checks if it has a route to the destination of the RREQ message.

If the node that receives the RREQ message does not know a route to the destination, it increments the value of the hop-count field and then forwards the RREQ message to its neighbors. If the RREQ message is received by a node that knows of a fresh route to the destination or if the RREQ is received by the destination itself, then the node responds to the originator of the RREQ request with a RREP message. The RREP is unicast towards the originator of the RREQ request via the path that was created during the propagation of the RREQ message. The RREP message also includes the hop count from the node that generates the RREP message to the destination (0 if the RREP is generated by the destination itself). The hop count in the RREP message is incremented by each node that forwards the RREP message towards the originator of the RREQ request. Since any node that knows of a fresh route to the destination may generate a RREP message, multiple RREP messages for the same destination might reach the source of the RREQ message. When the originator of the RREQ message receives multiple RREP messages it can compare them and select the route with the smallest hop count. Optionally the originator of the RREQ may also send a route reply acknowledgment (RREP-ACK) once it receives the RREP message. We illustrate the operation of the AODV protocol in Figure 4.2. In this figure we show node A as the source which desires to determine a route to node B, which is the destination. Node A creates and propagates the RREQ messages. Node B replies with a RREP message on receiving such a RREQ query.

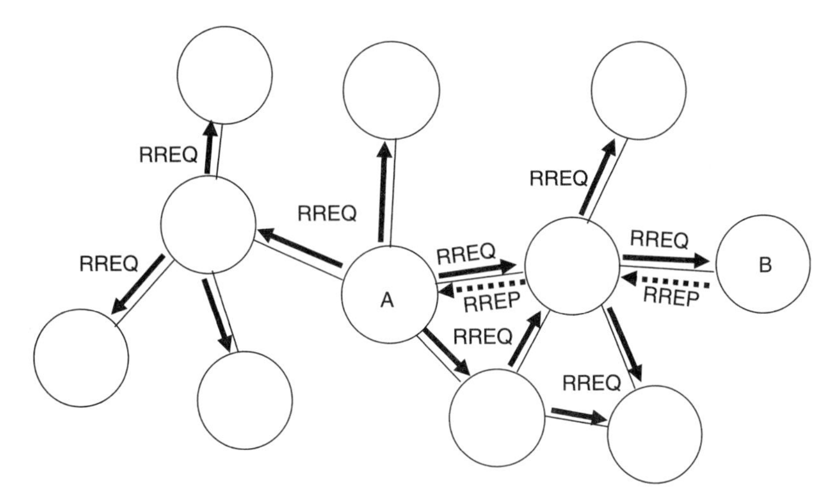

Figure 4.2. Illustration of the AODV routing protocol. In this case node A desires to find a route to node B.

AODV also supports route repair using the route error (RERR) message. This message is used by nodes for invalidating routes they are aware of when links supporting those routes break. This could happen due to node movements, increased link error rate, and so on. When a node loses connectivity to one of its neighbors, it sends a RERR message to other nodes that make use of the path through the "lost" neighbor. The RERR message contains the list of destinations that have become unreachable. The RERR may be broadcast or unicast to the affected nodes.

Based on the description of AODV provided so far, we see that AODV has no inherent security mechanisms. As a result, malicious nodes can perform many attacks. For example, a malicious node can impersonate some other node and initiate a forged RREQ message supposedly from the impersonated node. A malicious node can also decrease the hop-count field of a RREQ it is forwarding instead of increasing it. This could be intended to ensure that the path through the malicious node would become the shortest path. This would guarantee that all data packets will pass through the malicious node. An adversarial node could also pretend to be the destination or could pretend to know a route to the destination. The resulting RREP message can then be forged, ensuring that all messages to the destination will go to the malicious node. A malicious node can also choose not to forward the RREQ and RREP messages, resulting in suboptimal routes. In addition, a malicious node could also create forged RERR messages or choose not to forward genuine RERR messages. All these potential attacks demonstrate the need to enhance the security of AODV. We will describe some potential approaches for making AODV secure (at least against certain attacks).

4.2.1 Secure AODV

A secure version of AODV called Secure AODV (SAODV) has been proposed in [49, 50]. SAODV provides features such as integrity, authentication, and nonrepudiation of routing data. SAODV incorporates two schemes for securing AODV. The first scheme involves nodes signing the messages that they create (e.g. RREQ, RREP). This allows other nodes to verify the originator of the message. This scheme can be used for protecting the portion of the information in the RREQ and RREP messages that does not change once these messages are created. However, RREP and RREQ messages also contain a field (namely the hop count) that needs to be changed by every node. Such mutable information is ignored by the creator of the message when signing the message. The second scheme of SAODV is used for protecting such mutable information. This scheme leverages the idea of hash chains. We discuss the two schemes used by SAODV in detail later in this section. We would like to remark here that signing routing messages implies that the various nodes need to possess a key pair that makes use of an asymmetric cipher. In addition, nodes in the network also need to be aware of the authentic public keys of the other nodes. Therefore, a key management scheme is required and any of the schemes in Chapter 3 can be used for this purpose.

SAODV defines extensions to the basic AODV message in order to meet its objectives of security. This extension has the format shown in Figure 4.3. The type field specifies the type of message. For example, a secure RREQ message may set the type field to 64, indicating an RREQ message (see [50] for a complete list of values). The length field defines the length of the extension, excluding the type and length field. The hash function field specifies the hash function selected [denoted as $h(x)$ in the explanation following]. The max hop count field is set by the originator of the RREQ to the TTL value. The top hash value is calculated by applying the hash function to a randomly selected number, called the seed, max hop count times. Defining $h^i(x)$ as the result of applying the hash function i times to x, top hash can be described as follows:

$$\text{Top hash} = h^{\text{Max Hop Count}}(\text{seed})$$

The signature field contains the signature of the node originating the SAODV message on the complete SAODV message, excluding the hop count (in the basic AODV message) and the hash value field described next. The last field, which is the hash value field, is initially set to be equal to the seed. We will discuss next how the SAODV message extensions can be used for supporting the two SAODV security schemes.

4.2.1.1 Digital Signatures for SAODV As mentioned earlier, the first scheme in SAODV provides authenticity of the nonmutable information in the routing messages. Originators of routing messages (e.g. RREQ, RREP) digitally sign each message (excluding the hop-count field in the AODV message and the hash field in the SAODV extension), which ensures that nodes do not impersonate other nodes. Thus, digital signatures are used

<table>
<tr><td>Type</td><td>Length</td><td>Hash function</td><td>Max hop count</td></tr>
<tr><td colspan="4">Top hash</td></tr>
<tr><td colspan="4">Signature</td></tr>
<tr><td colspan="4">Hash</td></tr>
</table>

Figure 4.3. SAODV message extension.

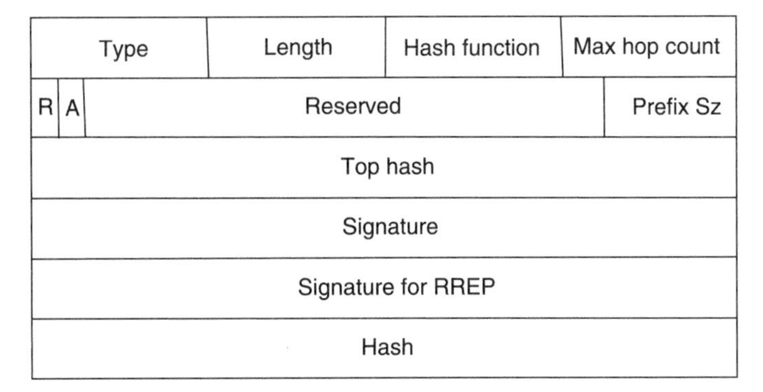

Figure 4.4. RREQ double signature extension.

to protect the integrity of the nonmutable data in RREQ and RREP messages. A node that receives either of these messages verifies that the signature on the message is correct before taking any other action.

This approach of requiring nodes to sign each routing message works well in general for AODV, though it has one major limitation. Note that AODV allows nodes receiving RREQ messages to respond to them if they know of fresh routes to the destination. This feature of AODV improves the efficiency of the protocol and reduces overhead. Unfortunately, this feature does not work with the simple signature scheme discussed so far. This is because intermediate nodes cannot generate RREP messages signed by the destination. To address this limitation [50] proposes a double signature scheme. The idea behind it is as follows. When a node (say node A) originates an RREQ message, it also includes in the RREQ message an extension to allow RREP messages to be created by the intermediate nodes on behalf of node A. Such an RREP message is in reply to a separate RREQ message with node A as the destination. This extension contains the flags for an RREP message, the prefix, and a signature that intermediary nodes can use to produce an RREP (for node A as the destination) message and sign it on behalf of node A.

The format for the RREQ message extension with the double signature is shown in Figure 4.4. The R, A, reserved, and prefix size fields are fields required in RREP AODV messages. The signature for the RREP field is provided by the originator of this

Type	Length	Hash function	Max hop count
Top hash			
Signature			
Old lifetime			
Signature of the new lifetime			
Hash			

Figure 4.5. RREP double signature extension.

RREQ (i.e. node A) and contains the signature that can be used by an intermediary node for creating a RREP in response to an RREQ that has node A as the destination.

Figure 4.5 shows the format for the RREP message for the double signature scheme. This RREP is created by the intermediary node. The signature field contains the signature for the RREP message provided by the destination of the RREQ message (i.e. node A). Since the lifetime of the route could have changed (since node A created the signed RREP fields earlier), the RREP double signature extension includes the old lifetime that was used by node A for creating the signature. The RREP double signature extension also includes the signature of the intermediary node for the RREP message with the actual lifetime (shown as signature of the new lifetime in Fig. 4.5). This way, when nodes receive a RREP message with the double signature extension, they can verify that the message is authentic since it has been signed by the intermediary node but also represents accurate information about the destination since it contains the signature of that node also.

4.2.1.2 Hash Chains for SAODV The second scheme in SAODV protects the mutable information such as the hop-count field in the RREQ and RREP messages. It does this by making use of the concept of hash chains. This assures nodes receiving AODV messages that the hop-count values provided are accurate and have not been decremented by an adversary on the path. Hash chains are created by applying a hash function (see Chapter 2) repeatedly to a seed number. We describe this scheme in detail next.

When a node originates an SAODV message (e.g. RREQ, RREP) it selects a random number (i.e. seed) and a hash function to use. The random seed is stored in the hash field. The max hop count field is set to the TTL value of the IP packet. The top Hash field is also populated as described earlier. The originator thereby completes the extension SAODV message.

Whenever a node receives an SAODV message it first checks the digital signature on the message. If the digital signature matches, then the node checks that the hop count (which has been incremented by nodes en route) has not been maliciously altered. This is achieved as follows. The node applies the hash function max hop count minus the value in the hop count field times to the value of the hash field. The resulting value must equal the value of top hash as shown below:

$$\begin{aligned} & h^{(\text{max hop count}-\text{hop count})}(\text{hash}) \\ & = h^{(\text{max hop count}-\text{hop count})}[h^{\text{hop count}}(\text{seed})] \\ & = h^{(\text{max hop count}-\text{hop count}+\text{hop count})}(\text{seed}) \\ & = h^{(\text{max hop count})}(\text{seed}) = \text{top hash} \end{aligned}$$

If the message is verified to have an unaltered hop count and the node needs to forward the SAODV message, it applies the hash function to the value in the hash field of the message it received. The resulting value is used to update the hash field in the message. Thus, hash $= h$(hash). The node also increments the hop count in the AODV portion of the message as appropriate. The node then forwards the message.

This scheme provides protection against nodes manipulating (more precisely decreasing) the hop count when forwarding AODV routing messages assuming a strong hash function. Several attacks are possible, however. One possible attack is when a node does not increment the hop count when it forwards a routing message. The value for the hash field remains the same and therefore other nodes will not be able to detect this malicious behavior. This SAODV scheme cannot protect against such an attack. Another possible attack involves a malicious node applying the hash multiple times. Note, though, that this attack might not have a significant impact. This is because it ends up causing nodes to avoid using routes through the malicious node since those routes appear to be longer (and therefore less desirable) than they should be. SAODV also does not deal with wormhole attacks [50].

AODV also provides for the transmission of RERR messages for invalidating broken routes. These messages also need to be protected. Route errors have some mutable information. In such a case SAODV proposes that the nodes originating such RERR messages digitally sign the whole message. Any node that receives the RERR message is then expected to verify the signature before using the information in such messages. This ensures that a node attempting to transmit a false RERR message will not succeed in doing so without being detected.

SAODV is able to handle many attacks leveraging modification, fabrication, or impersonation. The focus of SAODV is on ensuring that nodes do not impersonate other nodes and that nodes forwarding routing messages do not alter them while those messages are in transit. This covers a large number of attacks. For example, black hole attacks (see Chapter 1) are prevented because a node (node A) cannot pretend to be a neighbor of another node, say node B. This is because node B itself will have to respond to RREQ messages via RREP messages; node A cannot alter those messages or forge them.

There are certain attacks in addition to the ones mentioned earlier that cannot be prevented through SAODV. For example, nodes may drop routing messages (e.g. RREQ, RREP messages), impacting the effectiveness of the routing protocol. Nodes may also send very frequent RREQ requests even if they know the route to the destination increasing network overhead and draining CPU and batteries of other nodes. SAODV does not deal with such attacks. IDS systems (described in Chapter 5) can be used in conjunction with SAODV to deal with such attacks. SAODV also depends on the use of public key cryptography, which might be problematic in resource constrained ad-hoc wireless networks.

4.2.2 Authenticated Routing for Ad Hoc Networks (ARAN)

Another approach similar to SAODV based on the use of digital signatures has been proposed in [51]. The focus here is on securing on demand routing protocols in general. The authors in [51] assume that there is a trusted certificate server, T, which creates and distributes a certificate for each node before the node joins the network. All the nodes in the network have access to the true public key of T which they can use to verify the authenticity of the certificates. The certificate of node A has the following format $cert_A =$ [IP_A, PuK_A, t, e] PrK_T, where IP_A is the IP address of Node A, PuK_A is the public key of node A, t is the time the certificate was created, and e is the time the certificate expires. [...] PrK_T denotes that the certificate was signed with T's private key.

When a node, say node A, wishes to find a path to node X, it broadcasts a message called the route discovery packet (RDP), which is similar to the AODV route request message. The RDP message has the following format:

$$[RDP, IPx, cert_A, N_A, t]PrK_A$$

The RDP field specifies that this is an RDP message. The IPx field specifies the IP address of node X for which node A is trying to find a route. The RDP message also includes the certificate of node A, a nonce N_A that increases every time node A sends an RDP message, and the current time t. The message is then signed by node A's private key.

When another node (node B) receives the message it first verifies that the message is authentic by extracting node A's public key from the certificate included in the message and then verifying that the message has the proper signature. It also needs to check that the certificate has not expired. Node B then sets up a route (in the direction opposite to that traversed by the RDP message) towards the originator of the RDP message through the neighbor it received the message from (node A in this case). This is similar to the way AODV works (as discussed earlier in this section). If node B has not seen the message before and node B is not the destination, it signs the message with its own private key, attaches its own certificate and then rebroadcasts the RDP message to its neighbors as follows:

$$[[\mathrm{RDP}, \mathrm{IPx}, \mathrm{cert}_A, N_A, t]\mathrm{PrK}_A]\ \mathrm{PrK}_B, \mathrm{cert}_B$$

Note that node B has not changed the RDP message as created by the original node but merely added its certificate and signature to it. When the next node (node C) receives the message, it follows the same process as B, although it removes node B's certificate and signature first and replaces this with its own. The resulting message has the following format:

$$[[\mathrm{RDP}, \mathrm{IPx}, \mathrm{cert}_A, N_A, t]\mathrm{PrK}_A]\ \mathrm{PrK}_C, \mathrm{cert}_C$$

Eventually node X receives the message and responds to it with a Reply (REP) message. Node X is expected to respond to the first message it receives. ARAN does not attempt to verify the number of hops in the path (such as AODV) because a malicious node may attempt to manipulate the number of hops. The authors also suggest that it is preferable to send packets through the route that the RDP arrived to the destination first than through a route that may appear to be the shortest path. The reason is that the packets (RDP messages in this case) in the shortest path may have taken longer to arrive because of potentially malicious nodes delaying the packets or because of congestion in the route. Nodes unicast REP messages towards the requestor (node A in this case) through the path that was discovered by the RDP messages traveling towards the destination.

Assuming that node C was the last hop before node X, then node X sends to node C the following message:

$$[\mathrm{REP}, \mathrm{IP}_A, \mathrm{cert}_X, N_A, t]\ \mathrm{PrK}_X$$

The field REP specifies that this is an REP message. The message includes the IP address of node A, the certificate of node X, the nonce that node A sent, and timestamp that node A sent. Node X also signs the message with its private key. Once node C receives the REP message, it verifies its authenticity and then signs the message and attaches its certificate to message and sends it to the previous hope (i.e. node B in this case). As in case of RDP messages, intermediate nodes remove the signature and certificate of the previous intermediate node. Eventually node A receives the REP; it verifies the authenticity of the response, and can start communicating with node X. The authors in [51] do not discuss whether it is possible for intermediate nodes to respond to RDP messages if they have a fresh route to the destination. It seems possible to extend the scheme in a manner similar to the double signature approach of SAODV discussed earlier in this section. This could increase the efficiency of the protocol, but may also increase the security risk and complexity.

ARAN also supports a route maintenance scheme similar to the approach used in AODV. If, for example, node B discovers that the link to node C is broken, it sends an error message (ERR) towards the source of the route. The ERR message has the following format:

$$[\mathrm{ERR},\ \mathrm{IP_A},\ \mathrm{IP_X},\ \mathrm{cert_B},\ \mathrm{N_B},\ t]\ \mathrm{PrK_B}$$

Since it is in general difficult to distinguish malicious ERR message from correct ERR messages, especially in very volatile networks, it may be useful to maintain a count of the number of ERR messages that each node generates. If a node generates an abnormally high number of ERR messages (compared with other nodes), it is likely that this node is malicious (since ERR are signed and it can be verified that such a node actually generated those messages). Hence such a node must be avoided during routing.

ARAN also supports key revocation, presumably for nodes that have been verified to be behaving maliciously. Node T (the trusted certificate server) is responsible for generating such messages. The revocation messages have the following format:

$$[\mathrm{revoke},\ \mathrm{cert_R}]\ \mathrm{PrK_T}$$

Each node receiving this message broadcasts the message to its neighbors. Nodes keep track of the certificates that have been revoked until those certificates expire. When nodes receive the revocation message they also check to see whether any routes go through the untrusted node. If they do discover such routes, they initiate the route discovery process for those routes. Thus, we see that the approach in ARAN is to make use of digital signatures.

4.2.3 Security-Aware Ad Hoc Routing

The authors in [52] propose a different approach for securing routing in ad hoc networks. The key concept of the security-aware ad hoc routing (SAR) protocol is that, for applications that require security, it is preferable to route packets through trusted nodes rather than route the packets through the shortest path that may include untrusted nodes. SAR assumes that there is a trust hierarchy. Nodes lower in the hierarchy may be trusted less while nodes higher in the hierarchy may be trusted more. Trust may be derived for example through the organizational structure. In a military environment, nodes belonging to generals may be more trusted (since they are typically well protected) than sensor nodes out in enemy territory.

The approach is quite general and can be applied to several routing protocols, although the authors explain this in the context of AODV. In order to support routing messages based on the trust hierarchy, Yi *et al.* [52] propose modifying AODV slightly. Nodes sending a RREQ message add a field to each message called RQ_SEC_REQUIREMENT, indicating the level of security required for the route the node is trying to discover. When intermediate nodes receive the RREQ and cannot support the level of security required they drop the request. If they can support the security requirements of the route, they forward the RREQ to their neighbors in a way similar to AODV. They add an additional field to the RREQ message called RQ_SEC_GUARANTEE, indicating the maximum level of security that can be supported by the path. Once the field reaches the destination that node generates a RREP message similar to AODV. The RREP message includes a field RP_SEC_GUARANTEE copied from the RREQ RQ_SEC_GUARANTEE, indicating the maximum level of security that can be supported by the path. Intermediate nodes forward the RREP message as in AODV but cache the value of the RP_SEC_GUARANTEE value. This helps nodes respond to subsequent RREQ messages.

The authors in [52] propose the use of cryptography for ensuring the correct behavior of nodes. For example, SAR messages can be encrypted through a key that is shared among all nodes at the correct security/trust level. A shared key can be used for each trust level. Nodes that do not have the appropriate trust level will drop the packet which assures the expected routing behavior.

Additional security properties can be incorporated in the routing protocol depending on the needs of the application using the network. Note that each of these properties comes with a cost (e.g. additional bandwidth overhead, CPU processing overhead):

- timeliness can be achieved through the use of timestamps;
- ordering can be achieved through the use of sequence numbers;
- authentication can be achieved by using a public key infrastructure and requiring each node to sign the messages they generate and attach their signature to the message;
- authorization can be achieved through the use of credentials;
- integrity of routing messages can also be achieved using digital signatures;
- confidentiality can be achieved using encryption, as discussed earlier;
- nonrepudiation can be achieved by requiring each node to sign the routing message and appending the signature to the message.

4.3 DYNAMIC SOURCE ROUTING PROTOCOL

In this section we focus on the dynamic source routing (DSR) described in [53]. DSR is a reactive routing protocol. Thus, routes get created only when they are needed and there is no periodic routing traffic for creating or maintaining routes. DSR also makes use of source routing. In source routing, when a node originates a data packet it puts in the header of the packet all the hops that the packet needs to traverse to get to the destination.

DSR has two main components: route discovery and route maintenance. Route discovery is used by nodes for discovering new routes when they do not have routes to destinations they need to communicate with. Route maintenance is used for discovering routes that have been broken due to node movements or failures. When a route is broken, either a different existing (less optimal) route may be used or the node may initiate route discovery for discovering a new route. We now discuss each of these two components of DSR in more detail.

When a node needs a new route to a destination it initiates the *route discovery* process by sending a route request message. The route request is broadcast by the originator and contains the address of the originator and the destination (similar to AODV). The route request also has a unique identity associated with it. When a node receives the route request, it checks the unique identity to determine whether it has seen this request before. If it has not seen the request before, it appends its address in the route request message and then broadcasts the message to its neighbors. If the node has seen this request before, it just ignores it. Once the destination receives the route request message, it sends back a route reply message that contains the route information accumulated in the route request message.

If the wireless transmission scheme utilizes bidirectional links, the route reply message is sent (via unicast) through the route that was discovered via the route request message. That route is included in the header of the route reply message, since (as mentioned earlier) DSR is a source routing protocol. If the transmission scheme allows unidirectional links then the route may not be valid in the opposite direction and therefore the destination needs to discover an independent route to the originator of the route request message. This can be accomplished by sending a route request message towards the originator of the original message, which contains inside it the route reply message to improve the efficiency of the protocol. Once the originator receives the route reply message it caches the route. Thus, information about multiple routes can be available at the source. One of these routes could be used at a given time while the others could be used in the future if necessary.

Once a node has discovered the route to the destination it can start sending packets to the destination. Each of the packets contains all the hops in the route and therefore each node can easily route a packet by finding which node is next using the path information from the packet itself.

Since in the wireless ad hoc networking environment nodes move around and links break often, DSR has incorporated a *routing maintenance* mechanism. To support this mechanism nodes try to ensure that packets forwarded by them are actually received by the next hop. DSR accomplishes that through acknowledgment mechanisms contained in the MAC layer that acknowledge each frame transmitted by nodes. If such an acknowledgment scheme is not available, nodes may require that each packet forwarded is acknowledged by the next hop (this may be changed to acknowledgement only every few packets in order to reduce overhead). When nodes do not receive acknowledgments (even after an adequate number of retransmissions), they declare the link broken and send a route error message to the originator of the packet and other nodes that may be using this link for routing packets. The originator may then use a different route from its cache (if available) or reinitiate the route discovery process.

DSR has only been described at a high level here. Several optimizations to the protocol can be used to improve performance and the reader is referred to [53] for more details.

4.3.1 Secure Routing Protocol

Secure routing protocol (SRP) [54] is a protocol that can be used as an extension of a variety of routing protocols in order to add various security related attributes to these protocols. In our discussion we focus on the use of SRP for making DSR more secure.

SRP assumes that there is a bi-directional security association (SA) between nodes that wish to exchange routing messages. The two nodes can share a secret key which they use for protecting the routing messages they exchange. SRP focuses on the nonmutable fields of the routing messages and uses the keys that the two nodes share for signing the routing messages and hence ensuring that those portions of the messages are not tampered with. As will be explained later, signing only the nonmutable fields of the DSR messages is sufficient for ensuring acceptable operation of the routing protocol.

SRP defines a header that is added to each routing message of the underlying protocol (e.g. DSR). When a node initiates the route discovery process it generates a route request message and attaches the SRP header to it. The header has the format shown in Figure 4.6. The type field defines the type of message, that is, route request in this case. The query identifier is a random 32-bit identifier. The query sequence number is a 32-bit sequence number that increases monotonically for each route request generated by the originator for the specific route destination. The sequence number is initialized when the SA between the two nodes is established. If the sequence numbers are exhausted then a new SA needs to be established between the two nodes (although this should be very infrequent due to the size of the sequence number). The message authentication code (MAC) is a 96-bit long field generated by a hash function (e.g. SHA-1, MD5). The hash function takes as an input the entire IP header, routing message, the SRP extension, and the shared key between the originator and destination of the routing message. The mutable fields are excluded from this calculation. The mutable fields include any IP-header mutable fields and the list of hops on the route that are added to the route request as it travels towards the destination.

When intermediate nodes receive the route request they check whether they have seen the message before. If they have not, they add their address in the route request message and rebroadcast the message (just as in case of DSR). Otherwise, they drop the message. Nodes keep track of the number of route requests their neighbors generate and tend to ignore (or serve at a lower rate) requests from nodes that generate an excessive number of requests. This approach helps reduce denial of service attacks by nodes that generate too many routing requests trying to overload other nodes and consume their available resources.

Once the destination receives the route request message it checks to make sure that the request is valid. The sequence number is checked to make sure that the request is not old. The destination node then generates the hash of the route request (as computed by the originator) and compares it with the value in the SRP MAC field included in the route request

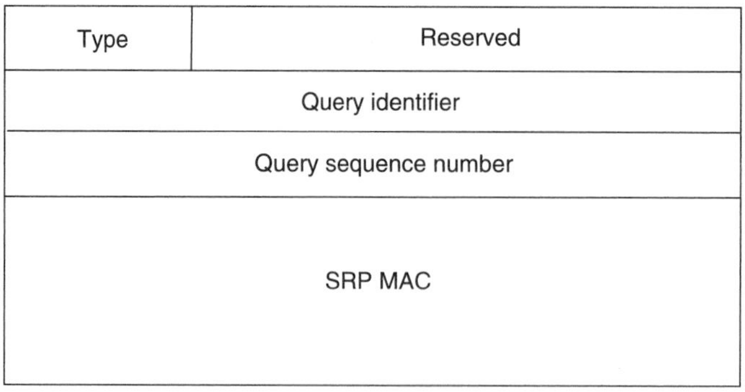

<table>
<tr><td>Type</td><td>Reserved</td></tr>
<tr><td colspan="2">Query identifier</td></tr>
<tr><td colspan="2">Query sequence number</td></tr>
<tr><td colspan="2">SRP MAC</td></tr>
</table>

Figure 4.6. SRP header format.

message. If the two values match, then the request is inferred to be valid. The destination usually receives multiple route requests with the same sequence number since the route requests may reach the destination through multiple paths. The destination responds to several of them so that the originator can be notified about the existence of multiple paths. Thus, if a malicious node disrupts some of the routes, it is still possible for the communication between the two nodes to continue.

The destination node then sends out the route reply message. The route reply includes the route determined through the route request, the query identifier and the query sequence number so that the originator can verify the freshness of the route. The MAC value in the response covers the message of the underlying routing protocol (e.g. DSR) and the remainder of the SRP header. The destination of the original route request message sends the route reply message to the originator of the route request message using source routing (assuming the underlying protocol is DSR) via the path that was discovered in the route request. When the originator receives the route reply it validates it and stores it in its route cache. The originator will probably receive multiple responses as discussed earlier and therefore will store multiple of those paths.

A major difference between this protocol and other protocols is that SRP does not attempt to protect the mutable fields in the routing messages. Those mutable fields contain among other things the complete route determined through the route discovery process. The rationale is that, if a malicious node is in the middle of a route, the only thing it can do is to change the hops in the route. The route will be detected pretty quickly as being incorrect since it is a source routing protocol and the protocol will recover through the route maintenance process. Since the node is malicious, it can disrupt the route anyway by just dropping packets routed through it, and therefore protecting the route information does not provide a lot of value. The protocol ensures that multiple paths are stored for each destination pair and, therefore, a small number of malicious nodes should not severely disrupt the operation of the network.

An optimization to the protocol is to allow nodes forwarding routing requests and responses to observe those packets and determine paths that are useful to them (and therefore store them in the routing cache). Such an optimization increases the security risk of the protocol, however, because a single malicious message could impact the routes used by many nodes. Another possible optimization is for an intermediate node that has a fresh route to the destination to respond to a route request with a route reply (assuming it has a security association with the originator of the request). In this case the route request will not travel all the way to the destination of the request. In order to support this process the SRP header includes one more fields called the intermediate node reply token, which is the MAC calculated on the message using a group key. This group key is shared among a few nodes that the originator and the intermediate node belong to. If the destination of the route request belongs to the same group, then the header does not need to be extended and the original MAC field can be used to contain this value.

SRP provides minimal protection for route maintenance errors. Nodes that discover broken links send route error messages towards the previous hops as identified in the source route contained in the packet. However, nothing prevents nodes from sending malicious route error messages declaring a route broken even though it is not. The rationale in SRP for not providing protection for maintenance messages is that malicious nodes can disrupt paths they are on anyway as discussed earlier. Of course, malicious nodes can also send route error messages purportedly from nodes on other routes, and this is a risk that SRP does not protect against.

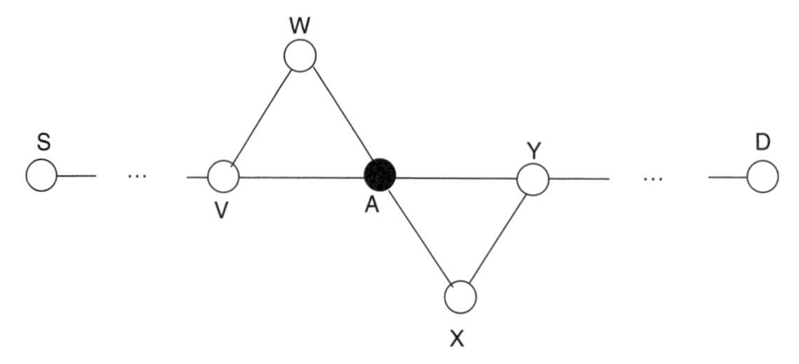

Figure 4.7. An attack against SRP.

An attack that SRP does not defend against is provided in [55]. Here the authors have used a formal model to investigate the security of on-demand source routing protocols. One of the protocols that they analyzed is SRP. To explain the attack, consider the network topology as given in Figure 4.7. Assume that source S needs to determine a route to the destination D. The attacker is shown as node A. The route request sent by source S reaches node V and is rebroadcast by node V. This transmission from V is received by A. The SRP header in this message is given by (rreq, id, sn, mac_S) while the DSR header contains information [S, D, (. . . , V)] where id is the query identifier, sn is the query sequence number and mac_S is the SRP HMAC from the source (see Figure 4.6).

The node A then broadcasts the following message [rreq, id, sn, mac_S, S, D, (. . . , V, W, θ, X)] where θ denotes an arbitrary sequence of identifiers. This message is picked up by node Y which processes the request and rebroadcasts it. Note that Y will process the message even if it is checking to filter messages not from its neighbors. If such filtering is being done, all that node A has to do is to rebroadcast the message while assuming the identity of node X. Thus the faulty path in the RREQ message will reach the destination and then make its way back in the RREP message. On the return path, when the corresponding RREP message is sent by node Y towards node X (since the path indicates that), node A overhears this transmission. Following this, node A can forward the message to node V while pretending to be node W. As a result, the RREP message will reach the source node S with a faulty path. SRP cannot protect against such an attack.

4.3.2 Ariadne

We next look at another secure routing protocol based on DSR, Ariadne [56]. Unlike SRP, Ariadne also ensures authenticity of information provided by intermediate nodes in the path between the source and the destination. Ariadne focuses on DSR and provides verification for much of the information provided by DSR. Ariadne provides similar security properties to SAODV, allowing nodes to authenticate routing messages and ensuring that those messages were not altered by nodes forwarding those messages. Securing DSR is more complex than securing AODV, though. The main reason for the increased complexity is that route request messages in DSR are modified by each forwarding node to include its own address. Therefore, the simple scheme used in SAODV (see Section 4.2.1) of originators signing the route request message does not work for DSR. We next explain Ariadne in detail.

Consider a network with bidirectional links. The network may drop, corrupt, reorder, or duplicate packets. Each node is assumed to be able to estimate the end-to-end transmission time to any other node in the network. Attacks at the physical layer and the medium access control layer are ignored. In Ariadne, nodes authenticate routing messages using one of three possible mechanisms:

- pairwise secret keys (between each pair of nodes);
- TESLA;
- digital signatures.

The first mechanism requires unique keys between every pair of nodes in the network. Thus, this requires $n(n-1)/2$ keys to be established in the network where n is the number of nodes. These keys may be installed by the key distribution center before deployment. Even though using pairwise keys is a fairly simple approach, it has both scalability and administrative challenges, as explained in Chapter 3. The second approach (use of TESLA) requires loose time synchronization, which could be challenging in a MANET environment. This approach also requires shared keys between all pairs of source–destination nodes. These keys could also be loaded on the nodes during initialization. In addition, one authentic key from the TESLA hash chain of every node also needs to be distributed to every other node in the network. The third approach (i.e. using digital signatures) on the other hand requires that each node has a set of asymmetric keys. The use of signatures has increased computational, overhead, and administrative requirements. For example, a 512 bit RSA signature generation has been found to take about 2.4–5.8 seconds, while signature verification has been found to take about 0.1–0.6 s. Hence this approach requires the use of nodes with significant computational capabilities.

The authors in their discussion consider the presence of active attackers in the network. These attackers can inject packets as well as eavesdrop on communications. There is no requirement for tamperproof modules in this scheme. We next describe Ariadne assuming the use of the TESLA approach. Use of either of the other approaches will require some minor but straightforward modifications.

Ariadne assumes that the source (node S) and destination (node D) of each communication share a MAC key for each direction denoted as K_{SD} and K_{DS}. Ariadne also assumes that every node has a one-way hash chain, which is required when using TESLA (see Chapter 2). Further, all nodes need to know an authentic key of the TESLA one-way key chain of every other node.

The authors of Ariadne assume that the sender of a route request message trusts the destination with which it communicates for authenticating nodes on the path between them. The approach in Ariadne is for the destination to authenticate route requests before sending the route reply. Each node on the way to the destination authenticates the new information that it adds to the route request. It does this by making use of one-way hash functions, as we explain later. This mechanism ensures that no intermediate node can remove any other node in the route request without being detected. The destination performs two checks and sends a route reply if both these tests are successful.

The destination then includes a HMAC in the route reply to certify that these conditions were met.

A route request in Ariadne contains the following eight fields:

- *route request*—this is a label that specifies that this is a route request message (rather than for example a route reply message);
- *initiator*—the address of the originator of the route request message;
- *target*—the address of the destination of the route request message;
- *id*—the unique id of the message so that nodes can eliminate duplicate requests;
- *time interval*—TESLA time interval of the pessimistic arrival time; it takes into account the expected delay for the message to reach the destination and the possible clock skew;
- *hash chain*—the sequence of hash values which are produced by repeatedly applying the hash function $h(x,y)$ to a number;
- *node list*—the sequence of nodes that have forwarded the route request message;
- *MAC list*—the list of HMAC values generated by each of the nodes forwarding the message.

A node that wishes to discover a route to a destination generates a route request message by populating the fields required in the message. The node initializes the hash chain field to $\text{MAC}_{K_{SD}}$ (route request, source, destination, id, time interval) and the node list and MAC list to empty.

When an intermediate node receives a route request message, it first makes sure that it has not seen the message before (by looking at the id field) and that the time interval is valid. The time interval must neither be too far in the future nor should it correspond to an interval of an already disclosed TESLA key. If the message passes those checks, the intermediate node adds its address to the node list and then replaces the hash chain value with h(address of current node, previous hash chain value). The node also appends the HMAC of the entire message to the MAC list. The HMAC is created using the TESLA key of the intermediate node for the time interval specified in the route request. The intermediate node then rebroadcasts the route request.

When the destination of the route request message receives the message, it performs two checks to validate the message. The first check is to make sure that the TESLA keys for the time interval specified in the message have not been disclosed yet. The second check is then to verify that the value of the hash chain field has been computed correctly by comparing it with the following value:

$$h[a_n, h(a_{n-1}, \text{h}\{\ldots\text{,[h(a}_1\text{,}\\ \text{MAC}_{K_{SD}}(\text{request, source, destination, id, time interval})]\ldots\})]$$

where a_i is the address of the ith node in the path and n is the total number of nodes in the list. The address of the nodes in the path can be obtained from the node list field in the route request message. This allows the target to verify the authenticity and freshness of the request.

Once the destination of the route request validates the message, it creates a route reply message. The route reply message contains the following fields:

- *route reply*—this is a label specifying that this is a route reply message;
- *target*—the address of the destination of the route request message;
- *initiator*—the address of the originator of the route request message;
- *time interval*—the TESLA time interval;
- *node list*—this includes the nodes on the path from the originator to the destination and obtained from the corresponding field in the route request message;
- *MAC list*—this list is copied from the route request message;
- *target MAC*—the HMAC of the destination computed on the above fields of the route reply message with K_{DS};
- *key list*—the list of disclosable TESLA keys of all nodes along the path. This list is initially empty. Each node that forwards the route reply message appends its key to the list.

The route reply message is sent across the route identified by the route request message. Each node on the route waits until it is able to disclose its TESLA key from the specified interval. It then appends that key to the key list and forwards the message towards the source of the route request message. This waiting does delay the return of the route reply message, but does not consume extra computational power.

When the source receives the route reply message, it verifies three things before accepting the route as valid. These three things are the values of the key list, the target MAC and the MAC list. The key list can be verified using the authentic element from the TESLA chain from each node on the path. The target MAC can be verified using the key shared between the source and the destination. Finally, each MAC value in the MAC list is verified using the TESLA keys. We illustrate the working of Ariadne in Figure 4.8.

We consider a source node denoted S that needs to determine a route to the destination node D. Node S creates a route request message, shown in step 1 above. Note that the node list and MAC list fields are empty initially. The hash chain is also initialized to $MAC_{K_{SD}}$ (route request, source, destination, id, time interval). Node S then broadcasts the message to all its neighbors. Consider node A to be one of the neighbors of S that receives the route

1. S $\rightarrow$*: [Request, S, D, id, ti, MAC_{KSD}(Request, S, D, id, ti), ( ), ( )]

2. A$\rightarrow$*: [Request, S, D, id, ti, h_A, (A), ($HMAC_A$)]
   $h_A = h$[A, $MAC_{K_{SD}}$ (Request, S, D, id, ti)]
   $HMAC_A = MAC_{K_A}$ [Request, S, D, id, ti,h_A, (A), 0]

3. B$\rightarrow$*: [Request, S, D, id, ti, h_B, (A,B), ($HMAC_A$, $HMAC_B$)]
   $h_B = h${B, h[A, $MAC_{K_{SD}}$(Request, S, D, id, ti)]},
   $HMAC_B = MAC_{K_A}$[Request, S, D, id, ti, h_B, (A,B), ($HMAC_A$)]

4. D$\rightarrow$B: [Reply, D, S, id, ti, (A,B), ($HMAC_A$, $HMAC_B$), $HMAC_D$, ( )]
   $HMAC_D = MAC_{K_{DS}}$ [Reply, D, S, id, ti, (A,B), ($HMAC_A$, $HMAC_B$)]

5. B$\rightarrow$A: [Reply, D, S, id, ti, (A,B), ($HMAC_A$, $HMAC_B$), $HMAC_D$, (K_B)]

6. A$\rightarrow$S: [Reply, D, S, id, ti, (A,B), ($HMAC_A$, $HMAC_B$), $HMAC_D$, (K_B,K_A)]

Figure 4.8. Route discovery in Ariadne.

request message transmitted by S. Node A on the receipt of this message needs to check that the time interval is valid. In addition, it also needs to check that it has not seen this message earlier. If both these checks pass, node A calculates the new value of the hash chain as shown in step 2 of Figure 4.8. In addition, node A also adds its identity to the node list and calculates a new MAC using the key K_A. The resulting message is then broadcast by node A to its neighbors and is shown as step 2 in the figure. Node B, which is assumed to receive the route request message transmitted by node A in step 2, also repeats the same procedure. This is shown in step 3. It is assumed that the destination node D receives the route request packet transmitted by node B. This concludes the route request phase.

The destination D, before transmitting the route reply, needs to perform two checks to validate the message as explained earlier. Following this, node D then creates the route reply message by copying many of the fields from the route request message received. In addition node D also creates the target MAC field as shown in step 4 above. Node D then transmits the message to node B. Node B after receiving this message, waits until it can release the TESLA key. It then appends the key to the key list and transmits the message to node A as shown in step 5. Node A repeats the same and transmits the message to the source in step 6. The source, on receiving the route reply message, performs the three checks and accepts the route if the checks are verified.

Ariadne also supports route maintenance as described in DSR. A node that determines that a link to the next hop is broken sends a route error message to the originator of the packets it is trying to forward. In order to prevent unauthorized nodes from sending such error messages, these messages also have to be authenticated by the sender.

The route error message contains the following fields:

- *route error*—label specifying that this is a route error message;
- *sending address*—the address of the node sending the route error message;
- *receiving address*—the address of the intended next hop the node generating the route error message is trying to forward packets to;
- *time interval*—the TESLA time interval;
- *error MAC*—the MAC of the preceding fields of the route error message computed using the TESLA key of the node originating the error message.
- *recent TESLA key*—the most recent disclosable TESLA key of the node sending the error message.

The route error message is routed as a regular data packet towards the source of the message that cannot be forwarded because of the broken link along the reverse path. The sender of the route error message can determine this path from the packet (since this is a source routing protocol). Each node forwarding the route error message and the destination of this message have to clear any routes from their cache that contain the failed link. They then use a different route if possible or use the route discovery process for identifying new routes. Before this, these nodes will have to authenticate the route error message. The authentication can be done once the node that originates the route error message discloses the TESLA key used in the error MAC field. The originating node does this once the time interval specified in the first route error message ends. Note that the nodes can continue to send data along the route until the error message is validated. Data using the erroneous link will result in multiple route error messages being created by the node that originates the first route error messages. All such route error messages created in the same TESLA interval would be identical. A route error message created after the TESLA interval would then contain the TESLA key needed for validation of the earlier route error messages.

Ariadne thus prevents adversaries from tampering with the routes that consist of uncompromised nodes. Use of any of the three different mechanisms discussed earlier to authenticate routing information implies that Ariadne can be used in a variety of different networks. Note that a node on the forwarding path of a route request cannot remove any nodes that are on the path. Doing so will cause a problem with the hash chain which will be detected. Similarly a malicious node on the forwarding path of a route request cannot add unnecessary nodes to the path unless this malicious node has access to the TESLA key chain of the nodes to be added.

Ariadne was analyzed using a formal model in [55] and was shown to have some security flaws. Specifically, the authors provided an attack on the routing path that Ariadne does not defend against. We explain this next for an Ariadne implementation using the digital signature option for simplicity. Consider the network topology shown in Figure 4.9. The attacker is denoted as node A while nodes S and D represent the source and the destination of the route. Node S sends a route request towards node D. This request reaches node V and is rebroadcast by node V. Both the nodes A and W receive the following route request message:

$$\mathrm{Msg}_1 = [\mathrm{rreq, S, D, id, h_V}, (\ldots.\mathrm{V}), (\ldots, \mathrm{sig_V})]$$

where id is the request identifier, h_v is the per-hop hash value generated by V and $\mathrm{sig_V}$ is the signature of node V. At this point, A does not rebroadcast this message while node W rebroadcasts it. As a result, later node A receives another route request from node X corresponding to the same source and destination. This message from node X has the following format:

$$\mathrm{Msg}_2 = [\mathrm{rreq, S, D, id}, h_X, (\ldots.\mathrm{V, W, X}), (\ldots, \mathrm{sig_V, sig_W, sig_X})]$$

From Msg_2 node A knows that node W is a neighbor of node V. Note that W need not be a neighbor of node A. Now node A can compute $h_A = \mathrm{H}[\mathrm{A}, \mathrm{H}(\mathrm{W}, h_V)]$ where H is the publicly known hash function and h_V is obtained from Msg_1. Node A can also obtain the signatures $\mathrm{sig_V}$, $\mathrm{sig_W}$ from Msg_2. As a result node A can generate and broadcast

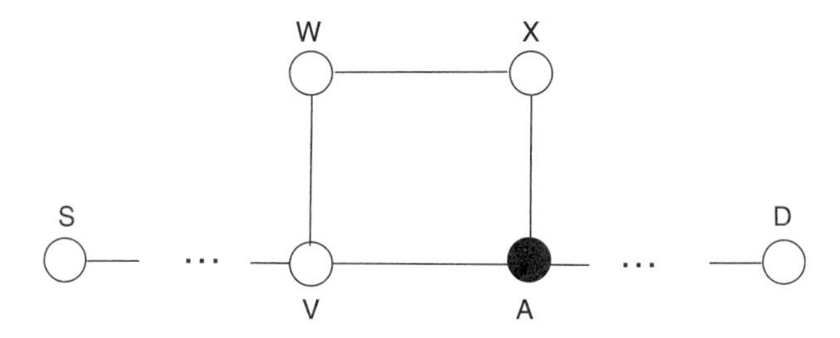

Figure 4.9. Configuration for an attack against Ariadne.

the following message:

$$\text{Msg}_3 = [\text{rreq, S, D, id, } h_\text{A}\text{, (....V, W, A), (..., sig}_\text{V}\text{, sig}_\text{W}\text{, sig}_\text{A}\text{)]}$$

This message passes all the tests at the destination, due to which node D generates the following route reply and sends it back towards the source node S:

$$\text{Msg}_4 = [\text{rrep, D, S, (....V, W, A,...), (..., sig}_\text{V}\text{, sig}_\text{W}\text{, sig}_\text{A}\text{,...) sig}_\text{D}]$$

When node A receives this route reply message, it can forward this to node V directly while pretending to be node W. As a result the source node S will receive the route (...V, W, A, ..., D) which is a nonexistent route.

4.3.3 EndairA: A Provably Secure Routing Protocol

Inspired by the formal analysis of SRP and Ariadne, the authors of [55] proposed a routing protocol that can be proven to be statistically secure. They call this protocol endairA, which is the reverse of Ariadne. They propose that each intermediate node should sign the route reply instead of signing the route request as in Ariadne. The operation of the protocol is shown in Figure 4.10. The initiator of the route request generates a message that contains the identifiers of the source and the destination as well as a randomly generated request identifier. Each intermediate node that receives this route request message appends its identifier to the route accumulated so far and rebroadcasts it. Route request messages received more than once are discarded. When the route request message arrives at the destination, it generates a route reply message. The route reply message contains the identifiers of the source and destination nodes, the accumulated route as obtained from the route request message and a digital signature of the target on the entire message. The route reply is then sent back on the reverse route as given by the accumulated route in the route request message. Each intermediate node on the reverse route verifies that its identifier is present in the accumulated route. In addition, it also verifies if the predecessor and successor nodes to it in the accumulated route are indeed its neighbors. If both these tests are valid, the intermediate node signs the route reply message and passes it to the next node in the path. As a result, the route reply reaches the source node. This node verifies whether it received the message from its neighbor and if this neighbor is the first node on the path. If so, it verifies the signatures of all the nodes in the reply. The path is then accepted to be valid if all the signatures are verified. We show the various steps assuming that source node S desires to determine a route to destination node D in Figure 4.10. The various steps are self-explanatory.

The authors have shown that the endairA is provably secure in a network that contains a single adversary and a single compromised node. The protocol is also less resource-intensive since nodes need to sign only the route reply messages. This is much

1. S→*: [RREQ, S, D, id, ( )]
2. V→*: [RREQ, S, D, id, (V)]
3. W→*: [RREQ, S, D, id, (V,W)]
4. D→W: [RREP, D, S, id, (V,W), (sig_D)]
5. W→V: [RREP, D, S, id, (V,W), (sig_D, sig_W)]
6. V→S: [RREP, D, S, id, (V,W), (sig_D, sig_W, sig_V)]

Figure 4.10. Operation of endairA.

more efficient than Ariadne, where a route request needs to be signed by a node, which might result in every node signing a message on account of the flooding of the route request. Further, in order to protect against DOS attacks resulting from flooding the network with dubious route request messages, the authors propose that the initiator of the route request signs the request. In that case the intermediate nodes will have to verify this signature before forwarding the route request message. Note that that intermediate nodes do not need to sign the route request message.

4.4 DESTINATION-SEQUENCED DISTANCE-VECTOR ROUTING PROTOCOL

The destination-sequenced distance-vector routing (DSDV) protocol [57] is a proactive distance vector routing protocol (see Section 4.1.1) that has been optimized for mobile networks. As in typical distance vector routing protocols, each node maintains a routing table that includes all possible destinations in the network, the number of hops required to reach the destination, and the next hop for reaching that destination. In DSDV, the routing table also includes a sequence number which is a number generated by the destination of the route and helps determine the freshness of routes. This helps prevent routing loops from routing updates received out of order.

Each node periodically transmits its routing information to its neighbors. This information may include the complete routing table or any updates in the table since the last transmission. Such transmissions may also be triggered when some significant event occurs (e.g. if a link breaks). In a variation of DSDV called DSDV-SQ (for sequence number) [58], events triggering a routing update include a change in a metric on a specific path or even just a new sequence number. Nodes may delay sending routing updates if they believe that other updates are expected to arrive soon (based on past experience). This reduces the number of updates and improves the protocol overall performance.

The information that is transmitted in the routing update broadcasts (whether triggered or periodic) includes the new sequence number of the routing update and for each route transmitted the destination's address, the number of hops to the destination, and the sequence number of the routing update generated by the destination of this route. If a particular link is broken then the metric (i.e. number of hops) is set to infinity.

When a node receives route advertisements from its neighbor, it adds one hop to each of the routes in the routing update (since it needs one more hop to get to the destination through its neighbor). It then replaces any of the routes in its own routing table if the route through its neighbor for that path has a larger sequence number or if the path is shorter than its current path and the path has the same sequence number as in its routing table. If the sequence number in the route update is smaller than the one in the routing table of the node, the route is ignored because it is out-of-date.

We can observe from this description that DSDV guarantees loop-free routes, but DSDV requires trustworthy nodes for its operation. A malicious node can disrupt the operation of the protocol. It can do this in several ways. It could provide an incorrect cost metric in the route update messages that it transmits, for example, pretending to be closer to some destinations than it actually is. It could manipulate the sequence numbers associated with each destination in the route update messages, eliminating fresher routes. Hence, it is necessary to enhance DSDV by securing the information propagated by the routing updates. The three protocols, namely SEAD, Super-SEAD, and S-DSDV, that we discuss in this section indeed attempt to do this.

4.4.1 Secure Efficient Distance Vector Routing (SEAD)

Attacks on the sequence number or on the metric could be addressed by authenticating the routing updates sent by DSDV. Each node could be required to digitally sign its routing updates, but a problem with the use of asymmetric cryptographic primitives is related to the resource constraints in ad hoc networks. Additionally, a compromised node can still modify the sequence numbers and the metrics in the routing updates it sends out. This requires that a detection mechanism be coupled with the routing protocol.

SEAD [59] attempts to address these problems while providing a secure version of DSDV. SEAD prevents a malicious node from increasing the sequence number or decreasing the metric associated with a destination in the routing updates the malicious node sends out. SEAD does this while making use of one-way hash chains (see Chapter 2).

SEAD assumes that a diameter for the network can be specified. This diameter is the maximum distance between any two nodes in the network. We denote the diameter by k_{m}. Note that this implies that this routing protocol would not scale to a large network, but this might be okay since a distance vector routing protocol is usually used for a small or medium-sized network.

SEAD provides two main mechanisms, namely a scheme to ensure the security of the metric and sequence number authentication as well as a scheme for neighbor authentication. We discuss these schemes next.

4.4.1.1 Metric and Sequence Number Authentication for SEAD SEAD uses one-way hash chains for authentication of metrics and sequence numbers in routing updates. This limits the ability of malicious nodes to modify these fields in routing updates. Each node during initialization selects a random seed and applies a hash function $h(x)$ repeatedly to that number. An element from this hash chain is then used for authentication. The detailed process is given below.

The first phase is system initialization. The following steps constitute this phase.

1 Every node in the network constructs its own hash chain by starting with a secret. Let s_{m} denote the maximum sequence number that the node expects to use and k_{m} denote the maximum number of hops in the network. Assume $N = s_{\mathrm{m}}\, k_{\mathrm{m}}$. Then the hash chain has a length of $N + 1$. The hash chain of a node V is arranged into s_{m} groups of k_{m} elements, as we show in Figure 4.11.

2 The last element, also referred to as the anchor of the hash chain of node V, h^{N+1} (secret), is not in any group, as shown in the figure. This value is used to authenticate routing update entries that specify this node as the destination. The groups are numbered from right to left, as shown in Figure 4.11. Thus, groups with a lower number are associated with a higher number of hashes of the basic secret. Further, hash elements within a group are numbered from left to right starting from 0 and ending at $k_{\mathrm{m}} - 1$. Thus, each element of the hash chain can be uniquely located given two numbers, namely the group id and the index of the element within the group. So if the group id is X (note that group id starts from right) and the element number is Y, then this corresponds to j hashes of the secret where $j = (s_{\mathrm{m}} - \mathrm{X}) \cdot k_{\mathrm{m}} + \mathrm{Y} + 1$. We show this in Figure 4.12. Let us denote this as hc[V, X, Y] to represent the hash element of node V corresponding to a sequence number of X (representing the group id) and number of hops Y (representing the

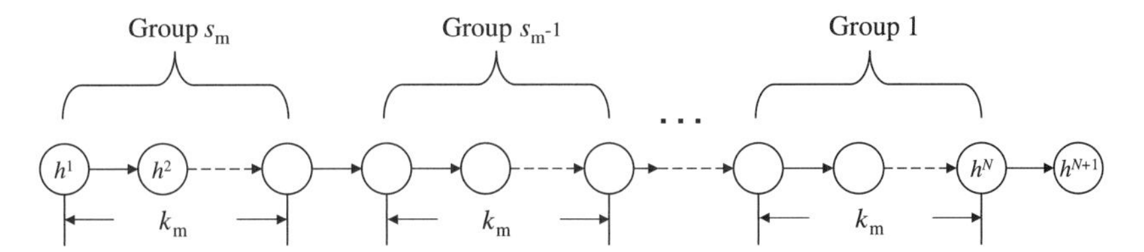

Figure 4.11. Hash chain used in SEAD.

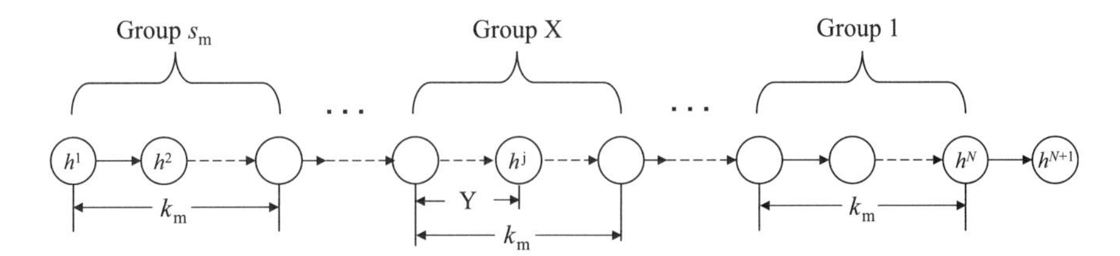

Figure 4.12. The element *j* of hash chain corresponding to X group and Y element.

element in the group). Since k_m is the maximum number of hops, there are enough elements in each group.

3 The node V then distributes the value h^{N+1}(secret) to all nodes in the network. This distribution can be achieved either through signed public key certificates or through symmetric cryptography or through other schemes [59]. h^{N+1}(secret) is used as the basis for the authentication of messages sent during network operation.

SEAD leverages the properties of a one-way hash chain. Due to this, a node cannot forge a routing update with a higher sequence number, nor can it forge a routing update with the same sequence number but with a lower metric,[6] as we explain below. Note that a routing update with a larger sequence number takes precedence over a routing update with a smaller sequence number.

We next describe the steps involved during these route updates.

1 *Transmission of Route Updates*—consider a node A advertising a route for node B with a distance of Y and a sequence number of X. This would be one record in the routing update that node A transmits. Note that there could be several records (entries) in the routing update message that node A transmits where every line corresponds to a record. SEAD requires that each record be supported by an authentication value.

   a The authentication value is given by hc[A,X,0] if A and B are the same node. This indicates that the node is at a distance of 0 from itself. This entry can be generated by the node itself since it has its entire hash chain.

   b Further, in case nodes A and B are distinct, then the authentication value is given by hc[B,X,Y]. Node A will not have access to this hash chain as it corresponds to Node B, but node A can obtain the value hc[B,X,Y] by applying the hash function to the authentication value for destination B it received in the routing update from its neighbor (which is hc[B,X,Y-1]).

[6]Since this would require traversing in the reverse direction on the hash chain. This would involve breaking the one way property of hash functions.

2 *Reception of Route Updates*—consider node C that receives the routing update from node A above. Node C has to first validate the routing update from node A that it has received and then revise the routing update.
   a *Validation*—consider a node F that is j hops away from node A. Then the authentication value that node C will receive in a routing update from node A will be hc[F, S, j] where the sequence number associated with destination F is S and the metric is $j - 1$. Node C can authenticate such an entry it received from its neighbor by applying the hash function $S^*k_m - j$ times to the hash value on the routing update (S, k_m, j are all known for node F). Thus, $h^{S^*k_m - j}$ (hc[F, S, j]) $= h^{S^*k_m - j}$ [$h^{(s_m - S)^*k_m + j + 1}$ (secret)] $= h^{sm^*km+1}$ (secret) $= h^{N+1}$ (secret) corresponding to node F. As was described earlier, h^{N+1}(secret) is known for each node (and therefore for node F) by every node in the system (hence by node C). Therefore the correct value can be verified. Applying the hash value can be optimized by maintaining the hash value for the previous sequence number and then calculating how many more times to apply the hash function to get to the right value. Thus, if the routing update entry corresponding to node F is verified, then node C will revise this entry as given next before propagating it further.
   b *Revision*—node C after verifying each entry in the routing update received from node A updates these entries. To do this, it adds one hop to the metric describing the distance to node A. It then applies the hash function to the hash value in the routing message record that node C has received. For example, consider node F. The revised entry for node F that node C will send out in its own routing update (assuming it is shorter than its existing entries) will contain the same sequence number S while the metric is increased by 1 to j. Additionally, the authentication value associated with this entry is given as hc[F, S, $j + 1$], which is obtained by h(hc[F, S, j]). Thus, node C can obtain the new entry in the hash chain value of node F.

3 The above process repeats at different nodes.

Thus, we see that the authentication approach is used to ensure that the sequence numbers cannot be increased, nor can the metric be reduced. Malicious nodes cannot affect the sequence number because they cannot create the right hash value and therefore cannot launch attacks that try to manipulate the sequence numbers. Similarly, a malicious node will not be able to reduce the metric associated with a destination that it has received in a routing update. Doing so would require the malicious nodes to either have access to the hash chain of a different node or would require the malicious node to violate the one-way property of hash chains. Both of these are not possible. Further, given a group of colluding nodes, the shortest distance that these nodes can claim to a destination corresponds to the shortest distance to the destination from any node in this colluding group.

This approach is open to some attacks from malicious nodes, though. Such nodes may increase the metric for the route more than once, let us say k times, and then apply the hash function the correct number of times (k times in this case). Other nodes will not be able to detect such malicious behavior (unless they use IDS techniques discussed in Chapter 5) because the behavior results in correct authentication. Such malicious behavior does not impact the operation of the network significantly. This is because such attacks will make routes through the malicious node appear longer than they actually are. This makes such routes less likely to be selected, which might not be what the adversary desires. Another possible malicious behavior is for a malicious node to not increase the metric of a route it has received from its neighbor and include the previous metric and the hash function it received from its neighbor in its own routing updates. Such behavior cannot be detected by SEAD because the metric and the hash value are consistent (they are unaltered from what the previous node advertised). Therefore in this case the route through the malicious node will appear to be one hop shorter than it actually is. Thus, we see that SEAD only ensures that the metric contained in routing updates that a node sends out is a lower bound. We next discuss another mechanism provided by SEAD, namely the neighbor authentication.

4.4.1.2 Neighbor Authentication for SEAD SEAD supports a scheme for authenticating the origin of routing update messages through a neighbor authentication scheme. In other words, neighbor authentication is used to ensure that the advertiser of the information is authentic. This is required due to the fact that a node selects the next-hop router towards a destination to be the source address of the routing update with the shortest distance towards this destination. This is also required to avoid routing loops by nodes spoofing the address of the originator of the message. If clock synchronization can be achieved, then a broadcast source authentication scheme such as TESLA (see Chapter 2) can be used.

Since clock synchronization is very challenging in a MANET environment, an alternative approach has also been proposed in [59]. That scheme assumes that there is a shared secret key between each pair of neighboring nodes. Whenever a node sends a routing update, it includes in that message the Message Authentication Code for that message for each neighbor using the shared secret key for that neighbor. This increases the size of the routing update (especially if there are many neighbors), but allows each of the neighbors to authenticate the source of the message. This approach assumes that each node knows its direct neighbors either through sensing at the medium access layer or by overhearing the broadcast routing updates its neighbors send out. Another simple way for a node to determine valid neighbors that leverages the earlier mechanism of SEAD is for the node to trust any routing update with a metric of zero that has a valid authenticator.

SEAD has other differences with DSDV. Some of them are given below:

- SEAD sends triggered updates immediately and does not wait to receive other further updates that may be imminent (based on past experience). This increases the overhead of the protocol but makes it more secure because malicious nodes may exploit such delays to preempt the propagation of the correct routing updates by injecting routing updates that force more traffic through it.
- SEAD does not allow nodes to increment sequence numbers to a destination when a link is broken. This again results in inferior routing performance but is a required change due to the security scheme used for authenticating the sequence numbers. Only the destination node has the ability to create a new valid sequence number that can be authenticated.

4.4.2 SuperSEAD

Another related protocol, called SuperSEAD, was proposed [59]. This enhances SEAD in ensuring that same-distance fraud cannot be perpetrated. Note that a malicious node in

SEAD can forward a routing update entry without increasing the metric or hashing the authentication value, as discussed earlier. SuperSEAD protects against such same-distance fraud. It does this by making use of the concept of hash tree chains.

The approach here is to tie the authenticator (the hash value) to the address of the node sending the route update. An adversary therefore cannot replay the same message as the one sent by the neighbor unless the adversary takes the identity of the neighbor. In order to achieve this, a special one-way chain called the hash tree chain is constructed. A hash tree chain is a hybrid between a hash tree and a hash chain. The hash chain is used in the same way as in SEAD to ensure that nodes cannot decrease the distance metric while the hash tree property is used to authenticate the identity of a node. We next describe the construction of the hash tree chain.

Consider the hash chain as constructed earlier and as shown in Figure 4.11. A hash tree is constructed between each pair of values in this hash chain. The hash tree is built as follows. Given each value in the chain, a set of values denoted $b0, b1, \ldots, bn$ is derived using a one-way hash function. The value bj is given as $bj = h(v \parallel j)$ where v is the value of the hash chain and j represents the identity of a node in the system. A hash tree is then built above these values as shown in Figure 4.13. In this figure we assume that there are four nodes in the system and hence the four vertices ($j = 0, \ldots, 3$) derived from the hash chain value. These vertices correspond to the leaf of the hash tree. The root of the tree becomes the next value in terms of the generation of the hash chain.

This is repeated for the entire hash chain giving a hash tree chain, as shown in Figure 4.14. The generation of the hash tree chain is from left to right while the usage as described earlier for SEAD is from right to left in groups of k_{m} each. Thus, we see that the identity of each node in the network is encoded into each element of the chain.

Now consider a node with the identity 1. Such a node corresponds to the node b1 in Figure 4.13. In SEAD this node sends the value of the hash chain as an authenticator, but with SuperSEAD, this node uses the elements in the hash tree to authenticate the routing updates. To do this, node with identity 1 forwards the values b_1, b_0' and b_{23} (the shaded nodes in Figure 4.13) along with the routing updates to its neighboring nodes. Each neighboring node can then use these values to verify the authenticity of the hash chain (since each neighboring node has the authentic hash chain anchor corresponding to the destination as in case of SEAD). Now consider a neighboring node with identity 3. Such a node cannot replay the hash values received from node with identity 1 since the node with identity 3 will have to forward the values b_3, b_2' and b_{01} along with the routing updates. In other words, each node has to specify a different path up the hash tree. Therefore a node cannot advertise the same metric as its neighbor and needs to follow the hash tree structure to the next level. Thus, the neighboring node is forced to increase the distance metric. A malicious node that does not do so can be easily detected. Packet leashes (see Section 4.7.1) can be used here to ensure that the node that sent it the packet is really its neighbor. Thus, packet leashes provide hop-by-hop authentication, thereby preventing an adversary from impersonating another node.

The hash chain as well as the hash-tree chain though have two problems namely:

- An adversarial node with fresh (higher) sequence number information can use this to fabricate lower metric authenticators for old (lower) sequence numbers.
- An attacker can flood a node in the network with route updates containing recent or very large sequence numbers with false authenticators.

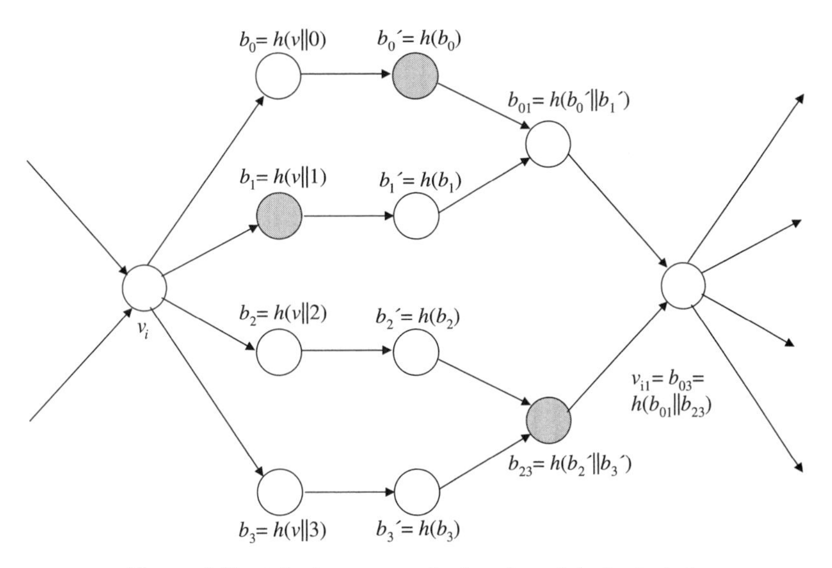

Figure 4.13. A hash tree on a single value of the hash chain.

In the first attack, the adversarial node can then combine this with other attacks such as rushing (see Section 4.7.2), resulting in valid nodes accepting false routing updates, even though this might be for a small time interval until the update with the higher sequence number reaches the valid node. Jamming messages with higher sequence number from reaching valid nodes can increase the time interval for which the false route updates are used by the valid nodes. The second attack forces nodes to perform a large number of hash operations before concluding that the authenticator and hence the route update is false. To address this, the authors propose a mechanism called as the tree-authenticated one-way chain in [59].

In this scheme, a new hash chain is used for each sequence number. Each node creates a tree-authenticated one way chain as shown in Figure 4.15. In this figure, we show four hash chains corresponding to four sequence numbers. The root (leftmost value) of each hash chain can be obtained as H(X,*s*) where H( ) is the hashing algorithm, X is the

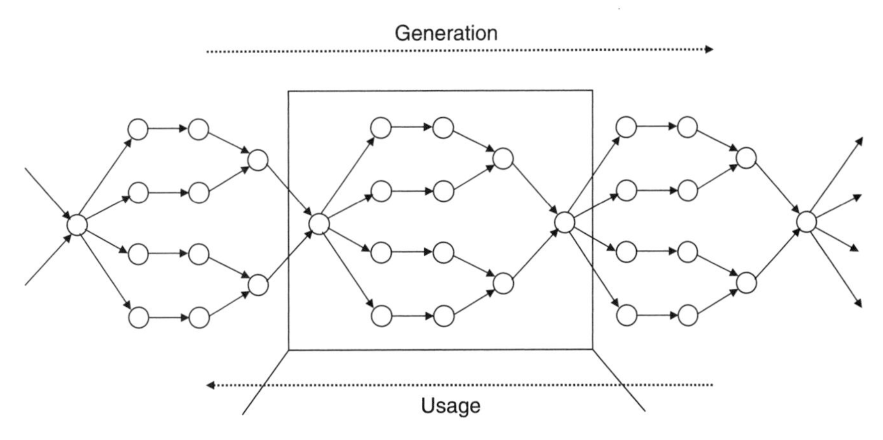

Figure 4.14. Hash tree chain.

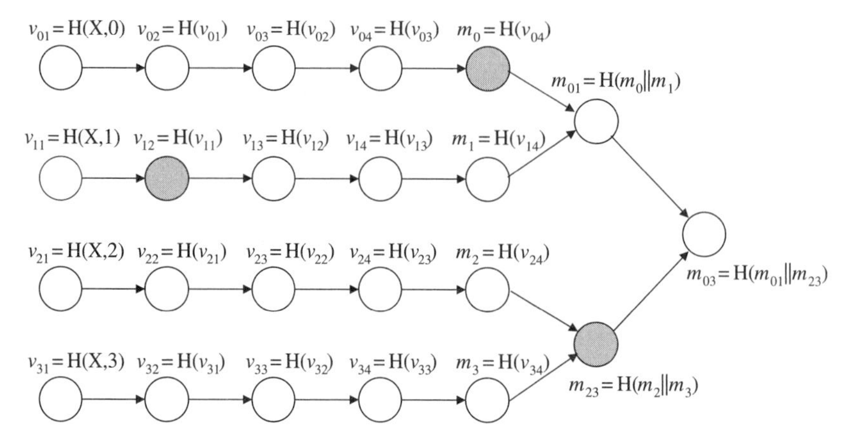

Figure 4.15. Tree-authenticated one-way chain (to authenticate a routing update with sequence number 1, metric 1, the shaded values have to be used).

random secret and s is the sequence number. The anchor of each hash chain is H^{k+1} (X,s) where k is the maximum value of the metric. To allow nodes to authenticate these anchors, each node also builds a hash tree using the anchors as leaves as shown in the figure. This node then needs to send the root of the hash tree (rightmost node in the figure) to every node in the network authentically. This hash tree root is then used to authenticate entries in route updates that have this node as the destination.

A node that sends a route update to itself with a new sequence number s includes the root of the hash chain H(X,s), the anchor of the hash chain and the path to the root of the hash tree. To authenticate the update, another node verifies the anchor for that sequence number by following the path to the root of the hash tree. It can then verify the authenticator corresponding to the sequence number and given metric by hashing it a fixed number of times and comparing with the anchor of that chain. If these tests are validated, such a node can then propagate the route update after updating the metric and the hash value. Any other node can repeat the same tests to validate the route updates. Thus, any node can verify the authenticator for sequence number s and metric m. Note that this approach addresses both the problems given earlier.

4.4.3 S-DSDV

S-DSDV [60] is another proposal to enhance the security of DSDV. S-DSDV ensures that nodes can neither increase nor decrease the distance metrics provided that no two nodes are colluding. S-DSDV assumes that any node in the system shares a pairwise secret key with every other node in the system.

S-DSDV classifies routes advertised by any node into two categories namely authoritative routes and unauthoritative routes. A route advertised by a node is authoritative if the route is for the node itself (therefore with a metric of 0) or the route it advertises is an unreachable destination (therefore with an infinite cost metric). Thus, routes to destinations other than the node itself, which will have a nonzero and finite cost metric, are all considered as unauthoritative routes.

Now when transmitting a route update to a neighbor, a node is expected to calculate the HMAC on the routing update using the key it shares with the neighbor. The neighbor node on receiving this route update first validates the HMAC value of the route update. If the HMAC value does not match, then the message is dropped. If the HMAC value matches, then the neighbor node classifies the entries into authoritative and non-authoritative entries.

1 The authoritative entries are accepted without any further check.
2 The nonauthoritative entries are accepted only after checking for their consistency.

To explain the consistency check, consider three nodes A, B, and C. Consider a routing update received by node C from node B, where node B advertises a route to the destination A with a metric of 1. Node C considers this as a nonauthoritative entry from node B. In order to check the consistency of such a nonauthoritative entry, node C needs to know the node from which node B received such a nonauthoritative entry. In this case, the node from which node B received the nonauthoritative entry is node A. Then node C sends a request to node A asking for the metric that node A has to the destination A and also information about the metric related to the path between nodes A and B. In addition, node A is also expected to send the sequence number corresponding to the destination of the nonauthoritative entry for which information is requested. Node A sends all this information to node C while ensuring the integrity of the message using the HMAC key shared between nodes A and C. When node C receives this message, it checks to see if the sequence numbers have been changed in the message received from node A and the routing update received from node B. Additionally, node C will also check if the value of the metric as received from the routing update from node B is the sum of the metric as received in the message from node A and the cost associated for routing between nodes A and B. The metric is accepted as reliable if it passes these tests.

Thus, S-DSDV ensures that the metrics associated with DSDV cannot be manipulated by malicious nodes acting along. A problem with S-DSDV, though, is the tremendous routing overhead associated with the consistency checks on the unauthoritative routes.

4.5 OPTIMIZED LINK-STATE ROUTING PROTOCOL

Optimized link-state routing protocol (OLSR) [61] is a typical proactive link state routing protocol that has been optimized for use in a MANET environment. As discussed in Section 4.1.1, in link-state routing protocols, nodes transmit routing advertisements listing their direct neighbors. These advertisements, called link-state advertisements (LSA), are flooded throughout the network. Since MANET networks usually have limited available bandwidth, OLSR incorporates a concept for efficient flooding of the routing messages throughout the network. This is based on the concept of multipoint relays (called MPRs), which will be described in this section.

The purpose of MPR is to optimize flooding of the link-state updates. In a typical routing protocol routing messages are flooded as follows. Each node that receives a message floods the routing message in all directions (to any nodes within its transmission range). Figure 4.16 shows how messages that node A generates would be flooded throughout the network.[7]

[7]The figure shows a simplified view of transmitted messages. Messages travel in all directions, including backward, i.e. towards the sender of the previous message. Those messages are omitted from this figure.

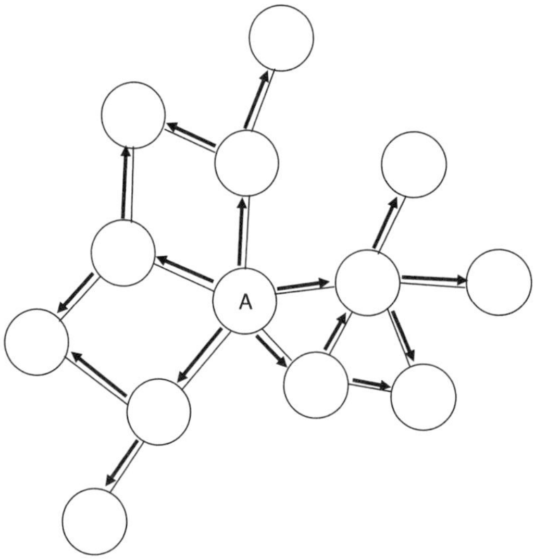

Figure 4.16. Flooding of routing messages.

As is clear from the figure, this flooding mechanism is not very efficient since various nodes can receive the same message multiple times. In OLSR a more efficient scheme for propagating routing information has been created. Here each node assigns the task of propagating its LSAs only to a few of its one-hop symmetric neighbors. These special nodes are selected in a way that ensures that the LSAs will reach all of its two-hop neighbors. Those nodes selected for relaying the LSAs are MPRs. For example, as shown in Figure 4.17, when node A transmits a routing update, it broadcasts it to each of its one-hop neighbors. All of those nodes receive and process the message, but only the nodes that are MPRs for node A retransmit the message. This reduces the number of retransmissions and hence the overhead generated by OLSR.

We next discuss the details of the OLSR routing protocol. In OLSR, each node transmits hello messages periodically (e.g. every second) over each of the node's interfaces. The main purpose of the hello message is to enable a node to discover its direct (one-hop) neighbors. Hello messages are only broadcast to the one-hop neighbors and do not get transmitted past the one-hop neighborhood of a node. The hello message contains the name of the originating node, the one-hop neighbors that the originating node has discovered so far, and the nodes that the originating node has selected to be its MPRs. Once a node hears a hello message it checks whether the message originated from a new neighbor, and if so, it updates its one-hop neighborhood list.

The hello message is also very important for supporting the MPR concept. Each node checks the hello messages received from its neighbors to see whether it has been selected to be an MPR by any of the neighbors. If so, the node needs to flood the routing updates generated from such neighbors that have selected it to be their MPR. Every node can also figure out which nodes are two hops away from it through the hello message, because each of its one-hop neighbors lists in the hello message all the nodes that are within one hop of them. Each node selects its MPRs based on the two-hop neighborhood so that each of its two-hop neighbors can be reached through the MPRs.

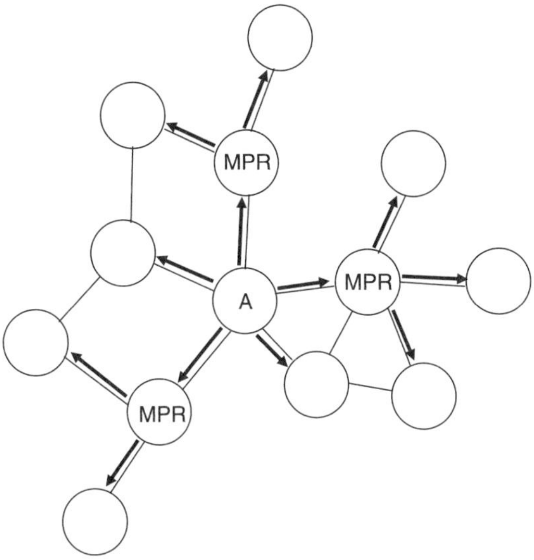

Figure 4.17. OLSR routing protocol.

Link-state updates are transmitted through the network via a message that is called a topology control (TC) message. TC messages are flooded throughout the network and every node can then recalculate its own routing table using this information. The flooding process is accomplished through the MPRs, as shown in Figure 4.17. In the case of OLSR it is not required for each node to advertise all its neighbors. It is adequate to advertise only the nodes that a node has selected as its MPRs.

OLSR also includes two additional message types: the host and network association (HNA) messages that are used by nodes for advertising connectivity to external networks, that is, networks that are not participating in the OLSR routing protocol, and the multiple interface declaration (MID) messages that are used only by nodes that have multiple interfaces that are participating in the OLSR routing protocol so that other nodes can associate the different interfaces with the same node. OLSR has no support for message authentication and is therefore vulnerable to a variety of attacks.

4.5.1 Secure Extension to OLSR

In the literature [62–64], schemes have been proposed for extending OLSR to make it secure against attacks. The main idea they propose is to use digital signatures for authenticating the OLSR routing messages. Such authentication may be done on a hop-by-hop basis or on an end-to-end basis. Halfslund *et al.* [62] focus on the hop-by-hop approach, in which each node signs OLSR packets as they are transmitted (such packets may contain multiple OLSR messages originated by a variety of nodes). The next hop verifies the authenticity of the message, strips the signature from the previous node, and adds its own signature. Therefore, the signature only verifies that the node that forwarded the traffic is the one that signed the message but does not verify the authenticity of the original message. Authentication of messages is based on symmetric keys that nodes share and the signature is created using some kind of a hash function such as SHA-1. Rafto *et al.* [63] and Adjih *et al.* [64] discuss schemes for authenticating OLSR messages on end-to-end

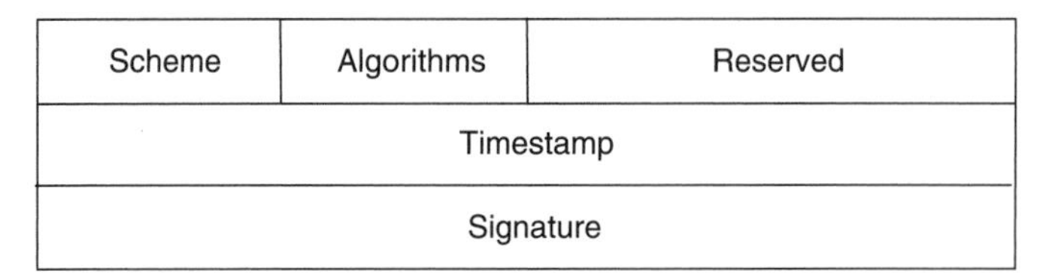

Figure 4.18. Basic signature extension.

basis so that nodes receiving OLSR message can authenticate the node that generated the original message rather than just the node forwarding the message. We will discuss both types of schemes in more detail next.

We first discuss the scheme given in [62]. The focus here is on authentication on a hop by hop basis. Figure 4.18 shows the basic signature extension that is attached to each OLSR packet. The scheme and algorithm fields shown in Figure 4.18 specify the algorithm that is used for creating the signature (Halfslund *et al.* [62] use SHA-1 as an example). The signature is generated by applying the hash function (based on the scheme and algorithm defined in the message) on the OLSR packet header, the OLSR routing messages contained in the OLSR packet, the fields in the signature extension shown in Figure 4.18 except for the signature field, and the shared secret key. The timestamp field is needed in this scheme so that malicious nodes cannot launch replay attacks by just moving to a different neighborhood and replaying messages they recorded earlier.

In order for this scheme to work, nodes need to know the current time of their neighbors. This does not require that nodes synchronize their clocks. It only requires that the nodes know the approximate time difference between themselves and their neighbors. It is also assumed that clocks move forward at about the same speed. In order for neighbors to discover the relative time of their neighbors, Halfslund *et al.* [62] propose the following scheme. When node A needs to discover the relative time of its neighbor, it initiates the timestamp exchange process by sending a challenge message as shown in Figure 4.19.

The destination field contains the IP address of the node (say node B) that node A is trying to get the time value for. The random value field contains a random number to avoid replay attacks and the signature is created by applying the hash function as described earlier. The destination node, that is, Node B, then verifies the authenticity of the challenge message and then responds with a challenge-response message with the format shown in Figure 4.20.

The challenge-response message contains the IP address of node A, the random number, and its timestamp. The response signature field is created by applying the hash function to the IP address of node B, the random challenge, and the shared key. The signature field is produced by applying the hash function to the entire message and the shared key. When node A receives the challenge-response from node B, it first verifies its

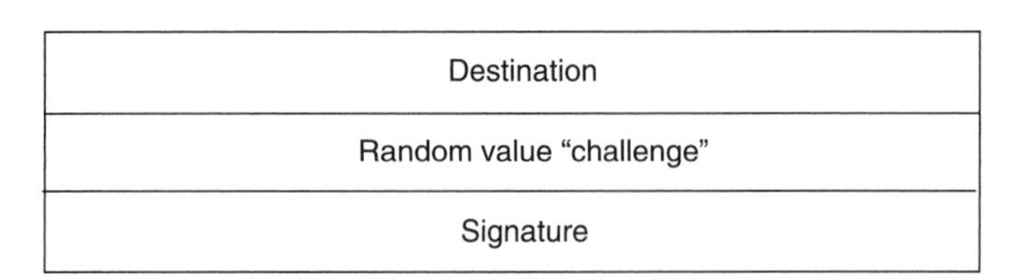

Figure 4.19. Challenge message.

Destination
Random value "challenge"
Timestamp
Response signature
Signature

Figure 4.20. Challenge-response message.

authenticity (by applying the hash functions as node B and comparing the results with what is contained in the message). Node A then sends its own timestamp to node B by creating a response message as shown in Figure 4.21.

Rafto *et al.* [63] propose a scheme for authenticating OLSR messages on end-to-end basis. The key idea of the scheme is as follows. When nodes send out OLSR messages they attach an ADVanced SIGnature (ADVSIG) message as defined in Figure 4.22. Nodes that advertise links sign the messages so that the origins of the message can be authenticated. Another key concept in this scheme is that, when nodes advertise a link to another node (e.g. through TC messages), they attach a proof that the link actually exists through a signed hello message from that node. This approach of securing OLSR requires a PKI scheme or some other key management scheme that can be used for verifying authenticity of messages. The scheme also requires a timestamp scheme [64] that allows nodes to have consistent timestamps as discussed earlier in this section.

In Figure 4.22, the signature method specifies the functions used for signing the messages. The MSN referrer field specifies the message sequence number the ADVSIG message refers to. The global timestamp message contains the timestamp. The global signature field contains the signature for the OLSR message and the ADVSIG message attached to it. The signatures of Certificates (#1–#n) are signatures of hello messages that are generated by the node sending the message and for which there are no proofs yet (because a signed hello message has not been received by the other side of the link). The timestamp and signature proof (#1–#n) contain the time when a signed hello message was received by the other side of the link advertised in the hello or TC message and the signature that was contained in that message (and was generated by the other side of the link).

4.5.2 Secure Link-State Routing Protocol

Secure link-state routing protocol (SLSP) [65] is a proposed scheme for securing link-state routing where security is achieved via the use of asymmetric primitives. SLSP assumes that each node in the network has a public–private key pair. Each node broadcasts its certified key to all nodes within its zone (i.e. all nodes that are within R hops from it). These broadcasts are periodic or when conditions require it (e.g. when there are substantial changes in the network topology), which enables new nodes entering the zone to discover the key. Certification of public keys can be achieved through a distributed certificate authority (see Chapter 2). We next describe how this scheme works.

Destination
Timestamp
Response signature
Signature

Figure 4.21. Response message.

Signature method	Reserved	MSN referrer
Global timestamp		
Global signature		
Signature of certificate #1		
...		
Signature of certificate #*n*		
Timestamp of proof #1		
Signature of proof #1		
...		
Timestamp of proof #*n*		
Signature of proof #*n*		

Figure 4.22. ADVSIG message format.

The first step in any link state routing protocol is neighbor discovery. Neighbor discovery in SLSP is achieved through signed hello messages that contain the medium access control MAC address and IP address of the node. This approach attempts to ensure that a single node cannot impersonate multiple nodes, and that duplicate IP addresses can be detected quickly.[8] Messages from nodes violating the expected behavior may be discarded. This approach also allows nodes running SLSP to calculate the rates at which other nodes generate routing messages. Messages from nodes that are generating too many routing messages can be given lower priority, preventing those nodes from launching DoS attacks by generating too many routing messages.

As is typical in link-state routing protocols, once nodes discover their neighbors they advertise their direct neighbors through link-state update (LSU) messages. In SLSP, LSU messages have the header shown in Figure 4.23. The originator of the LSU specifies

[8]A node can still take on a different identity by changing both the MAC and IP addresses while using a different signature. Thus, the node has to have multiple pairs of public–private keys.

Type	R_{LSU}	Reserved
Zone_radius		
SLSP_LSU_SEQ		
LSU_signature		
Hops_traversed		

Figure 4.23. Header of an LSU message.

the radius R of its zone in the R_{LSU} field. It then selects a random number x and applies a hash function on the number $h(x)$. $h(x)$ is put in the hops_traversed field and $h^R(x)$ in the zone_radius field. Subsequent nodes that forward the LSU apply the hash function on the hops_traversed field and replace that field with the new value. Thus we have hops_traversed = h(hops_traversed). The SLSP_LUS_SEQ contains a 32-bit sequence number that is increased monotonically with new LSUs generated by the node. The node also appends the signature into the LSU_signature field. The TTL value of the IP packet is set to $R - 1$ so that routing advertisements stay within the zone. It is also possible that the certified key is attached by the node to the LSU message itself (rather than in the periodic broadcast of the key). That ensures that the node receiving the LSU will be able to verify the message.

A node that is within the zone of the originator of an LSU most likely has the public key of the originator. When such a node receives the LSU message it can verify the authenticity of the message. Authenticity of the message is verified as follows. The node checks the hops_traversed field. The node knows the number of hops the message has already traveled which is the zone radius R minus the TTL value in the current packet (the TTL value is decreased by one at each hop). Therefore, the node applies the hash function TTL times on the hops_traversed field and compares it with the zone_radius field [which equals $h^R(x)$]. The two values should be equal:

$$h^{\mathrm{TTL}}(\text{hops\_traversed}) = h^{\mathrm{TTL}}[h^{R-\mathrm{TTL}}(x)] = h^R(x) = \text{zone\_radius}$$

If the LSU is validated, the node decrements the TTL, applies the hash function to the hops_traversed field and then rebroadcasts the LSU. The LSU is only kept until an LSU from the other side of the link is also received. When the LSU is confirmed it is used for updating routes on the node, otherwise the LSU is discarded. There is still a possibility that two nodes may collude to advertise a nonexisting link. SLSP can not prevent such attacks.

4.6 ANONYMOUS ROUTING PROTOCOLS

The routing protocols that we have discussed so far in this chapter focus on ensuring that the information propagated by the routing protocol is authentic. A facet of routing that has not been considered by these routing protocols is traffic analysis. Traffic analysis is a passive attack where an adversary observes the network traffic in order to obtain sensitive information about the communicating entities. The information could be related to the identities of the communicating parties, or to the network traffic patterns or even to the changes in the traffic pattern. For example, in a network formed on the battlefield, such information could be used to infer locations of command centers or could be used to infer impending activity. Routing protocols such as SAODV or Ariadne use the fixed node identifiers in plaintext during the routing process. As a result of this, adversaries can easily perform traffic analysis.

Prevention of traffic analysis requires a routing protocol that maintains the anonymity of nodes in the network. Such anonymous routing protocols are orthogonal to the secure routing protocols that we have seen earlier. The next two routing protocols, namely ANODR and MASK, are such anonymous routing protocols designed to prevent traffic analysis in ad hoc networks. These protocols do this by hiding the sender and/or the recipient's identity from an outside observer. This makes it impossible in some cases and harder in others for an adversary to correlate eavesdropped traffic information to network traffic patterns. The approach here is to remove the identities of the nodes present in the packet. Instead, the packets make use of link identifiers. These link identifiers are created between two neighbors and hence an intended receiver can easily recognize a packet destined for it. On the other hand, the link identifier appears as a random number to other nodes. Further, these link identifiers change for every link due to which an adversary will not be able to trace the path of a packet. We explain these protocols next, starting with ANODR.

4.6.1 ANODR

ANODR [66] is an on-demand routing protocol. As a result, there are two important phases of the protocol, namely route discovery and data forwarding. The source initiates route discovery by locally broadcasting a RREQ packet. The RREQ packet contains four fields as follows:

- *RREQ*—this is a label that specifies that this is a route request message;
- *seqnum*—this is a globally unique sequence number that could be obtained using hash functions on proper information;
- *tr_dest*—this is a cryptographic trapdoor that can only be opened by the destination;
- *onion*—this is a cryptographic onion critical for route pseudonym establishment, which will be explained below.

The source could populate the seqnum field above using a hash function on a node's unique identity pseudonym. In order to populate the tr_dest field, the source needs to share secret trapdoor keys with the destination. This secret trapdoor key will ensure that only the destination can obtain the information. Any node on route to the destination will not be able to obtain this information since they do not possess the relevant keys. This ensures that the identity of the destination is kept secret. Finally, to populate the onion, the source has different choices. In one option the source randomly picks a symmetric key K_S, which it uses to encrypt its own address (or a special source tag) and populates the resulting ciphertext in this field. Following this the RREQ packet is broadcast by the source.

Every intermediate node will check the seqnum field of the incoming RREQ packet and will discard packets with previously seen sequence numbers. Otherwise, the intermediate node checks to see if it can open the cryptographic trapdoor. If it cannot open the cryptographic trapdoor, it is not the destination. In this case, it determines a random nonce $\overline{N}$ as well as a random symmetric key K′ (different from the random symmetric key chosen by the other nodes). It appends the onion in the RREQ packet received to the random nonce $\overline{N}$ and encrypts the result using the key K′. Note that both K′ and $\overline{N}$ are not known to any entity other than the node itself. These will be made use of later during the receipt of the RREP packets to ensure that it is the correct receiver for those packets. Thus an onion is built by an intermediate node. We show the steps involved in building the onion in Figure 4.24. Here node A is assumed to be the source which desires to determine a route to node E. Node A creates the onion in step 1 and the other intermediate nodes keep on appending to this after performing the necessary checks.

This is one method of building the onion. Other methods using the public key of each node as well as using symmetric keys without the help of nonces are also explained in [66]. We do not explain those here.

On the other hand, if the node can open the cryptographic trapdoor, then it is indeed the destination and needs to create and send an RREP packet. The RREP packet contains the following fields:

- *RREP*—this is a label that specifies that this is a route reply message;
- *N*—this is a locally unique random route pseudonym;
- *Pr_dest*—this is a proof of global trapdoor opening;
- *onion*—this is the same cryptographic onion from the received RREQ packet.

The destination transmits the RREP packet after populating the above fields. Note that the locally unique random route pseudonym will be used for data forwarding. Any node that

1. $TBO_A=K_A(src)$
2. $TBO_B=K_B[N_B,K_A(src)]$
3. $TBO_C=K_C\{N_C,K_B[N_B,K_A(src)]\}$
4. $TBO_D=K_D(N_D,K_C\{N_C,K_B[N_B,K_A(src)]\})$

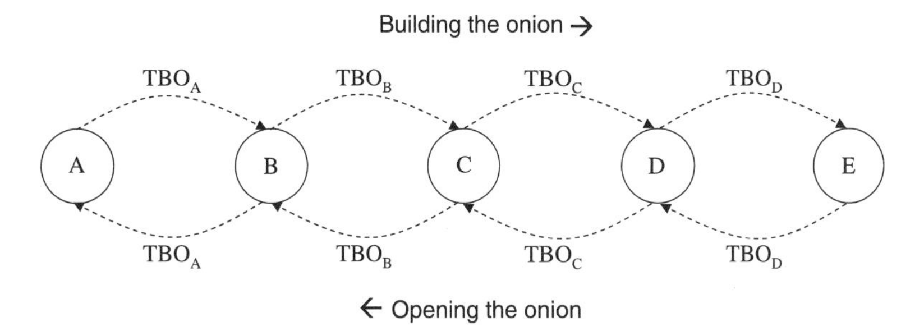

Figure 4.24. RREQ processing in ANODR.

receives the RREP packet will try to open the onion. Only the proper node (which is the previous hop during the RREQ phase) can correctly open the onion using the key K. The opening of the onion is also shown in Figure 4.24. Such a node (which can open the onion) then selects another random route pseudonym N′, peels off a layer of the onion, replaces N with N′ and then locally broadcasts the RREP packet. This node also stores the mapping between N and N′ in its forwarding table. These actions will be repeated by all nodes until the source receives the RREP packet. The source then verifies the proof of global trapdoor opening, and if this is found correct then the source is sure of communicating with the intended destination. We would like to remark here that the destination can initiate the same RREP procedure to realize multiple anonymous paths between itself and the source.

At this point, an anonymous route making use of locally unique pseudonyms is set up between the source and the destination. This route can then be used to forward data packets. The source uses the outgoing route pseudonym to encapsulate the data packet, which is then broadcast locally. This way neither the sender nor the local receiver can be identified. Each local receiving node will have to look up the route pseudonym in their forwarding tables. The packet is dropped by a node if no match is found. If a match is found in the forwarding tables of a node, it changes the route pseudonym in the packet to the matching route pseudonym and rebroadcasts the packet. This procedure is repeated until the packet arrives at the destination. We show this in Figure 4.25. We see from this figure that the nodes that are not on the data forwarding path drop the received packet. The nodes that are on the data forwarding path translate the pseudonyms and forward the packet. An anonymous route maintenance procedure based on the use of route pseudonyms is also specified in ANODR.

ANODR is robust against node intrusion. No node identity is revealed nor does it have a single point of failure. Of course, ANODR has a strong assumption that each source shares a pairwise secret with each potential destination that is used in the trapdoor construction. In addition, ANODR requires that every intermediate node try to open the trapdoor in the RREQ to see if it is the intended destination. An intermediate node also needs to try to decrypt the RREP packet to determine if it is the correct forwarder. This implies that the route discovery process in ANODR is computationally expensive. ANODR is also sensitive to node mobility. Finally, note that ANODR allows only the destination to initiate

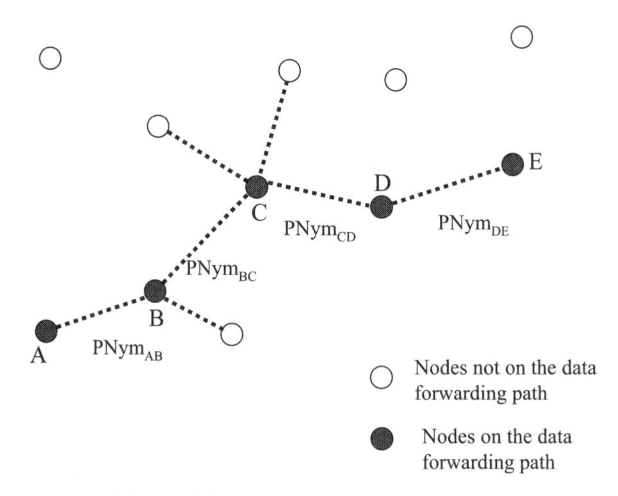

Figure 4.25. Data forwarding in ANODR.

route replies. Optimizations whereby intermediate nodes send the route replies if they have the path to the destination are not possible under ANODR.

4.6.2 MASK

MASK [67, 68] is another routing protocol intended to prevent traffic analysis. As in case of ANODR, this protocol provides countermeasures against traffic analysis from passive adversaries. MASK facilitates routing without disclosing the identity (and hence the location) of the mobile nodes. MASK is similar to ANODR in the sense that packets are routed based on link identifiers instead of real node identities.

In order to facilitate the working of MASK, a node needs to run its MAC interface in the promiscuous mode. This allows the node to receive all MAC frames broadcast in its neighborhood. In addition, every node also needs to be able to manipulate the source addresses of the outgoing MAC frames (as was done in ANODR).

MASK uses the following phases to reach its objective:

- anonymous neighborhood authentication;
- anonymous route discovery; and
- data forwarding.

The first phase of neighborhood authentication is used to authenticate nodes anonymously. In addition, this phase also results in setting up of dynamically changing pseudonyms (also called linkIDs) and the corresponding keys between neighboring nodes. Hence, the nodes will not have to use their real identifiers such as network layer addresses or MAC addresses for communication. During the anonymous route discovery phase the routes are set up using these linkIDs and the corresponding keys. The data is then routed between the source and the destination during the data forwarding phase. We explain these phases in detail next.

The authors use the technique of bilinear pairing [69, 70] in their approach.[9] Every node before deployment is assumed to be provided with a set of pseudonyms and the corresponding secret points by a trusted authority (TA). Each secret point is connected to the pseudonym using the system secret. The system secret is known only to the TA. In addition, each node also knows of various system parameters such as a system wide function f ( ) with certain special properties [specifically, f( ) is a bilinear map] as well as two different hash functions H1 and H2.

The anonymous neighborhood authentication phase is shown in Figure 4.26. Consider two parties, Alice and Bob, who need to authenticate each other. Each of these parties has a set of pseudonyms and the corresponding secret point, as shown in the figure. Note that H_1 () is a hash function. The secret point of each party is only known to the party and the TA. The TA calculates the secret point for each party using the pseudonym and a system parameter denoted as g. Assume that Alice initiates the authentication. In this case, Alice picks one of its pseudonyms and sends this information to Bob along with a nonce n_1 as shown in Figure 4.26. The secret point corresponding to Alice is known only to Alice. Bob on receiving this information computes a common key K_{BA} by using his own secret point and a hash of the pseudonym used by Alice as f[Bob's secret point, H_1(Alice's pseudonym)]. Note that Bob is only aware of the pseudonym of Alice and not her identity.

[9]We do not explain the details of bilinear pairing since we explain the principles at a high level without getting into the mathematical details.

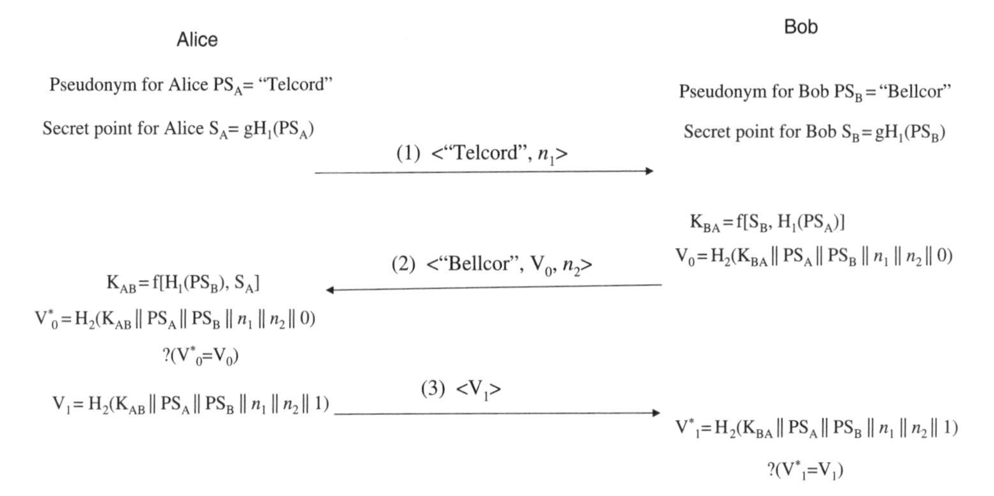

Figure 4.26. Anonymous neighborhood authentication.

Following this Bob determines its own nonce n_2. He then calculates a ticket (denoted as ticketB) as a hash of the common shared key, pseudonym of Alice (PS_A), Bob's own pseudonym (PS_B), as well as the two nonces n_1 and n_2. The ticket is calculated as $H_2(K_{BA} \|PS_A\|PS_B\|n_1\|n_2\|0)$, where H_2 is another hash function.

Bob then transmits a message to Alice. The message includes Bob's own pseudonym, which is "Bellcor" in this case, the ticket and the nonce n_2. Alice on receiving this message can also calculate the common key K_{AB} as f[H_1(Bob's pseudonym], Alice's secret point). The properties of function f() ensure that this key is the same as K_{BA} calculated by Bob (assuming that both Bob and Alice are given the corresponding pseudonyms and secret points by the TA accurately). Alice also calculates if ticketB provided by Bob is valid by computing $H_2(K_{AB}\|PS_A\|PS_B\|n_1\|n_2\|0)$. If the ticket is valid, then Alice calculates another ticket (denoted as ticketA below) as $H_2(K_{AB}\|PS_A\|PS_B\|n_1\|n_2\|1)$. Alice then transmits a message with this ticket to Bob, who can then verify it. As a result of the verification of these tickets, both Alice and Bob are convinced that they both belong to the same group. Note that the identity of either party is not known to the other. This completes the anonymous neighborhood authentication phase.

To summarize, the following message exchange is used in this phase:

1 Alice → Bob: $<$Alice's pseudonym (PS_A), $n_1>$
2 Bob → Alice: $<$Bob's pseudonym (PS_B), ticketB, $n_2>$
3 Alice → Bob: $<$ticketA$>$

Alice and Bob then calculate L pairs of link identifiers to be used for the link between them and the shared session keys to be used with each link identifier. Each pair of neighboring nodes thereby shares multiple link identifiers uniquely, that is, no two links have the same link identifier, while the same link has multiple link identifiers. Hence, when one node broadcasts a packet identified by the link identifier, only the neighboring node which shares this link identifier can interpret and accept the packet. These link identifiers (linkIDs) and the shared session keys are used in the routing process.

The anonymous route discovery in MASK is similar to that in AODV. The difference, though, is in the use of the anonymity property. Each node maintains four tables:

- *Neighbor Table*—each entry in this table contains the pseudonym of a neighbor, the pairwise shared key pairs (namely the linkIDs and corresponding keys) used on the link with this neighbor as well as the index of the pair currently in use.
- *Forwarding Route Table*—this table contains entries of format <dest_id, destSeq, pre-link-list, next-link-list>. The dest_id is the real identifier of the destination node, destSeq is the corresponding sequence number, pre-link-list is the set of pre-hop link identifiers from which packets destined for dest_id may arrive and next-link-list is the set of next-hop link identifiers to which packets destined for dest_id are to be forwarded.
- *Reverse Route Table*—this contains entries of the format <dest_id, destSeq, pre-hop-pseudonym> where pre-hop-pseudonym is the pseudonym of the node from which the RREQ packet for dest_id was received. This is then used to send route replies back to the source.
- *Target Link Table*—this contains the list of linkIDS for which the current node is the final destination. Thus packets bearing these linkIDs are intended for this node.

The source sends out a RREQ packet. Each RREQ packet has the following fields:

- *ARREQ*—this is a label that specifies that this is a route request message;
- *ARREQ_id*—this is a globally unique identifier for the ARREQ;
- *Dest_id*—this is an identifier of the destination;
- *Destseq*—this is the last known sequence number for the destination or an unknown flag if this information is not known;
- *PS*—active pseudonym.

The source populates the above fields with PS denoting its own pseudonym and transmits the RREQ message. Note that MASK, unlike ANODR, does not hide the destination address. The address of the sender though is hidden. We show the route discovery process in Figure 4.27.

When an intermediate node receives the ARREQ message, it checks whether it has seen the corresponding ARREQ_id previously. If so, it discards the message. Otherwise, it inserts an entry into its reverse route table with information about the identity of the destination, the pseudonym of the node from which this ARREQ message was received as well as the sequence number associated with the destination. It then changes the pseudonym field of the ARREQ message to contain its own pseudonym and rebroadcasts the message. This process continues until either the ARREQ reaches the destination or it reaches an intermediate node that has a valid routing entry to the destination. Note also that, in order to avoid correlation attacks, each node including the destination or any intermediate node with a valid path to the destination is expected to rebroadcast the ARREQ once. If not, the adversary can look at the fact that the ARREQ has not been forwarded and infer the identity of the node with the destination address in some cases.

An ARREP message is generated by either the destination or by an intermediate node with a path to the destination. The ARREP packet contains the following fields:

- *LinkID*—this is the link identifier shared by this node and the pre-hop-pseudonym node;
- *Enc(ARREP, dest_id, destSeq)*—the ciphertext obtained by encrypting the information using the key corresponding to the link identifier being used.

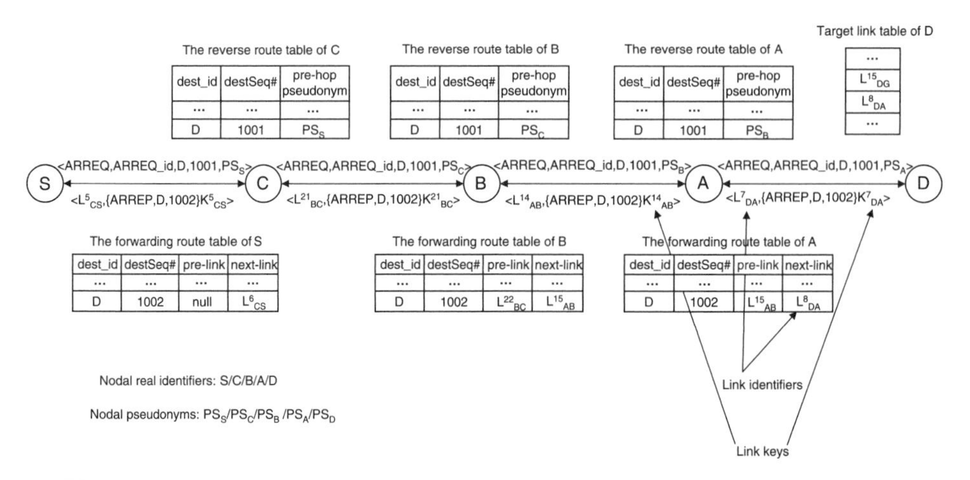

Figure 4.27. Anonymous route discovery with a route reply generated by destination D.

Thus, note that the ARREP packet is confidential. The ARREP packet is identified only by the LinkID which can be recognized only by the intended receiver by looking at its neighbor table. If the destination is generating the ARREP packet, then it has to modify its target link table by entering the corresponding linkIDs. An intermediate node will not have to do this.

Any intermediate node that receives the ARREP packet will check if it is the intended recipient. If so, it decrypts the relevant portion of the ARREP packet and verifies if the destSeq is larger than that in its reverse route table. If so, the node will form and transmit a new ARREP packet. This packet includes the linkID shared with the pre-hop-pseudonym node corresponding to the ARREQ. This information is obtained from the reverse route table. In addition, this packet contains the information encrypted using the key corresponding to the used linkID. The node will also have to update its forwarding route table. This process continues until ARREP reaches the source. We show the steps involved during this process in Figure 4.27.

Data forwarding then is akin to a virtual circuit switching process. When the source has to send a packet to the destination, the source picks a linkID from the next-link-list field of the forwarding table. A packet is then formed and transmitted. Note that the source and destination address of the data MAC frames are set to the value of the linkID, as for the ARREP packets. The key corresponding to the linkID can be used to encrypt communication on this link if needed. Any intermediate node that receives this data packet will check if it is an intended recipient. If so, it changes the linkID field in the packet to a value selected from the next-link-list field of its own forwarding table. The packet is then transmitted. This process continues until the packet reaches the destination. We illustrate this process in Figure 4.28. Here we see that different packets between the same source and destination can take different paths. The linkID on each hop is also shown in the figure. Note that, during data forwarding, packets are identified by secure link identifiers between two neighbors. Additionally, the link identifiers change over every hop. Hence, an adversary will not be able to track a packet.

MASK uses pseudonyms only for neighborhood authentication and during the route discovery process. Even here, since every node has many pseudonyms, it is secure against traffic analysis. If the node had a single pseudonym, then it could have been analyzed the same way as a real identifier. MASK, however, uses the real identity of the destination during the route discovery process. This is intended to improve the efficiency of

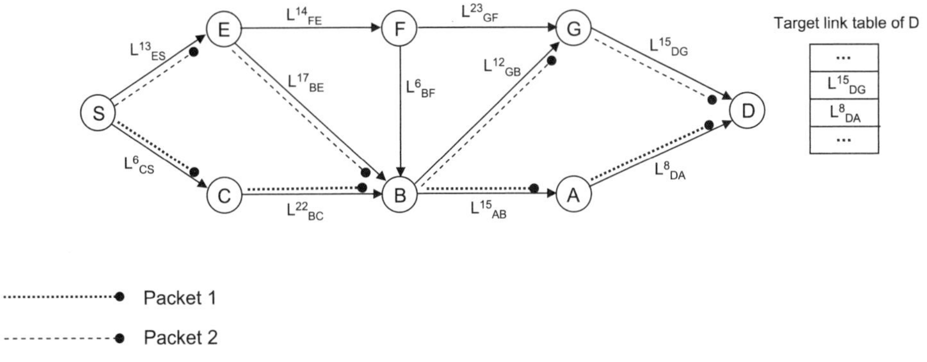

Figure 4.28. Anonymous packet forwarding.

the routing protocol. The subsequent data traffic though makes use of linkIDs, due to which the data traffic cannot be tied to the route discovery process.

MASK has the following properties: it conceals the identities of the senders as well as the relationship between the sender and the receiver. In addition, an adversary will not be able to trace the intended recipient of a data packet. MASK also ensures that adversaries cannot trace a packet back to its source nor trace it forward to its destination. MASK also achieves neighborhood authentication while ensuring that the nodes do not have to disclose their real identifiers. Note also that MASK can provide a better packet delivery ratio than ANODR. This is because in ANODR, data packets are locally broadcast which, combined with the hidden terminal problem, can lead to many packet collisions. On the other hand, MASK allows anonymous RTS/CTS exchange, leading to reduced packet collisions.

4.7 GENERIC ATTACKS AGAINST ROUTING

We have so far considered specific routing protocols which have several security attributes. In this section we focus on some attacks and discuss the impact of these attacks on routing in ad hoc networks. We consider three types of attacks:

- wormhole;
- rushing;
- sybil.

We explain each of these attacks and summarize briefly the various proposals to protect against such attacks.

4.7.1 Wormhole Attacks

A wormhole attack typically requires the presence of at least two colluding nodes in an ad hoc network.[10] The malicious nodes need to be geographically separated in order for the

[10]We ignore a variant that depends on a single malicious node being a common neighbor of two nonneighboring nodes. The malicious node can relay packets between the nonneighboring nodes to convince them that they are neighbors.

attack to be effective. In this attack, a malicious node captures packets from one location and "tunnels" these packets to the other malicious node, which is assumed to be located at some distance. The second malicious node is then expected to replay the "tunneled" packets locally. There are several ways in which this tunnel can be established. We consider two possible methods below.

In the first method for establishing the tunnel shown in Figure 4.29, a malicious node denoted X in the figure, encapsulates a packet received from its neighboring node A. Node X then sends the encapsulated packet to the colluding malicious node Y. Node Y then replays the packet in its neighborhood after decapsulating the packet. Thus, the original packet transmitted by node A in its neighborhood is replayed by node Y in its neighborhood, which includes node B. For example, if the original packet transmitted by node A (and tunneled by node X) was a hello packet, then node B on receiving this packet would assume that node A is its neighbor, which is not true. As another example, if node A transmits a route request packet for node B, then node X can "tunnel" such a packet to node Y by encapsulating the packet. As a result, this route request packet will arrive at the destination node B with a lower hop count than the other Route Request packet going through the other route. This happens in spite of using any secure routing protocol such as the ones given earlier. Note that nodes between X and Y that relay the packet cannot interpret the packet as it is encapsulated. Therefore, they cannot increment the hop count.

In the second method for establishing the tunnel shown in Figure 4.30, the two malicious nodes X and Y are assumed to have access to an out-of-band high bandwidth channel. This could be achieved for example by having a wired link between the two nodes or by having a long range high bandwidth wireless link operating at a different frequency. Thus, this method requires specialized hardware capability and hence is more difficult than the previous method. In this case also, a hello packet transmitted by node A can be retransmitted in the vicinity of node B. As a result node B infers that node A is its neighbor. Similarly, a route request packet, from node A for node B, can also reach node B (which is the destination for the route request packets) faster and possibly with fewer hops, since a high-bandwidth direct link is being used between the two malicious nodes. As a result, the two endpoints of the tunnel can appear to be very close to each other. To see this, consider Figure 4.30. Here node B receives three route requests. It is clear that the route request received through the wormhole will have the least hops.

It seems as if the malicious nodes are performing a useful service by tunneling the packets. This would be so if the nodes were performing this service without any malicious

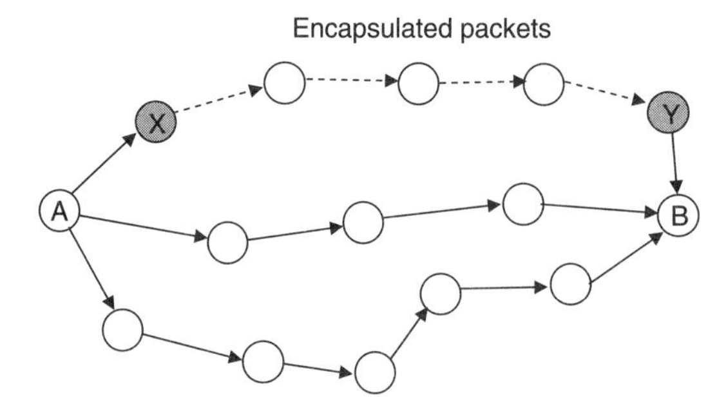

Figure 4.29. Wormhole attack (encapsulated packets).

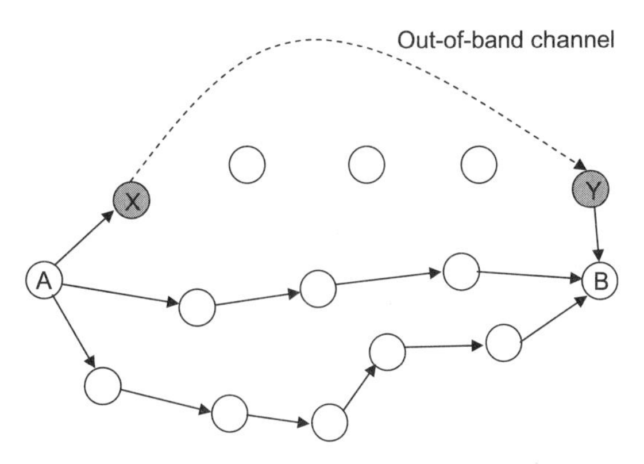

Figure 4.30. Wormhole attacks (out-of-band channel).

intent, but malicious nodes could use this attack to undermine the correct operation of various protocols in ad hoc networks. The most important protocol that is impacted is the routing protocol, as we can see from the examples given earlier. Data aggregation, protocols that depend on location information, data delivery, and so on, are some other examples of services that can be impacted. Note that the wormhole attack can be successful even without access to any cryptographic material on the nodes. For example, in the above figures (Figures 4.29 and 4.30), the wormhole attack can be successful even without knowledge of the keys used by the valid nodes in the system (such as nodes A and B). In addition, nodes in the network do not have to be compromised. Thus, in the same figures, node X and node Y could be outsider nodes which are not part of the regular network.

There have been some proposals recently to protect networks from wormhole attacks by detecting such attacks [71–73]. In [71], the authors introduce the concept of leashes to detect wormhole attacks. A leash is any information added to a packet in order to restrict the distance that the packet is allowed to travel. A leash is associated with each hop. Thus, each transmission of a packet requires a new leash. Two types of leashes are considered, namely geographical leashes and temporal leashes. A geographical leash is intended to limit the distance between the transmitter and the receiver of a packet. A temporal leash provides an upper bound on the lifetime of a packet. As a result, the packet can only travel a limited distance. A receiver of the packet can use these leashes to check if the packet has traveled farther than the leash allows and if so can drop the packet.

When using geographic leashes, a node is expected to know its own location. Each node transmitting a packet will include its own location in the packet and also the time at which the packet is transmitted. The node receiving this packet uses this information (about location and time) to calculate if the packet has traveled more than the allowable distance and if so the packet is dropped. With temporal leashes, each node transmitting a packet includes the time at which the packet was sent. The receiving node notes the time at which the packet was received and uses this to infer if the packet has traveled too far. In an alternate formulation, a packet can contain the expiration time after which a receiver should not accept the packet. The transmitter node decides on this expiration time as an offset from the time of transmission. Note that in both these cases, geographical leashes and temporal leashes, the receiver needs to authenticate the information about the location and time included by a transmitter in the packet. This authentication can be achieved by a mechanism such as a digital signature.

Geographic leashes require loose time synchronization among the nodes in the network, as opposed to temporal leashes, where the nodes in the network need to have tightly synchronized clocks with nanosecond precision. In addition, temporal leashes might not be practical in networks (such as those based on the use of 802.11) that make use of a contention-based MAC protocol. Temporal leashes will also be effective only against the first and easier mode of tunneling, where a packet is encapsulated. Of course, nodes in systems dependent on geographic leashes need to be able to determine their locations securely (see Chapter 7).

Another approach for detecting wormhole attacks is given in [72]. In this case the authors assume the presence of directional antennae. The approach here is based on the use of packet arrival direction to detect that packets are arriving from the proper neighbors. Such information is possible due to the use of directional antennae. This information about the direction of packet arrival is expected to lead to accurate information about the set of neighbors of a node. As a result, wormhole attacks can be detected since such attacks emanate from false neighbors.

To illustrate this idea consider Figure 4.31. Here we show two nodes, A and B. We show the directional antenna with six zones explicitly for both nodes. Each node is assumed to have knowledge of the zone from where a packet is received. Given this, the basic idea that is used to determine the set of authentic neighbors is that, if a node is in a given direction of another node, then the latter node is in the opposite direction of the former node. For example, in Figure 4.31, node B is in zone 6 of node A while node A is in the opposite zone, which is zone 3 of node B. An implicit assumption here is that the directional antennae on the various nodes are perfectly aligned.

Now consider nodes A, B, C, D, and E as shown in Figure 4.32. Assume that B transmits a message (call it a Hello message for simplicity) in its neighborhood which includes C. Then node C replies back to B informing node B of the zone from which node B's hello message was received. If node B receives this reply in the opposite zone to what zone C reports, then node C can possibly be an authentic neighbor. Thus, in this case node C replies back to node B that the hello message was received in zone 1 of node C. This reply is received by node B in its zone 4, which is the zone opposite to zone 1. Thus, B can infer that node C is an authentic neighbor. Every node can repeat this and form the list of authentic neighbors. Any message that does not emanate from an authentic neighbor is rejected.

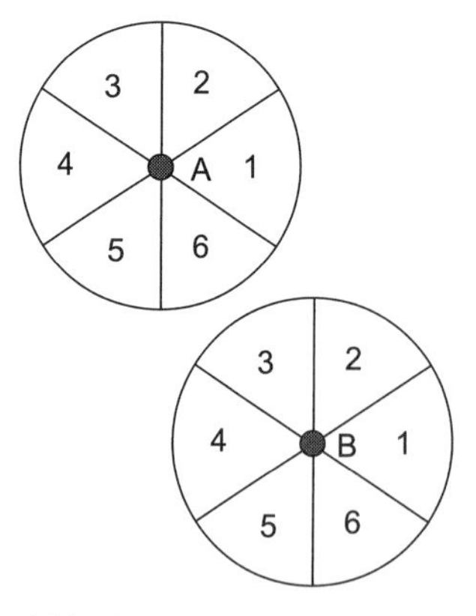

Figure 4.31. Zones on a directional antennae.

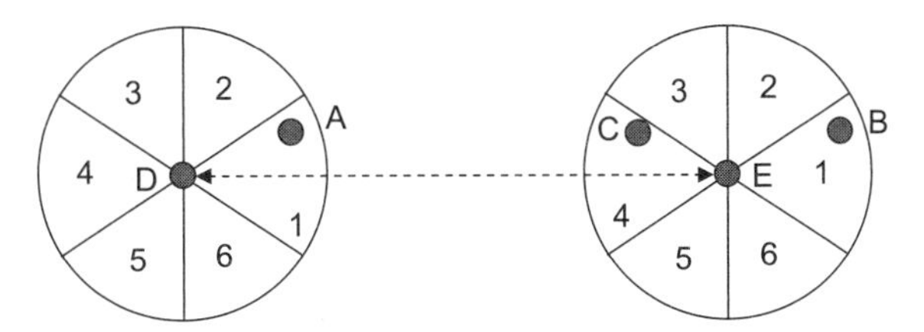

Figure 4.32. Detecting wormholes using directional antennae.

We next show how this mechanism can be used to detect wormholes. Assume that nodes D and E are malicious and they intend to form a wormhole between nodes A and B. Consider hello messages being transmitted by node A in order to determine the set of real neighbors. In this case, node D receives the hello message of A and tunnels it to node E, which then retransmits the Hello message in its neighborhood. Node B receives this message and reports back using a reply message that the Hello message of A was received in zone 4 of node B. If this reply of node B is replayed back by node D, then node A receives this reply in its zone 4. Thus, node A can infer that there is an inconsistency and reject this reply from node B. Note, though, that the same mechanism does not protect a wormhole from being established between nodes A and C since they are in the opposite zones as compared with the wormhole endpoints.

Thus, we see that the simple mechanism given earlier is not sufficient to detect all wormholes, although it does detect many of them. In order to address this shortcoming of the simple mechanism, the authors enhance the basic scheme. The enhancement requires that nodes cooperate with their neighbors by having other nodes called verifiers validate the legitimacy of the various packets being transmitted. Restrictions are placed on the nodes that can act as verifiers. This prevents the adversary from leveraging the weakness earlier whereby nodes that are in opposite directions to the wormhole endpoints are tricked into accepting each other as neighbors. In this case, after the messages of the basic protocol, the verifier is requested to corroborate the zones from which the messages were received by the verifier. As pointed out earlier, however, this approach requires perfectly aligned directional antennae.

In [73], the authors present a graph theoretic framework for modeling the wormhole attack. They provide a necessary and sufficient condition that any solution to the wormhole problem needs to satisfy. In addition, the authors also propose the use of local broadcast keys whereby the keys in different geographic regions are different. As a result, an encrypted message replayed via the wormhole in a different location cannot be decrypted by the receivers in that region.

4.7.2 Rushing Attacks

We next describe the rushing attack [74]. This attack impacts a reactive routing protocol. Note that, in the case of a reactive routing protocol, a node that needs a path to a destination floods the network with route request packets. Such route request packets are flooded in a controlled fashion in the network. Thus, every node only forwards the first route recovery packet that it receives and drops the rest. The adversary can exploit this feature of reactive routing protocols. It does so by "rushing" the route request packets towards the destination. As a result, the nodes that receive this "rushed" request forward it and discard the other route requests that arrive later. The resulting routes would then include the adversary, which places the adversary in an advantageous position.

We see that this attack is not very difficult to launch. All it requires is that the adversary be able to forward route request packets faster than can be done by valid nodes. The adversary can do this by forming wormholes. The adversary can also do this by choosing to ignore the delay between the receipt and forwarding of route request packets. Such delays are specified by the routing protocols to avoid collisions of route request packets. The adversary can also choose to ignore delays specified by the protocols to access the wireless channel. Thus, in all these cases the route request packet can be rushed by the adversary.

A simple way to address the rushing attack is to allow for randomized selection of route request messages. Thus, every node is expected to collect a threshold number of route requests. Following this, the node can randomly choose to forward a route request from among the received requests. Note that a timeout also needs to be associated here so that, if a node does not receive the threshold number of route requests within the timeout, it can choose from among the received route requests. This mechanism does add to the latency of route discovery, though, and hence both the threshold value and the timeout value have to be chosen with care. Note, though, that the mechanism is general enough and can be easily integrated with any reactive routing protocol to protect against rushing attacks.

4.7.3 Sybil Attacks

The Sybil attack [2] consists of a node assuming several node identities while using only one physical device. The additional identities can be obtained either by impersonating other nodes or by making use of false identities. These identities can all be used simultaneously or over a period of time. This attack can impact several services in ad hoc networks. For example it can impact multipath routing [75], where a set of supposedly disjoint paths can all be passing through the same malicious node which is using several Sybil identities. This attack can also impact data aggregation where the same node can contribute multiple readings each using a different identity [76]. Fair resource allocation mechanisms will also be affected since a node can claim more than its fair share by using the various Sybil identities. Mechanisms based on trust, voting, and so on, are also affected by this attack.

A simple approach for detecting the Sybil identities could be to issue a public-key certificate to each of the identities. The problem, though, is the need for a central authority that can distribute the certificates. In addition, this solution requires higher computational capabilities than might be available on resource constrained devices such as sensor nodes. Resource testing was suggested as an approach to prevent Sybil attacks in [2]. This approach makes use of the fact that each node is limited in some resources. The test then verifies if each identity has the proper amount of the tested resource. Thus, the implicit assumption here is that every identity has the same quantity of the tested resource. The resources suggested in [2] are computation, storage, and communication, but these resources might not be the right ones to test for in ad hoc and sensor networks because of the heterogeneity possible in such networks due to which different physical devices have different values for the quantity of the tested resource.

Another resource suggested in [76] is the radio. Here the assumption is that every physical device has only one radio that is incapable of simultaneously transmitting and receiving on more than one channel. Thus, a node that desires to verify if any of its neighbors are

Sybil identities will allocate a channel to each of its neighbors. The neighboring node is expected to transmit a message on the allocated channel. The verifier node then picks random channels to listen to. If no message is heard on the channel selected, then the corresponding node identity is assumed to be a Sybil identity. Another approach [76] to detecting Sybil identities leverages the random key predistribution technique (explained in Chapter 3). When using random key management schemes, a node is loaded with a set of random keys before deployment. This scheme can be leveraged to detect Sybil identities if the set of keys assigned to a node is related to the identity of the node. As a result, a node claiming any identity will have to prove the claimed identity by also demonstrating that it has the keys corresponding to that identity. The node does so by participating in encryption or decryption operations with the key. In this case, an adversary will first have to compromise many nodes, thereby getting access to the keys corresponding to each identity. Following this, the adversary might be able to create fake identities. Some other approaches proposed are based on verifying the location for each of the node identities or on registering the identities with a base station.

Another approach for detecting Sybil attacks is given in [77]. In this scheme each node is provided a unique secret by a central authority. A hash chain is derived from this secret by the node. The node is also provided an identity certificate by the central authority binding the node's identity to the secret information. A node that claims a given identity is then expected to present the identity certificate and prove that it possesses the unique information certified in the identity certificate. The proof might be needed during every interaction between two nodes. The hashing property of pre-image resistance can be used to create the proof. Thus, the identity certificate might contain the authenticated value of the root of the hash chain, while every proof instance might require the node to expose the predecessor value in the hash chain.

4.8 SUMMARY

We conclude this chapter by summarizing the various secure routing protocols that we studied in this chapter in Figure 4.33. We consider several attributes and comment on these attributes with respect to each of the protocols.

Performance parameters	ARAN	ARIADNE	SAODV	SAR	SEAD	SLSP	SRP	ANODR	MASK
Type	Reactive	Reactive	Reactive	Reactive	Proactive	Proactive	Reactive	Reactive	Reactive
Encryption algorithm	Asymmetric	Symmetric	Asymmetric	Symmetric/ asymmetric	Symmetric	Asymmetric	Symmetric	Symmetric	Asymmetric
MANET protocol	AODV/ DSR	DSR	AODV	AODV	DSDV	ZHLS	DSR/ZRP	NA	NA
Synchronization	No	Yes	No	No	Yes	No	No	No	No
Central trust authority	CA required	KDC required	CA required	CA/KDC required	CA required	CA/KDC required	CA required	KDC required	CA required
Authentication	Yes	Yes	Yes	Yes	Yes	Yes	Yes	Yes	Yes
Confidentiality	Yes	No	No	Yes	No	No	No	Yes	Yes
Integrity	Yes	Yes	Yes	Yes	No	No	Yes	Yes	Yes
Nonrepudiation	Yes	No	Yes	Yes	No	Yes	No	No	No
Antispoofing	Yes	Yes	Yes	Yes	No	Yes	Yes	Yes	Yes
DoS attacks	No	Yes	No	No	Yes	Yes	Yes	No	No
Anrnymity	No	No	No	No	No	Yes	Yes	Yes	Yes

Figure 4.33. Comparison of various protocols.

5 Intrusion Detection Systems

5.1 INTRODUCTION

Security in most enterprise environments today is based on the concept of defense-in-depth, where multiple layers of defenses are used to prevent adversaries from violating the security policies of the enterprise. The defense-in-depth approach is based on the premise that, even if an adversary penetrates one of the layers of defense, he will not be able to cause much harm because the other layers will provide an adequate level of protection. Typically technologies attempting to prevent attacks provide the first layer of defense. Our discussion so far has focused on such technologies. In traditional wireline networks, the key technology that is used as a first layer of defense at the perimeter of the network is a firewall. Firewalls (see Chapter 6) are used to prevent outsiders from penetrating the enterprise network. Cryptography is another technology being used as a preventive layer to keep outsiders from accessing critical resources. We have seen this in the earlier chapters. Authentication through password, biometrics, or other means is typically used to prevent unauthorized access to enterprise systems and networks. Encryption is often used to prevent eavesdropping by unauthorized users.

Even though the preventive mechanisms discussed earlier present a formidable barrier to malicious users, an additional layer of defense called intrusion detection is often used to protect networks. Intrusion detection technologies focus on detecting malicious activity typically from attackers that have successfully penetrated the perimeter defenses. Penetration of preventive mechanisms may be achieved because of vulnerabilities of deployed protocols or because of vulnerabilities of the preventive barriers themselves. These vulnerabilities can then be used by the adversary to launch attacks. In addition to exploiting vulnerabilities of preventive technologies, outsiders can also potentially penetrate the preventive security barriers by obtaining information (such as passwords) through social engineering (for example obtaining passwords from legitimate users by pretending to be a security administrator) or by gaining information about how to penetrate the defenses through espionage. In other cases it may be possible that a trusted insider is exhibiting malicious behavior either because the insider is trying to obtain financial gain through unauthorized activities or because he/she is a disgruntled employee and is attempting to cause harm to the enterprise.

There are two ways of protecting networks from such intrusions. One way is to remove the vulnerabilities from the system or protocol itself by redesigning these entities. Good examples of such an approach include the design of attack resistant protocols like stream control transmission protocol (SCTP) [78], which can resist TCP SYN flood attacks, patching the operating systems, and so on. However, implementation of such redesigns may not

Security for Wireless Ad Hoc Networks, by Farooq Anjum and Petros Mouchtaris

always be possible for various reasons, including the significant costs associated with replacing existing systems with more attack resistant solutions, the diverse capabilities of different devices (e.g. CPU, data storage, bandwidth), the inability of devices to support attack resistant solutions (which is often more of a problem for wireless networks), the inefficient configuration (e.g. users do not change default security settings or apply patches), and so on. For example, SCTP is rarely used today due to the cost of changing all of the existing systems using TCP. In addition, processor intensive preventive mechanisms (such as some cryptographic operations) intended to protect the enterprise systems might also be used against the system. For example, such technologies may be used for launching DoS attacks by overloading the available CPU resources.

The second way of protecting enterprises from such intrusions is to monitor events occurring in a computer system and detect attempts to either leverage the vulnerabilities of preventive mechanisms or misuse preventive mechanisms. This is called intrusion detection and is the focus of this chapter. Intrusion detection *per se* does not prevent users from obtaining unauthorized access to data or systems but rather focuses on monitoring the behavior of users. Following detection, efforts might be made to stop such attempts from succeeding. It might also be necessary to have the system recover from the effects of any attack. Some aspects of recovery will be the focus of the next chapter.

There are several advantages to implementing intrusion detection systems (IDS) in a network. An IDS acts as a deterrent for attackers. It also can detect attacks that bypass other security measures. It further provides information about the actual intrusions that can be used for responding to such attacks. An IDS is also helpful in performing quality control of security design and administration. In this chapter we will focus on IDS. In Section 5.1.1 we focus on discussing the IDS concept and how it is applied today in a traditional enterprise environment. We then focus on the challenges of adapting existing IDS techniques to MANET and discuss IDS approaches that are well suited to MANET.

5.1.1 Traditional IDS Systems

As discussed earlier, intrusion detection systems were developed for detecting malicious behavior of users in an enterprise environment. Malicious behavior includes users trying to obtain unauthorized access to systems or data, users trying to disrupt availability of services to legitimate users through launching DoS attacks, and so on. In order to detect such behavior, intrusion detection systems typically contain two types of components:

- data collection components;
- data analysis components.

Data collection components consist of entities that are responsible for monitoring and collecting data about user and application activities. The collected data is then used by the second type of components, called the analysis components. Data analysis components are responsible for analyzing the data collected and detecting any malicious activity. We now discuss each type of component in more detail.

5.1.1.1 Data Collection Components The first step in intrusion detection is to collect data from the system being monitored. Two main approaches for data collection have traditionally been used in large enterprises. These approaches also result in two types of intrusion detection systems:

- host-based IDS (HIDS) that runs on a host and focuses on collecting data on each host usually through operating system audit logs;
- network-based IDS (NIDS) that runs in the network and focuses on collecting data by monitoring traffic flowing through the network.

The advantage of HIDS is that it is not affected by the use of end-to-end cryptography or by changes in topology, and routing that may be triggered by node mobility in case of ad hoc networks. However, HIDS has several serious drawbacks. First, an intruder can avoid detection and produce maximum damage by exploiting the knowledge that only the destination of the traffic analyzes the packets. For example if the intruder knows that the destination is using a particular version of the operating system, then it can try to transmit a packet that immobilizes the machine as soon as the destination's network layer assembles the packet and before the HIDS at the application layer analyzes the packet. Second, an attack against a host may affect the HIDS itself. Often this makes it impossible for the IDS to detect and report the attack since the detection mechanism uses the computation resources and network interfaces at the end host under attack. For example, a DoS attack on the target may simultaneously exhaust the bandwidth and CPU on the target. These same resources might be needed for detecting and reporting the attack by the HIDS, however. HIDS is also not suited for detecting activity such as network scans that could be the precursor to the launch of attacks because such attacks cannot be detected by a single host and require that the IDS looks at traffic across the whole network in order to detect the malicious activity.

Some limitations of HIDS are more relevant for ad hoc networks than for enterprise networks. In ad hoc networks, many of the destination nodes may not be able to execute the IDS due to limited computational resource and low residual energy. Further, if only the end hosts execute the IDS, then the malicious packets will not be dropped until they reach the destination. This is of concern in ad hoc networks where several nodes may expend their limited energy and bandwidth in relaying malicious packets.

The other approach to intrusion detection in NIDS has several advantages. First, an intruder can no longer be certain that only the destination is executing the IDS. Moreover, the IDS active nodes can be selected so that they have different characteristics, thus distributing the IDS load among devices with appropriate capabilities and making it more difficult for the intruders to devise attacks to bypass the IDS. Second, the NIDS approach results in fewer entities (such entitites are also loosely referred to as sensors[11]), protecting multiple hosts. This reduces the cost of running the IDS. Also, the sensors themselves are more easily secured in this approach since they can be designed to focus on monitoring in the stealth mode. Third, it is easy to retrofit existing networks with minimal effort since the IDS only needs to be installed in a subset of the nodes. Fourth, the IDS active nodes can be selected only among those that have the required capability. Finally, NIDS captures malicious packets in transit and thus limits the wastage of bandwidth and energy in relaying them as discussed earlier. The last two advantages in particular are very significant when considering ad hoc networks.

There are also several disadvantages associated with NIDS. NIDS might not always be able to work with encrypted information. NIDS can analyze encrypted traffic when

[11]Such sensors, however might not be as resource-constrained as nodes in a sensor network.

encryption is not at a layer at which it operates. For example, when traffic is encrypted at the application layer, IDS modules can detect attacks at transport and lower layers, for example, ping-of-death, TCP SYN flood, smurf, or bubonic [79]. If encryption is used at all layers, which is sometimes the case in battlefield networks, then schemes have to be designed to distribute the keys securely to the IDS active nodes so that the IDS can decrypt the traffic and analyze it for detecting attacks. The NIDS approach in a wireless ad hoc network has another limitation. It may be possible to devise attacks based on packet fragments and multiple routes that send fragments of the attacks through different paths. NIDS sensors do not get the chance to observe all the fragments of the attack and therefore may not be able to detect the attack. Such attacks are more effective in a wireless ad hoc network since nodes move often and routes change frequently. This makes it more likely that a single NIDS entity may not observe all relevant traffic for detecting the attack.

It is clear from the earlier discussion that both the HIDS and the NIDS approaches have their own advantages and disadvantages. Therefore, both approaches are often used together in complementary roles. In a typical enterprise environment today, the activity that could be monitored by an IDS is incredibly large and it is often impractical to monitor all of it. Enterprises usually select the key points in the enterprise to monitor depending on what is important to protect and where potential attacks may be originating from. Hence, it is typical for enterprises to monitor activities on critical servers and applications through HIDS. It is also typical for enterprises to position NIDSs right behind firewalls protecting the enterprise from external traffic since most attacks are expected to originate outside the enterprise and enter through the firewall. A few well-placed NIDSs in an enterprise network can monitor most of the traffic in the network and detect malicious activities. This architectural approach is shown in Figure 5.1. In the remainder of this chapter we will focus mostly on NIDS because they are impacted more by the unique features of wireless ad hoc networks.

5.1.1.2 Data Analysis Components Once the data is collected by IDS sensors, it has to be analyzed so that malicious activity can be detected. IDSs typically incorporate analysis engines that automatically analyze the data collected by the various sensors to detect

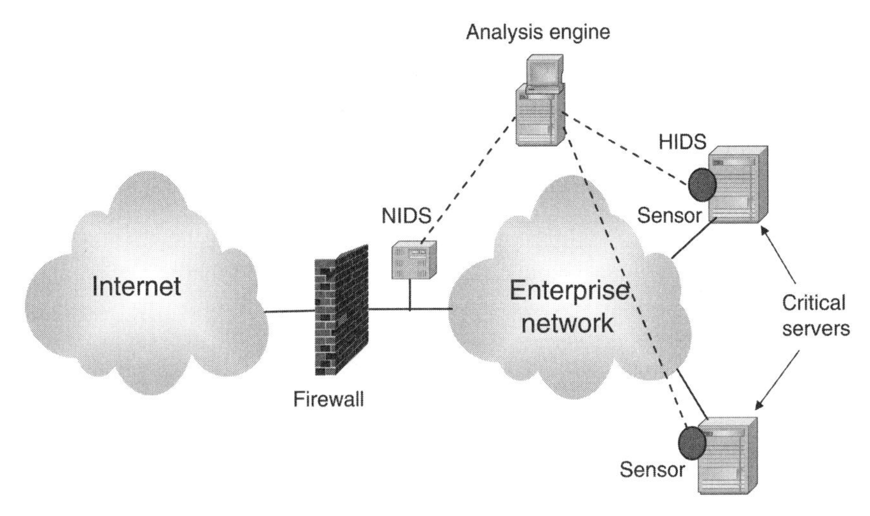

Figure 5.1. IDS placement in an enterprise network.

malicious activities. Since the amount of data usually collected by sensors is very large for humans to go through, analysis engines that can examine all of the available data collected by NIDS and HIDS are very important. Analysis of IDS data involves consolidating the data possibly at a central location and identifying malicious activities automatically to the extent possible. In some cases the HIDS and NIDS sensors themselves contain some preliminary analysis capabilities. Such a capability allows the IDS to detect the attack locally at the sensor, increasing detection speed and allowing faster reaction. Such a capability also minimizes the need to transfer all of the collected data to a centrally located analysis engine, reducing the amount of overhead data that is generated.

Analysis engines may utilize a variety of techniques for detecting malicious behavior. The two most widely used techniques are:

- misuse detection (often also called signature based detection); and
- anomaly detection.

A third less widely used technique is called specification-based detection. We now discuss each of these techniques in detail.

The misuse detection technique involves analyzing the collected data for specific behavior patterns known to be consistent with specific attacks. These behavior patterns are called signatures. For example, a UDP packet destined to port 0 can crash some machines [79]. The signature of a ping-of-death attack is a very large ping packet; the signature of an RPC locator attack is a packet intended for port 135 that contains a command that the system is not expecting; the signature of a Bubonic attack is made from various values, such as a TTL of 255, a TOS field value of $0 \times C9$, exactly 20 byte payload in the IP datagram, and the fragment ID value with consistent increments of 256 [79]. The misuse detection technique is being used widely today by commercial systems because it allows for very precise detection of attacks, thereby resulting in low false alarm rates. When a signature for an attack has been identified (typically a bit-by-bit precise specification of an attack packet), it is straightforward to detect the attack by comparing a packet against the precise attack signature. The problem with misuse detection intrusion detection techniques is that these techniques can only detect previously known attacks with well-defined patterns of malicious behavior. Novel attacks that have not been seen before cannot be detected until a new signature for the attack is created.

The anomaly detection technique involves looking for behavior that is outside the normal expected behavior. This is usually done by utilizing statistical techniques that compare the observed behavior against the statistics of the expected normal behavior. These techniques often utilize thresholds (such as amount of excessive loading of CPU, counts of traffic of specific type, etc.) that if exceeded indicate an attack. Therefore, detectors utilizing this technique require prior training so that the thresholds used for detecting abnormal behavior are set at the appropriate values. The advantage of anomaly-based detection is that this technique does not require the existence of precise signatures and can therefore be used for detecting attacks not previously seen. The disadvantage of anomaly-based detection is that, due to its statistical nature, it is more prone to false positives depending on how the thresholds are set. There is always a trade-off between setting very tight thresholds (close to normal behavior), thus triggering a lot of false positives when users deviate even slightly from expected behavior, and setting very loose thresholds that minimize false positives but allow attackers to evade detection. Given the disadvantages, this technique is not used widely in commercial systems.

Another technique for detecting malicious activity has been introduced recently although it is not commonly used yet in commercial systems. This technique, called specification-based detection, assumes the existence of a precise protocol specification. Malicious behavior is detected by comparing the protocol traffic with the protocol specification. The detectors typically build precise models of expected behavior (e.g. by using state machines) based on the protocol specifications and then compare the observed behavior in the network against the model. The advantage of specification-based detection techniques is that, given accurate specifications of normal behavior (e.g. protocols specifications), malicious behavior can be detected with a high degree of certainty. This eliminates the possibility that a detector will classify a normal behavior as malicious, which reduces false positives. Such detectors may also spot novel attacks since these detectors do not depend on the existence of specific attack signatures. On the other hand, developing models of normal behavior for each protocol is often a fairly complex task. Further, this approach requires models of normal behavior for all the protocols being used in the network in order to detect a wide range of attacks. These models need to be executed for each node in the network. As a result, these detection schemes require significant CPU resources for a large enterprise network. Another limitation of specification-based detection is that detectors using this technique do not detect attacks that do not violate the specification, but exploit permitted behavior to launch an attack. For example, in a flooding attack or a TCP SYN flood attack, a node's behavior is acceptable by the protocol specification, but the behavior is actually harmful to the operation of the enterprise and is therefore malicious.

Detection of attacks is typically not sufficient for protecting an enterprise because the attack may continue to cause harm to the network. Therefore, intrusion detection systems are usually coupled with attack response systems. Once an attack has been identified by the IDS, the response system is responsible for stopping the attack. It could do this by isolating the malicious behavior or by cutting off the attacker from the network and if possible restoring the damage caused by the attacker. Responses are of two types:

- active response; and
- passive response.

In the case of active response, actions are taken automatically when certain types of intrusions are detected. These might involve activities such as collecting additional information or changing the environment in order to stop the attack. The purpose of collecting additional information could be to aid in the apprehension of the attacker. Changing the environment on the other hand will result in halting an attack in progress and blocking subsequent access by the attacker. We will discuss some potential approaches in this area in the next chapter. This might involve actions such as injecting TCP reset packets, reconfiguring routers and firewalls, revoking the credentials of malicious users, patching vulnerable machines, and so on. Another possible action is also to respond back towards the machine inferred to be the attacker. Note, though, that there are legal issues associated with this line of action and hence this is not generally recommended. In the case of passive response, the objective is to trigger additional monitoring that can more closely observe the malicious behavior and provide information to system users by sending alarms and notifications. It is expected that subsequent action will be taken by humans.

5.2 UNIQUE IDS CHALLENGES IN MANET

Although IDSs for enterprise environments have been studied extensively and several products/tools have been developed and are available today, these systems are not readily usable in a MANET environment. In this section we will focus on discussing why existing IDSs are not appropriate for a MANET environment. That discussion will also help us elucidate the unique characteristics that an IDS needs to have to operate in a MANET environment. We will also discuss the potential intrusion detection approaches that may be effective in a MANET environment.

Defending MANET networks is much more challenging than defending traditional enterprise networks for a variety of reasons. Characteristics such as volatility, mobility, as well as the ease of listening to wireless transmissions make the network inherently less secure. Existing tools usually assume a well-structured and static network and therefore cannot be used as they are.

The nature of MANET networks makes it easier for malicious users to disrupt the network because by definition MANETs are flexible and lack a fixed infrastructure. It is possible for a malicious node to join the network and become responsible for providing key services. This implies that, when securing the network we cannot assume that the threat is mostly from outside the network. Therefore, the network needs to be protected from all nodes, both external and internal.

Another key difference between MANET and enterprise networks is that MANET networks do not have an established perimeter like traditional enterprise networks. This plus the ability of nodes to move implies that there are no well-defined chokepoints where the IDS systems can be placed to monitor the traffic. This lack of chokepoints implies that nodes may only be able to observe a small portion of the network traffic and many attacks might escape detection by a single IDS. Therefore, a much more cooperative approach to intrusion detection is necessary. In this approach, multiple nodes have to work together to detect attacks. Nodes monitor traffic around them and then exchange information with other nodes. Nodes then use the exchanged information to detect attacks that cannot be detected by local information. As a result more attacks can be detected.

MANET networks also introduce a large number of complex protocols. New protocols for performing important functions such as routing, automated configuration, and mobility management, have been introduced. These protocols create new opportunities for malicious users to identify protocol vulnerabilities and use them for launching attacks. New techniques for detecting attacks exploiting the behavior of these protocols are needed. MANET protocols by their nature depend on nodes cooperating with each other in performing the functions of the network. One such critical function is routing. In static enterprise networks, routers are usually specific dedicated nodes. Routers are usually also located in well-protected areas. In a MANET, every node can be a router and therefore it is possible for a single malicious node to cause significant disruption in the operation of the network. For example, node A may begin advertising that it has direct connectivity to a large number of nodes. This may result in node A receiving a large number of packets which it can then inspect, drop, or loop around. Such attacks against routing protocols can be very disruptive in a MANET environment. Design of secure routing protocols (see Chapter 4) is one approach to solve this problem. The other approach is to detect routing attacks using IDS. The latter approach will be discussed extensively in this chapter.

Another unique challenge of MANET is the limited bandwidth. Traditional IDS systems require a large amount of sensor information to be sent around for analysis, which is typically not a problem in large enterprise networks. Transmitting large amounts of intrusion detection data is often not an acceptable solution for a MANET. Data reduction, which involves compressing the captured data being transmitted among nodes, is then a required feature for IDS in a MANET environment.

In a MANET, nodes tend to move a lot and therefore the connectivity of nodes changes dramatically. This makes it challenging to collect data from several sensors to a central entity for analysis and correlation. For example, the node that is responsible for analysis and correlation may lose connectivity to most of the sensors and gain connectivity to several new nodes due to node movements. A potential way of overcoming this challenge may be to develop a much more flexible architecture for intrusion detection where there is a loose association between sensors and analysis/correlation components. These constraints also rule out the use of a centralized architecture for intrusion detection. In addition, the detection algorithms must be able to seamlessly adapt to topological changes due to node mobility and intermittent connectivity due to the wireless environment.

MANET networks are also much more dynamic and unpredictable because connectivity depends on the movements of nodes, terrain, changes in the mission (e.g. for a military application or a first responder application), node failures, weather, and other factors. As a result, it is difficult to accurately characterize normal behavior. Hence, it is often difficult to distinguish malicious behavior from normal but unexpected events. For example, a network may be seen connecting and disconnecting from the rest of the network. This may be symptomatic of an attack but can also be due to the fact that a node is moving in and out of range of the network. Existing detection tools (anomaly detection tools in particular) may not be effective in such an environment because they have been developed with a much more static and predictable environment in mind and cannot deal with the dynamism and unpredictability of MANETs.

Finally, unlike in wireline networks, nodes in an ad hoc network have limited energy. Hence, often only computationally simple, energy-efficient detection strategies can be used by such nodes. The detection algorithms must also be distributed as communication with a central computing unit will consume significant energy and bandwidth.

All these differences with wireline networks motivate the design of detection strategies specifically geared towards ad hoc networks. We summarize this discussion in Table 5.1.

5.3 THREAT MODEL

The threat model in a MANET environment is also different from the threat model in traditional static enterprise networks. In this section we discuss the type of threats that we consider in this chapter and discuss potential attacks that are possible in such an environment. Later in this chapter we discuss specific schemes for detecting some of these attacks.

In this chapter, we focus on both outsider and insider threats against MANET. The threats from outsiders can result from users that have penetrated the preventive security layers protecting the MANET. Such users are not authorized to be a part of the

TABLE 5.1. IDS Challenges in a MANET.

Feature of MANET	IMPACT of the Feature on the Design of IDS Used Traditionally
Lack of fixed infrastructure	A distributed detection approach would be necessary
No established perimeter. Mobility of nodes	Lack of well-defined chokepoints at which to place the IDS. Hence a distributed and cooperative detection approach is necessary
MANET protocols are based on node cooperation	The IDS has to focus on areas that might traditionally be ignored, such as routing
Limited bandwidth availability	Data reduction might be a required feature
Dramatic changes in the connectivity of nodes	A requirement for a flexible architecture. Also points to the need for detection algorithms that can seamlessly adapt to topological changes
Large number of new and complex protocols	Requires new techniques to detect attacks leveraging vulnerabilities in these protocols
Unpredictability of the network topology and connectivity patterns	Static techniques may not be effective
Limited energy of nodes	Energy-efficient and computationally simple detection strategies required

MANET. The threats due to insiders can manifest in the form of nodes that are authorized to be a part of the MANET but start exhibiting malicious behavior. Attacks resulting from such threats include attacks launched through insider nodes that have been captured by a malicious user or through nodes that can successfully pretend to be a part of the network since they have access to compromised keys. Such attacks are also possible in a traditional wired network, but are much more likely in a MANET because of the features of MANETs as described earlier. Such attacks may also have a potentially bigger impact in a MANET because, while critical servers can be physically protected in a wired network, that may not always be possible in a typical MANET.

An area of particular vulnerability is routing as mentioned earlier. Several attacks on routing are possible as described in Chapter 1. In addition to routing attacks, many other attacks that can be launched in a MANET. Below are some examples of such attacks:

- *Packet Dropping*—in this type of attack a node (let us say node A) may drop packets destined for other nodes that are being routed through node A. Node A may drop all of the packets, packets destined only for a specific node, or randomly drop a percentage of the traffic. These attacks are often difficult to detect in a MANET because packet dropping occurs in MANET even in the absence of malicious users due to the wireless nature of the links. It is often possible for the packet loss rate to be high due to nodes moving in and out of range, due to a terrain, or weather that increases the bit error rate of transmissions, and other factors. There are also similar attacks that nodes may launch that focus on just delaying packets, reordering them, modifying them, or sending them to the wrong destination, rather than dropping them. Such attacks cause severe problems with protocols such as TCP and SCTP.
- *Denial of Service Attacks*—in this type of attack a node launches an attack with the goal of affecting some services and denying access to those services. A malicious

node may launch such attacks by flooding the network with excessive traffic so that other nodes cannot use the network, flooding a specific node so that other nodes cannot communicate with that node or get access to services hosted by that node, jamming the wireless spectrum, and so on.

- Other types of attacks that are extensions of traditional attacks that have appeared in wireline networks such as scanning, relaying packets with malicious load, and so on.

In this chapter we discuss a variety of schemes for detecting these attacks. We will focus in particular on attacks against routing protocols and packet dropping attacks because these represent two very broad classes of attacks. The techniques that we will discuss are general and can be used for detecting many of the other attacks discussed here.

When we discuss the various detection schemes we make a simplifying assumption. We assume the existence of an authentication scheme in which each node has an individual key and uses it to sign all individual messages it sends (or at least the ones we use for detecting the attack such as routing messages). This scheme will help us ensure that, when an attack is detected, we can easily identify the malicious node leveraging the authentication scheme. If such an authentication scheme is not available, it is typically difficult if not impossible to identify the malicious nodes because any node may have created an offending message pretending to be any other node. If there is no way to trace back malicious traffic to a specific node, then the detection schemes discussed later in this section will only be able to detect that there is an attack by observing the malicious traffic.

5.4 ARCHITECTURE FOR INTRUSION DETECTION IN MANET

It is evident from the previous discussion that there are significant challenges in utilizing existing IDS tools in a MANET environment. A new intrusion detection architecture that is much more distributed and cooperative needs to be introduced. Such an approach will be able to detect more attacks reliably. In order to create such an architecture there are some fundamental questions that need to be considered.

A key question that will have to be answered first is related to the roles that a node can have in a MANET intrusion detection architecture. These roles include the following:

- *Self-Detection*—a node may run a detector that focuses on monitoring the node itself to see whether it is behaving as expected. It could do this for example by monitoring the messages that the node itself is sending to other nodes. Since this detector has perfect knowledge of the state of the node, it can detect malicious behavior without false positives. The limitation of this approach is that, if a node has been penetrated or taken over by an attacker, it is likely that the detector performing the self-detection may be penetrated, turned off, or completely by-passed. Therefore, this approach can only provide for reliable detection of attacks by unsophisticated attackers. This is also inefficient since every node will have to perform the monitoring function itself. This could result in redundant execution of the same function on different nodes.
- *Local Detection*—a node may run a detector that detects attacks based on evidence available locally. This may include evidence from packets received by the node, forwarded by the node (as part of the routing process), or packets that have been

observed by the node going through the wireless link (by eavesdropping on the link). The advantage of this approach is that, since the evidence is collected locally, it is dependable. Therefore the node can be fairly confident that when an attack is detected it is indeed an attack. Some false positives may occur due to the fact that monitoring of packets over the wireless link has limitations, as discussed in Section 5.5.2.

- *Data Collection*—since a number of attacks in a MANET environment cannot be detected locally with a high degree of certainty by a single node, it is necessary for nodes to collect intrusion detection evidence and share it with other nodes. This data is then shared either with everybody or with some subset of the nodes, depending on the specific IDS architecture used.

The next key question is related to the number of nodes that need to be part of the intrusion detection architecture as well as the role of each. If our goal is to detect all the attacks for all possible mobility scenarios, it is likely that all nodes in the system will have to be part of the intrusion detection process because there is always a possibility that a malicious node A will be directly connected to a single node B, which is the only node with direct visibility to the behavior of the malicious node. In order to detect node A's malicious behavior, node B will either need to have the ability to analyze A's behavior or will need to have a sensor that collects sufficient evidence and then sends the evidence to some other node C for further analysis.

Requiring every node to participate in the intrusion detection process may not be desirable, however, due to inefficiencies associated with this approach. It is also likely that in practice it will not be possible to require that all nodes devote resources to the intrusion detection process given the resource constraints and the need for the node to execute other applications. Sensor networks are good examples of networks with nodes that have severe resource constraints and most likely cannot support IDS running on each node in the network. If we are willing to tolerate less than 100 percent detection accuracy, which is probably the case in a lot of applications, it may not be necessary to run IDS on each node. If we can utilize only a subset of the total number of nodes in the intrusion detection process, it is important to decide what role each node needs to play in the intrusion detection process. There is no general answer to this question that is well suited for all situations. The answer depends on several parameters, such as:

- The degree of certainty required for the detection. If we need to detect intrusions under almost all conditions with minimal false positives then a very large percentage of nodes will need to execute the IDS.
- The mobility scenario expected for the specific application and mission. For example, in a military environment, units tend to move together and be close to each other for a long period of time. In such cases a small number of nodes running IDS placed within a unit may be able to detect most intrusions.
- Mode of monitoring. If nodes are capable of monitoring in the promiscuous mode whereby each node can receive and analyze any packet transmitted by any of its neighbors, then each node can monitor a larger portion of the traffic. If this is not possible, then a node might be restricted to monitoring traffic that it relays. In this case, each node can monitor a relatively smaller portion of the traffic. A disadvantage of the promiscuous mode of operation, though, is the increased

power consumption associated with it and the possibility of false positives (see Section 5.5.2).

- The connectivity environment. In a flat area where there are few obstructions, most nodes usually have multiple neighbors that can observe their behavior. In that case, a small percentage of nodes may have visibility of most of the traffic and therefore be able to detect most attacks.
- The capabilities of nodes. It may only be possible for a small subset of the nodes to run IDS due to limitations on resources such as power, CPU processing, storage, and bandwidth. In that case there may not be a choice and the nodes that have available resources and the necessary capabilities will have to execute the IDS functions.
- Number of nodes that need protection. It may be critical to only protect a small subset of the nodes that either store important data or run critical servers. The specific mission or application may be willing to tolerate attacks against a number of other nodes. In that case it is probably more important to place nodes running IDS around the important nodes. A few additional nodes running IDS scattered throughout the rest of the network might then suffice.
- Percentage of compromised nodes that can be tolerated. Certain applications can continue functioning at an acceptable level even after some percentage of the nodes has been compromised. If that is the case then a smaller percentage of IDS nodes may be sufficient to ensure that no more than the acceptable percentage of nodes has been compromised.

Unfortunately, since there are so many parameters in deciding how many and which nodes need to execute the IDS functions, there is no clear-cut answer to this question. The answer depends to a large degree on the specifics of the application and the mission. We will discuss some possible schemes that attempt to address this issue later in this section.

The most important factor in deciding which nodes are part of the intrusion detection system and the role of each is the architecture of the system. Typically, we can have two different types of architecture:

- noncollaborative intrusion detection architecture;
- cooperative intrusion detection architecture.

We next look at each of these architectures and also address the problem of the role of nodes in the intrusion detection system.

5.4.1 Noncollaborative Intrusion Detection System

Noncollaborative intrusion detection systems are those in which the nodes that are part of the intrusion detection system act individually in performing the various roles associated with detection. Thus, such nodes will not only have to collect data but will also have to analyze the collected data themselves. This would lead to lower communication overhead amongst the nodes. In addition, this is also a less complex approach since each node acts individually. An attempt at addressing the question of the role of nodes in a noncollaborative intrusion detection system has been made in [80–83].

5.4.1.1 IDS Module Placement for Noncollaborative Intrusion Detection In [80, 83] the authors address the problem of placement of intrusion detection modules considering ad hoc networks with multiple mobile intruders. They consider networks in which each node is acting individually. Cooperation amongst nodes is limited to sharing the recommendations. Traffic capturing and analysis is done individually by every node. In this work, the authors assume the use of misuse detection techniques, although some of the ideas are also applicable when the network is using the anomaly detection techniques. The objective in this work is to maximize the detection probability given a constraint on the resources consumed for executing the IDS functions. The proposed approach assumes that the nodes can monitor traffic promiscuously. A node, after capturing traffic promiscuously, analyzes the captured traffic for any malicious signatures. It is obvious that this leads to a significant wastage of network resources if every node has to execute the capture and analysis processes. A more efficient solution is to let a smaller subset of nodes monitor the traffic while ensuring that all of the network traffic is analyzed.

The question then is how to select this smaller subset of nodes that have exposure to all of the network traffic. A concept that can be used to select this smaller subset of nodes is that of dominating sets. A dominating set of nodes is the set of nodes such that these nodes along with the neighbors that each node can hear make up all the nodes in the network. Identifying all the nodes that constitute the dominating set might not be sufficient for answering the question of where to place IDS capabilities. This is because some of the nodes on the dominating set may not be able to execute IDS due to resource constraints. Hence, the challenge is in selecting the set of nodes that constitute the dominating set while satisfying the resource constraints. In addition, constraints on the capability of nodes to execute IDS imply that some of the traffic might not be analyzed for intrusions. This implies that perfect detection might not be possible because it may not be possible to identify a complete dominating set where all nodes in the dominating set are IDS capable. Hence, the authors seek to maximize the detection rate.

To understand these concepts, consider Figure 5.2. Here we consider five nodes. A node is denoted by a vertex and an edge represents the "neighbor" relation. Thus an edge between two nodes indicates that the nodes are neighbors. We would like to remark here that a node is able to receive all packets transmitted by its neighbor although it may not receive all packets received by its neighbor. Thus a dominating set in this context is the set of nodes such that these nodes can hear all the traffic transmitted by their neighbors. One such example of a dominating set is the set of nodes [v_1, v_3, v_5] since these nodes along with their neighbors constitute the entire network. But note that this is not the minimum dominating set, that is, a dominating set with the least number of members. [v_2, v_4] corresponds to a minimum dominating set. In such a case, executing the IDS on the nodes that constitute the minimum dominating set is sufficient to capture all the network traffic and hence achieve perfect detection.

Now assume that the shaded nodes v_2, v_3, v_4 are incapable of executing IDS. In this case the set of nodes that can execute IDS is [v_1, v_5], but as is obvious, this set of nodes will not be able to detect any malicious traffic originating from v_3 to either of its neighbors. This implies that it is not possible to achieve perfect detection. Note that a

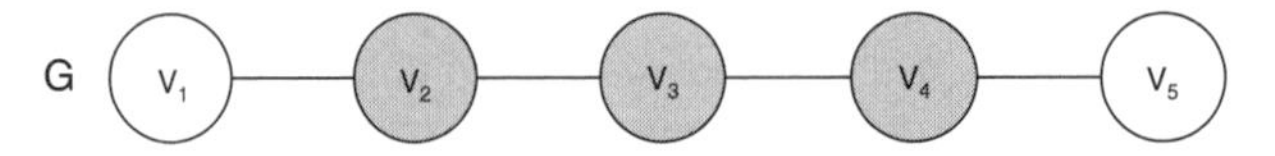

Figure 5.2. Concept of dominating set.

node is said to be covered if it is the neighbor of any node that executes IDS and is hence IDS active. Further a node covers itself. Thus, in this example, node v_3 is not covered.

The authors formulate this problem mathematically by aiming to maximize detection while satisfying the resource constraints as well as constraints on the nodes that can execute IDS. The optimal solution to the mathematical formulation is shown to be an NP-hard problem. Hence the authors propose two polynomial complexity algorithms [80] called APPROX and maximum unsatisfied neighbors in extended neighborhood for maximizing the coverage (MUNEN-MC).

APPROX iteratively selects the nodes that have to execute IDS. During each iteration, APPROX selects a node in the entire network which has the maximum number of neighbors that are not covered by any IDS active node. After the selection, every node in the selected node's two-hop neighborhood recomputes its priority. The selection process terminates when a given number of insider nodes are selected or the IDS capable insider nodes have zero priority, that is, no nodes are left uncovered. MUNEN-MC on the other hand also selects the same set of nodes as APPROX, but has significantly lower communication complexity than APPROX. It achieves this by considering a two-hop neighborhood and selecting nodes which have the maximum number of uncovered neighbors in this neighborhood.

A problem with these approximation algorithms is that the set of IDS active nodes needs to be recomputed every time there is a neighborhood change. The computations and related message exchanges also consume significant resources. Thus, these algorithms cannot be used when the network is highly dynamic. To address this, the authors propose two heuristics. The first heuristic proposes that every node execute the IDS with a probability selected to regulate the resources consumed and detection rate. The second heuristic is called the geometric dominating set algorithm, GO-DOM.

GO-DOM uses geometric information to select the set of nodes that have to execute IDS. The geographic area of the network is covered by the minimum possible number of circles each with radius r, where r is the transmission range of a node. The center of the outermost circle has a distance r from the closest network boundary. Each node is assumed to know or is assumed to be able to compute the coordinate of the centers of the circles. Note that this is a one-time computation or message exchange for each IDS capable insider node. Each node is also assumed to know its coordinates using GPS or other techniques. Each IDS capable node broadcasts its distance from the center of each circle it resides in to its neighbors. It sends this broadcast packet when it joins the system, and thereafter each time it moves. Each node then selects an IDS capable neighbor which is the nearest to the center of a circle it currently resides in to execute the IDS.

Figure 5.3 provides an illustrative example of the GO-DOM algorithm. The circles in solid lines are some of the circles that cover the geographic area of the network. For the current positions of the nodes, two nodes in circle 3, namely nodes v_2 and v_5, execute the IDS because both of them are nearest to the center of the circle they reside in (i.e. circle 3) in their neighborhood. Nodes v_1, v_3, and v_4 do not execute IDS because they are within v_2's neighborhood and v_2 is closest to the center of the circle. v_5 is not within v_2's neighborhood (shown with a dotted line) and therefore is executing IDS.

The authors in [80, 83] also compare the performance of the optimal algorithm with the approximations and the heuristics. They also investigate the fraction of nodes needed to execute the IDS and show that, even for high detection rates, only a modest fraction of the nodes needs to execute IDS. These results are quite useful in the design of effective intrusion detection systems.

We would like to remark here that our description above is simplified. Specifically, the authors in [80, 83] consider special types of ad hoc networks where there are two types of

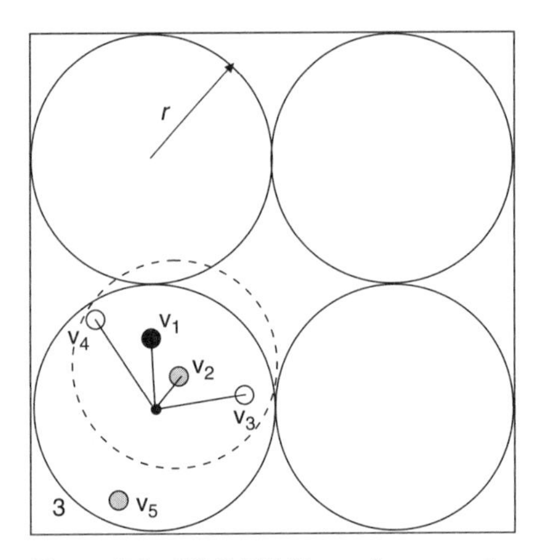

Figure 5.3. GO-DOM illustrative example.

nodes that they call insider and outsider nodes. The insider nodes are assumed to be trusted and are the ones which execute the IDS. Thus all the explanation above applies to insider nodes. The authors had to consider these types of special ad hoc networks since the proposed IDS placement algorithms (such as random placement, GO-DOM, approximation to optimal) are not secure to attacks from untrusted nodes. Rather the security of these algorithms is derived from the fact that the nodes executing these algorithms are trusted. The same ideas can be applied to general ad hoc networks provided secure versions of the IDS placement algorithms can be designed. This was the reason why we have described the algorithms more generally here.

An important assumption in [80, 83] is that the nodes that execute IDS in the promiscuous mode operate perfectly, but this is not always true. A node may fail to capture and analyze the packets perfectly due to many reasons. This failure may happen due to faults in the node, power-saving operations that the node might use or if the attack has been designed to evade the IDS, or if the attacker successfully launches a DoS attack on the node, or if the node does not receive the packet due to the vagaries of promiscuous mode monitoring such as collisions (see Section 5.5.2), poor transmission quality in wireless links, and so on.

Such a failure model is considered in [81], where the authors extend the algorithms proposed in [80, 83]. The authors, however, assume that a node is not compromised and hence never reports falsely on the occurrence of malicious packets. To address such failure, the authors propose to make use of redundancy whereby more than one node in a neighborhood monitors packets promiscuously. We show this in Figure 5.4. In this figure the shaded nodes indicate the nodes that are executing IDS. We see that node A is protected by four other nodes that execute IDS. As a result, traffic emanating from node A will always be analyzed if at least three of the four nodes (C, D, E, and F) have not failed at any point in time. A question is related to the number of nodes in a neighborhood, denoted as k, that have to execute IDS to reach the desirable level of detection. The authors address this by relating the number k to the desired detection probability. Using this number k, the authors provide extensions to the mathematical models used in the previous techniques to ensure that every node is covered by k other nodes that execute IDS. In addition, the approximation algorithms are also extended by considering a node as satisfied if it is covered by more than k nodes. In each iteration, the approximation algorithms then select the node with the maximum number of unsatisfied neighbors to execute the IDS.

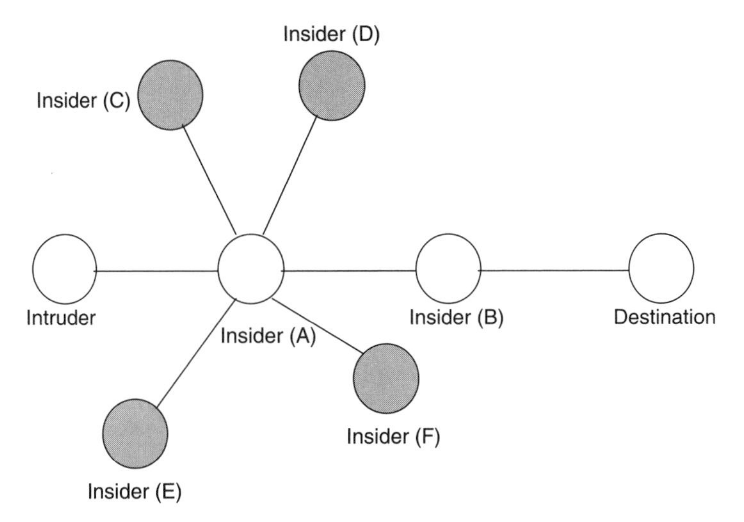

Figure 5.4. Use of redundancy to address node failures.

Other schemes have also been proposed for dealing with the problem of detecting uncooperative nodes. Marti *et al.* [84] proposed a simple scheme in which each node utilizes a *watchdog* for monitoring its neighbors and identifying misbehaving nodes (e.g. nodes that do not forward packets). Each node utilizes a *pathrater* that ensures that routes used for routing do not contain the misbehaving nodes. The effectiveness of this scheme is limited because each node acts independently and there is no cooperation among nodes in exchanging information about misbehaving nodes. This scheme was shown through simulations to increase throughput up to 27 percent in the presence of misbehaving nodes while increasing overhead by a few percentage points.

So far we have been assuming that each node acts independently in order to capture and analyze traffic, but as discussed earlier, cooperation amongst the nodes executing the IDS functions might result in more efficient monitoring of the network. We discuss this aspect next.

5.4.2 Cooperative Intrusion Detection

Cooperative intrusion detection is expected to result in better detection of malicious behavior in MANETs. This is particularly true in cases where a single node may not have enough data to detect an attack. On the other hand multiple nodes may each have a portion of the data required for detecting the attack and would need to work together to detect the attack. For example, an attack may consist of multiple packets that take different routes and may not all be observable by a single node. It is therefore necessary for nodes to cooperate in the intrusion detection process for sharing data that is required for detecting an attack by multiple cooperating nodes. Cooperation among the nodes might also result in sharing information about the attacks detected.

There are two main approaches that can be used by nodes to coordinate the detection process:

- peer-to-peer; and
- hierarchical.

In the first approach each node is acting completely autonomously. Cooperation is achieved by each node sharing with all (or a subset of) other nodes its intrusion detection data or alerts. The second approach is the hierarchical approach in which the responsibilities of nodes are structured through a hierarchy. Specific nodes may be responsible for data collection and other nodes may be responsible for correlation and initiating responses. Further, intrusion detection data and alerts may only be shared through the hierarchy, thus minimizing the overhead traffic generated by the IDS scheme. We will now discuss each of these two approaches in more detail.

5.4.2.1 Peer-to-Peer IDS Architecture In the peer-to-peer approach every node tends to act independently of other nodes. Each node has its own local detectors that are used for detecting attacks. To improve the detection accuracy at each node, nodes share information with other nodes in the network on what they are observing. Each node monitors for suspicious activities by monitoring traffic terminating on that node (e.g. port scans). Each node also tends to monitor its neighbors either through promiscuous monitoring or by inspecting traffic destined for other nodes that is routed through it. When malicious activity is detected by the detectors on a node, its detectors generate alerts that are sent to all other nodes in the network. The other nodes can then use that information to confirm detection of attacks within those nodes and increase the confidence of their own detection. Based on the local intrusion detection and alerts from others, nodes can also determine appropriate responses to the attacks. Examples of such responses include not routing traffic through a malicious node and not forwarding malicious traffic generated by a malicious node. In order to reduce the potential harm from malicious nodes wrongly accusing well-behaved nodes of launching attacks, each node should trust local information much more than alerts received from other nodes.

In the peer-to-peer architecture, in addition to alerts, nodes may also share with other nodes local data about observed events. This may help the IDS modules in other nodes detect attacks that could not have been discovered by them based exclusively on their local data. For example, for detecting malicious port scanning activities it is usually beneficial to collect statistics across multiple nodes. If the raw data needed by the intrusion detection algorithms is shared with every node in the network then the detector running on each node will have the ability to perform much more accurate detection of stealthy attacks (attacks that do not exhibit sufficient suspicious activity on a single node). However, the amount of IDS data that can be shared across all nodes needs to be limited because otherwise the level of overhead in terms of bandwidth used and delay incurred could be very large. Figure 5.5 illustrates the peer-to-peer IDS architecture at a high level. Here each circle represents a node executing the IDS functions. As shown in the figure, different nodes are able to exchange data with each other.

The main advantage of the peer-to-peer approach is that the IDS architecture is fully distributed. A fully distributed approach makes the IDS architecture much more resilient. Every node acts independently of other nodes and does not rely on any other node for detecting attacks. In a MANET environment nodes move around and lose connectivity to other nodes, and networks join and split. All this makes reliance on other nodes and any centralized components problematic.

The other key advantage of the peer-to-peer approach and its distributed nature is its tolerance to attacks against the IDS architecture from malicious nodes. A malicious IDS node can only have a limited impact on the overall intrusion detection process because each node trusts mainly its own evidence. The data received from other nodes

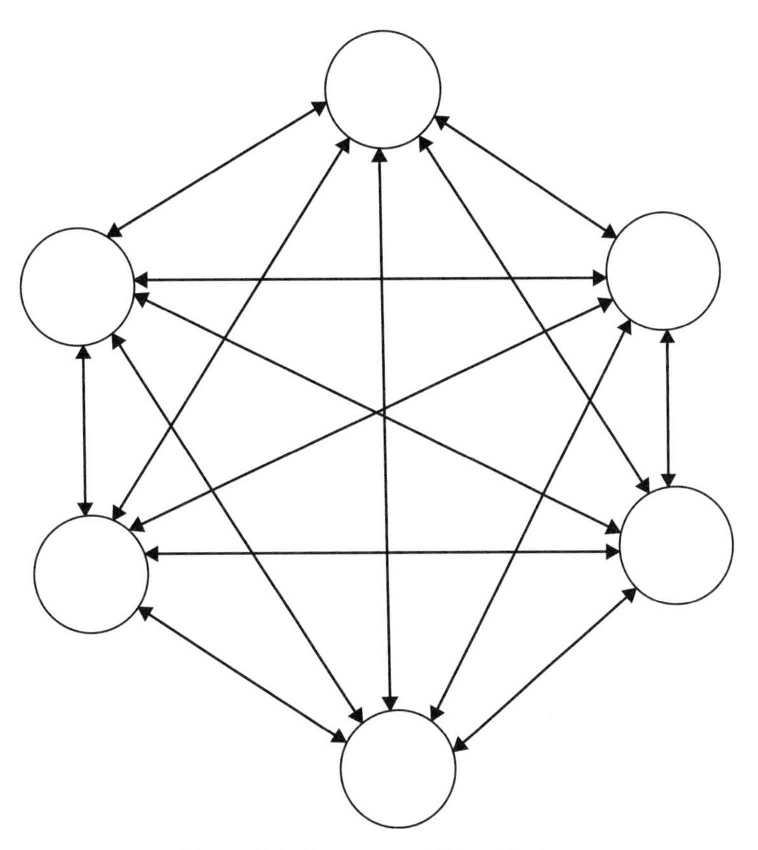

Figure 5.5. Peer-to-peer IDS architecture.

can be used only to improve the detection process. For example, if a single node alerts every node in the network that a node, let us say node X, is malicious, other nodes may not believe that claim (or rely on that claim in their detection process). That is because typically a node has multiple neighbors and therefore multiple nodes should observe node X's malicious behavior and report that to others. Alerts from multiple nodes would be expected and therefore a single alert from a single node may be considered with caution. This approach may also limit false positives since most nodes will not be influenced by a single outlier.

The main disadvantage of the peer-to-peer approach is that it tends to generate a lot of overhead. Even if we assume that IDS messages are multicast, it is still necessary for every single node to communicate with every other node in the network. Given *N* nodes in the network, this approach may result in an unacceptable level of overhead. One potential solution for minimizing the overhead is to limit the distance that the IDS messages are propagated to. Since most IDS information is useful only in the neighborhood of the malicious node, it may only be necessary to propagate that information in the neighborhood of the malicious node rather than to the whole network. This introduces some level of risk since nodes tend to move around and therefore malicious nodes may be allowed to join the network again in a different neighborhood. The other limitation of the peer-to-peer approach is that it may not be effective for detecting attacks that require collection of large amounts of intrusion detection data across multiple nodes. It is infeasible to exchange a lot of raw data in the peer-to-peer approach among all nodes because of the $N \times N$ communication problem. Without such information, nodes can perform intrusion detection based only on local data and possibly limited data provided by other nodes. This may be insufficient for detection of certain types of attacks.

5.4.2.2 ***Hierarchical IDS Architecture*** The alternative to the peer-to-peer architecture is the use of a hierarchy [85, 86]. In this approach, every node plays a role in the intrusion detection process but some nodes have greater responsibilities. Nodes organize into small neighborhoods to perform intrusion detection. All nodes in the neighborhood are responsible for performing local intrusion detection for detecting attacks that can be detected locally. At the same time, nodes in the neighborhood elect a leader and all nodes in the neighborhood collect data relevant to the intrusion detection process. The collected data is sent to their leader. The leader is then responsible for correlating the data and detecting attacks that cannot be detected by a node in the neighborhood using local data. This is similar to the way attacks are detected in the peer-to-peer approach except that in that case the data is shared with every node in the network (not just the leader of the neighborhood). Each node in the peer-to-peer approach is then responsible for analyzing the data for attacks (not just the leader in the neighborhood).

Leaders of each neighborhood may also cooperate with leaders of other neighborhoods for correlating data across neighborhoods. This will help in detecting attacks spanning multiple neighborhoods. In the hierarchical architecture, alerts about malicious behavior within the neighborhood are sent to the leader. The leader is then responsible for sharing the alerts with other nodes in the neighborhood and with leaders from other neighborhoods. In a similar manner, responses to attacks can be either decided locally or driven by the leader once it determines that there is an attack by correlating events received from the neighborhood. Leaders can also coordinate responses across neighborhoods.

This concept can be extended so that a complete hierarchy that covers the whole network may be formed. Neighborhoods can be formed using clustering algorithms and leaders can be elected using the same algorithms that elect clusterheads. Such algorithms have been used in the past in routing protocols for MANET. We will discuss several potential clustering algorithms in Section 5.4.3.1. An example hierarchy is shown in Figure 5.6. Nodes in a neighborhood elect a leader (first-level clusterheads). First-level clusterheads may elect second-level clusterheads responsible for correlating data across neighborhoods and so on until there is a well-established hierarchy represented as a tree with a root node and leaf nodes, as shown in Figure 5.6.

The hierarchy formation needs to be designed in view of the unique requirements of the MANET environment. For example, the hierarchy needs to adapt well to movements of nodes. A static hierarchy where leaders are elected at the beginning of the operation of the network without any changes being allowed is not acceptable in a MANET environment. Nodes that move together as a group probably need to be part of the "neighborhood" in the hierarchy concept discussed earlier. Also, not every node may be elected a leader and take increased IDS responsibilities. Nodes with limited battery life probably should not be assigned leaders and take on those additional responsibilities. It is also preferable that nodes that are hardened from a security viewpoint and better protected than other nodes become leaders and take on more of the coordination responsibilities. This makes it more likely that the IDS functions will execute correctly. We will discuss in

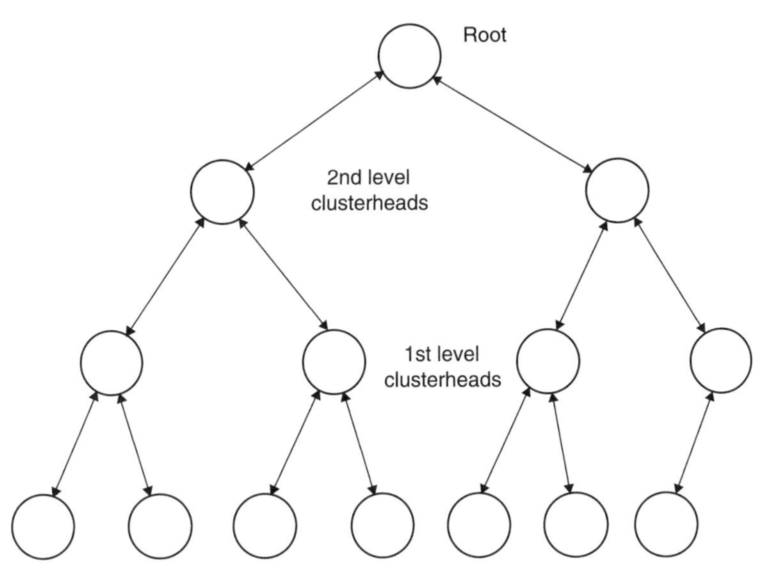

Figure 5.6. Hierarchical IDS.

Section 5.4.3.1 potential approaches for forming the hierarchy in a way that is well suited to the MANET environment.

Another potential feature of the hierarchical approach is that, if the hierarchy is formed in a way that nodes are just one hop away from their parent IDS, data and alerts can be shared through messages that are broadcast to its direct neighbors. This allows the IDS architecture to operate even in the absence of routing. If the attacker has completely disrupted the routing process the IDS may be able to detect the attacker and recover from it by isolating the intruder and allowing the routing protocol to recover and resume.

The key advantage of the hierarchical approach is it incurs much less overhead as compared with the peer-to-peer approach because nodes need to communicate with a smaller number of nodes. Intrusion detection data is shared in an orderly and efficient manner. This allows IDS information to propagate throughout the network very efficiently. At each level data aggregation and consolidation can further reduce the overhead. This is preferable to having every node share the same data with every other node as is done in the peer-to-peer architecture. The efficient transmission of IDS consolidated data in the hierarchical architecture also allows for raw IDS data to be shared among nodes in the network. This may make it possible for detection algorithms to detect attacks that cannot be detected by the peer-to-peer architecture.

The disadvantage of the hierarchical approach is that it is prone to Byzantine attacks from IDS nodes. Nodes that become a leader take on critical responsibility for the intrusion detection process because they are responsible for collecting and correlating intrusion detection information within neighborhoods and across neighborhoods and coordinating dissemination of alerts. If one of those nodes is taken over by a malicious user it can cause a lot of harm. One possible way to do so is by falsely accusing well-behaved nodes of malicious behavior. Such false accusations may lead nodes in the neighborhood to consider the falsely accused nodes as malicious. A leader may also attempt to convince leaders in other neighborhoods that a well-behaved node is malicious. If such an attempt is successful, the falsely accused nodes may end up getting completely isolated by the network. One solution to this problem is the election of back-up leaders that monitor the main leader and interfere if they detect malicious behavior from the leader. This requires some kind of a voting scheme to ensure that a single malicious leader cannot cause a significant amount of harm. This of course increases the overhead of the approach because IDS data and alerts need to be sent both to the primary and back-up leaders, limiting the benefits of the hierarchical approach. To resist the compromise of multiple leaders, a large number of back-up leaders is required. If every node ends up playing the role of a leader or a back-up leader then the hierarchical scheme becomes very resistant but ends up behaving as the peer-to-peer approach exhibiting the same advantages and disadvantages.

The other challenge of the hierarchical scheme is that, due to mobility, connectivity changes, failures, and so on, it may be challenging to maintain the hierarchy and ensure correct operation of the IDS hierarchy. For example, it may be necessary in several instances for a new leader to be elected for a neighborhood either to increase efficiency or just because a leader has moved out of range. The result is that, for a brief period of time, a neighborhood may operate without a leader, making the IDS unavailable in that neighborhood. This may introduce a window of vulnerability that a malicious user may attempt to exploit to launch an attack. Another related problem with the hierarchical approach is that, as the intrusion detection data and alerts flow through a potentially changing hierarchy, it may take a significant amount of time for the hierarchy to correlate the data and detect an attack. This results in increasing the detection latency of the architecture (especially compared with the peer-to-peer approach).

5.4.3 Key Novel Concepts for Cooperative Intrusion Detection in MANET

As discussed earlier, IDS in a MANET needs to be based on cooperative approaches in which most if not all nodes play some part in the intrusion detection process. Because of the importance of the cooperative approaches, we focus in this section on specific novel cooperative schemes that have been proposed. In particular we consider:

- clustering, which is a scheme that has been proposed for organizing nodes in a hierarchy;
- reputation schemes that allow good or bad reputation of nodes to be shared across the network;
- schemes that encourage node cooperation by providing incentives.

5.4.3.1 Clustering One of the key questions in a cooperative intrusion architecture is related to the organization of nodes. If every node is required to cooperate with every other node (as in a peer-to-peer architecture) without a structured organization, then overheads will grow proportionally to the square of the number of nodes. Rather than assuming a completely flat structure, it may be possible to organize nodes in a hierarchy by forming groups efficiently, leading to reduced communication overheads (as discussed in Section 5.4.2.2). We would like to remark here that it is critical to organize the groups appropriately in order to achieve a significant reduction in overhead.

This problem of designing groups efficiently is a fundamental problem that has been the focus of research in MANET and which has been considered especially in relation to routing. The same concepts can be adapted for the intrusion detection function. When designing the clustering schemes for detection, researchers have tried to exploit several characteristics of the network in order to minimize overheads. An intuitive approach for organizing into clusters is according to node mobility. Clusters are not static but change as nodes move around. As nodes move, they may drop out from their existing cluster and join a different cluster that is closer to them. Clusters in some cases may join together or a single cluster may split into two or more clusters. Typically a node might be elected in the cluster as a leader and such a node is typically called the clusterhead.

There are several algorithms that have been proposed in the literature for organizing nodes in clusters based on a variety of criteria, including how the clusters adapt to connectivity changes due to mobility, and how clusterheads for each cluster are elected (although certain cluster-based algorithms do not require the election of clusterheads). We will now discuss some representative clustering algorithms and will then consider how these concepts can be applied to intrusion detection. We consider clustering both with and without the use of clusterheads.

5.4.3.1.1 Clustering with Clusterheads One of the heuristics for selecting clusterheads that has been proposed in [87] is the "highest-degree" algorithm. In this algorithm, the nodes that have the largest number of direct neighbors become the clusterheads. In this case, neighbors are those nodes that are within each other's transmission ranges. The clusterhead selection can be implemented as follows. Each node discovers through periodic hello messages its direct neighbors and the number of neighbors that each of them have. Based on that information, a node selects itself as the clusterhead if it has more neighbors than its neighbors. This is a very simple algorithm that has a low rate of clusterhead change.

Another potential algorithm is the Graphcluster algorithm [88]. The first part of this algorithm involves forming a spanning tree by using the breadth-first-search (BFS) tree algorithm. In the BFS algorithm, a node initiates the algorithm and becomes the root. If multiple nodes initiate the algorithm, the one with the lowest ID wins. The root starts transmitting periodic tree discovery beacons and flooding them in the network. As each node receives a beacon it transmits its own beacon which includes its id, the root's id, the id of its parent (its neighbor that is closest to the root), and its distance from the root. Nodes that receive such beacons can discover the shortest path to the root and they send updated tree building beacons. Eventually the tree forms. One such possible tree is shown with bold lines in Figure 5.7.

In the Graphcluster algorithm, once the tree is formed through the BFS algorithm, clusters are then formed as follows. Each node counts the number of its children (it can discover those through the BFS advertisements). This information is then included in the subsequent tree building beacons. These cluster members then start to be aggregated from the leaf nodes towards the root. Once a node discovers that it has more than k children it forms either one cluster (if it has $<2k$ children) or multiple clusters by partitioning its children's subtrees into clusters. Figure 5.8 shows one example for the formation of clusters (assuming $k = 3$). Once nodes become part of a cluster they are not counted by their parents in the cluster formation algorithm. The algorithm terminates once the cluster formation is stable. After that the algorithm tries to maintain the clusters as they exist by making small changes for nodes that join or leave clusters due to mobility.

5.4.3.1.2 Clustering Without Clusterheads Algorithms have also been proposed for creating clusters that do not involve the election of an actual leader in the cluster. The advantage of these approaches is that they do not require that a node take on extra

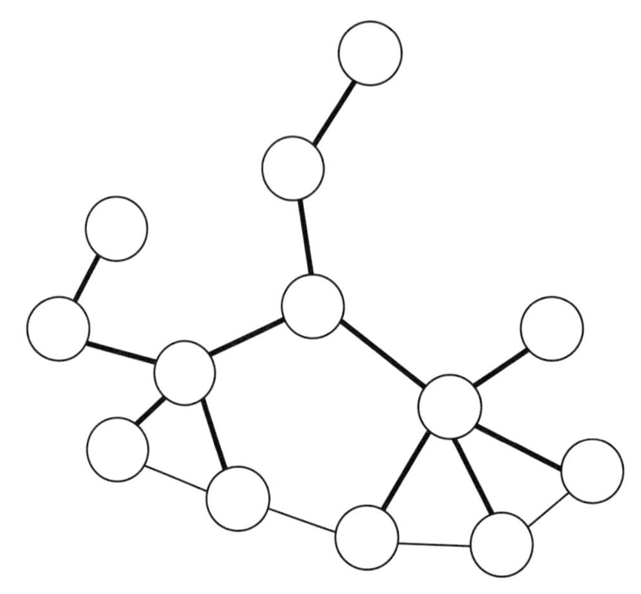

Figure 5.7. Breadth-first search example.

responsibilities. Such extra responsibilities could result in that node draining its power, and expending a lot of its CPU cycles and memory. We next look at some such algorithms.

Krishna *et al.* [89] have introduced a scheme for organizing nodes in clusters without a clusterhead. In this case, clusters are formed in a way that members of each cluster are always one hop away from each other. In other words, each member of the cluster has direct connectivity to every other member of the cluster. This type of a cluster is called a clique (see Fig. 5.9).

Krishna *et al.* [89] describe one approach for forming cliques. In this approach, each node maintains a list of its direct neighbors, a list of all clusters, and a list of nodes connecting clusters together. Whenever a new node appears on the network it gets from its

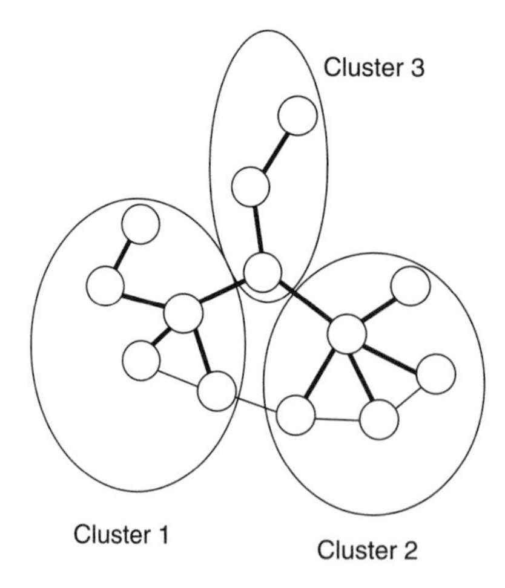

Figure 5.8. Graphcluster example.

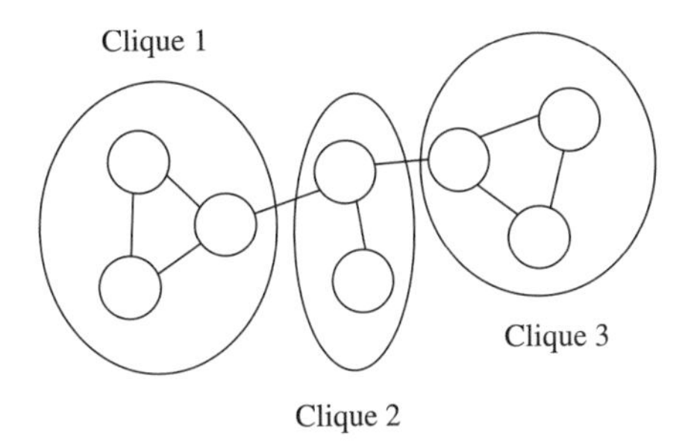

Figure 5.9. Clique example.

neighbor the clustering information each neighbor maintains and based on that it reorganizes the clusters so that every node is one hop away from every other and propagates the new information throughout the network. Whenever a node disappears, its neighbors initiate a similar process for reducing the number of clusters. This is a simple algorithm for forming cliques, although it is not very efficient.

5.4.3.1.3 Applications of Clustering for Intrusion Detection in MANET The clustering algorithms have been studied extensively for MANET and have been shown to improve the performance of routing and other functions in MANET. So a natural question here is whether it is possible to use the clustering concept for organizing the intrusion detection functions efficiently in a MANET. In general that seems to be possible and certain recent publications have suggested this [85, 86]. The clustering algorithms used for routing may not be quite applicable directly and might have to be modified because of the unique characteristics of the intrusion detection process.

One potential use of clustering in intrusion detection is for creating a dynamic hierarchy [85] for collecting and consolidating intrusion detection information. This would be in lieu of the peer-to-peer architecture that was described in Section 5.4.2.1. Another potential application of clustering is in distributing the function of collecting evidence in the network. For example, the authors of [86] propose the formation of clusters in which a clusterhead is one hop away from every member of the cluster. In that case a clusterhead can observe all communications in the cluster through promiscuous monitoring and therefore become the natural chokepoint in the neighborhood for placing the intrusion detection function. The clusterhead in this case can then be the single node in the cluster that needs to run the intrusion detection function. Such a clusterhead would focus on malicious behavior that can be observed from the network traffic. Multiple nodes in the cluster do not need to perform the same function, reducing the power consumption on those nodes.

The use of clustering algorithms for intrusion detection also has significant limitations specific to intrusion detection. The biggest limitation, as remarked earlier, is that these algorithms are prone to attacks themselves since they were not developed with security as a focus. They are not resistant to attacks such as clusterheads being taken over by an adversary in order to bypass the intrusion detection mechanism. A clusterhead that becomes malicious can introduce malicious intrusion detection reports and share these reports with the rest of the hierarchy. This could result in a well-behaved node being declared malicious and being expelled from the network. The higher the clusterhead is in the hierarchy, the more severe is the damage caused by a malicious clusterhead. The root of the hierarchy can potentially make the whole network inoperable if it becomes malicious.

Another limitation of the use of clustering for intrusion detection purposes is that clustering algorithms may be manipulated by a malicious node to cause that node to be elected as a clusterhead [86]. The election of a malicious node as a clusterhead can have a significant impact in bypassing the overall intrusion detection architecture. A potential solution to this problem has been proposed in the clustering scheme in [86]. Here cliques are created as described in Section 5.4.3.1.2. Cliques are clusters in which nodes are directly connected with each other and do not have clusterheads. Within each clique, a clusterhead is elected randomly in a way that cannot be manipulated by a single node. This is achieved by having each node select a random number and sending only the hash value of that number during the initiation of the election process. After every node has submitted the hash value, it also sends the actual random number used for the clusterhead election. This ensures that a malicious node cannot manipulate the election process to be elected as the clusterhead because, if the actual number was initially sent rather than the hashed number, a malicious node could manipulate the election process by waiting until every node has submitted its random number and then choosing the highest random number. The proposed scheme to address this vulnerability depends on the one-way property of hash functions, due to which, given a hash value, it is not possible to derive the input. Use of the hash function also commits a node to a random number on account of the collision properties of hash functions (see Chapter 2).

This approach proposed in [86] still has the earlier limitation that, if a clusterhead is taken over, it can launch attacks because it is not monitored by any node in the cluster. This is an open issue in this scheme, although it can be somewhat mitigated by requiring that clusterheads are reelected after a period of time. Another potential solution proposed is the use of tamper-proof hardware to run the IDS function on the node so that it is protected. The use of tamper-proof hardware, though, is not a magic bullet. Not only does this increase the cost, but in some cases this approach might also not be effective [90].

5.4.3.2 Reputation Schemes MANET networks depend on the cooperation of the network nodes. Each node depends on other nodes for forwarding packets, for obtaining routing information, for accessing the wireless medium, and so on. Thus, a misbehaving node has many opportunities to interfere with the working of the network. To address this, each node can monitor the behavior of its neighbors (e.g. through promiscuous monitoring) and determine over time whether nodes are behaving as expected or are misbehaving by dropping packets, advertising wrong routes, trying to monopolize the wireless medium, and so on. Unfortunately, it takes time to realize whether a node is well-behaved or not. This is even more challenging in a MANET environment given the node mobility, link dynamics, and other changes because it is difficult to determine whether suspicious behavior such as a packet being dropped is due to the dynamic nature of the network (e.g. connectivity changes) or due to malicious behavior. Therefore, a node needs to cooperate with other nodes to increase detection confidence either through the peer-to-peer or hierarchical cooperative approaches, as discussed earlier. In both models, as nodes move around and network connectivity changes, a node may depend on the cooperation of other nodes that it has not communicated with before. Hence a node might have to also deal with untrusted nodes whose behavior cannot be concluded to be malicious.

One approach for making it possible to detect malicious behavior more quickly is to have nodes proactively communicate with each other about the behavior of other nodes. Each node is also expected to maintain information about the "reputation" of other nodes. Nodes inform others about the good and bad reputation of nodes that they have observed.

Therefore, when a node sees a new neighbor it may have already heard about it from others. As a result, it may already have enough information to determine whether the new node can be trusted. Such a system is called a reputation system [91]. This scheme has been introduced on the internet before and has been very successful. For example, eBay allows customers to rate merchants after a transaction. New customers can see the ratings given by previous customers. Based on this the trustworthiness of the merchants can be decided by the customers. The main strength of the reputation system is that the merchants (provider of services) are encouraged to behave well. If they do not behave well, they will get a bad reputation, will not get new customers, and therefore will go out of business.

Reputations systems have also been proposed for use in MANET. In these schemes, each node monitors all its neighbors. When a node observes a neighbor behave well, it lets other nodes know about this. This increases the reputation of the well-behaved node. When a node sees its neighbor behave maliciously (e.g. dropping a packet it was supposed to forward), it lets other nodes know of this behavior also. This decreases the reputation of the malicious node. In general each node maintains a table with the reputation of every node it has come in contact with or has heard about from other nodes. A node updates this table whenever it obtains direct evidence locally of good or bad node behavior and whenever it receives reputation updates from other nodes. Each node may put more weight on direct evidence it obtains than on indirect observations or reputation updates received from other nodes. This approach can be applied both to the hierarchical and the peer-to-peer cooperative architecture.

A protocol called CONFIDANT (cooperation of nodes: fairness in dynamic ad hoc networks) has been proposed in [92]. This scheme takes into account only the negative reputation. Good behavior is not considered in deciding the reputation. The advantage of this approach is that it responds quickly to bad behavior as compared with schemes where good behavior is rewarded. This is because, in schemes that reward good behavior, a node can behave well for a period of time to boost its reputation before launching an attack. The CONFIDANT protocol is not prone to such attacks and malicious behavior has an immediate effect on the node's reputation. However, the scheme is sensitive to false accusations. The authors propose to deal with this by periodically reinstating malicious nodes. The problem with this solution is that the nodes might resort to the same malicious behavior right after being reinstated.

Another scheme called CORE (a collaborative reputation mechanism to enforce node cooperation) is proposed in [93]. Unlike the CONFIDANT scheme, the CORE scheme allows nodes to have both a good reputation and a bad reputation. Each node utilizes a watchdog for monitoring the behavior of its neighbors. If the watchdog detects that one of the neighbors is not cooperating in the network functions (e.g. routing or forwarding functions), the reputation of that node is decreased. Negative reputation is not shared with other nodes to prevent denial of service attacks. When the watchdog detects that a node performs the expected function, then the reputation of the node is increased locally. Positive reputation is shared with other nodes. The nodes with a bad reputation are not allowed to use the network services. For example their route requests are denied and their packets are not forwarded by the network. Michiardi and Molva [93] also discuss the application of this scheme to DSR by basically providing a layer on top of DSR to make DSR more resistant to misbehaving nodes.

The advantage of a scheme such as CORE is that it takes into account the limitations of promiscuous monitoring (as discussed in Section 5.5.2) on account of which a node might be perceived to be misbehaving even though it is not. In this scheme such perceived misbehavior of a node for a short period of time will not adversely impact the node's reputation because it will merely decrease the good reputation while the node will still be considered to have good reputation. Unfortunately, because of the difficulty of distinguishing between perceived and real misbehavior, such an approach also makes it possible for malicious nodes to attack other nodes for short periods of time, making sure that their reputation is still good and therefore no actions are taken against them.

The reputation of each node can also be taken into account for responding to malicious behavior. For example, a node may exclude nodes with bad reputation from its routing table so that its packets are not routed through such nodes. Nodes may also refuse to provide route updates to a node with bad reputation or they may refuse to forward packets from such nodes. The fear of retribution hopefully encourages nodes to be well behaved and cooperate in providing the network services so critical to making a MANET operate. Even if there are nodes that continue misbehaving due to their knowledge of the working of the reputation system, the reputation systems still result in ad hoc networks that are much more resistant to the presence of adversarial nodes.

There are challenges associated with the design of reputation systems in a MANET environment because reputation systems are prone to attacks from malicious nodes. A malicious node (which is otherwise well behaved and has a good reputation) may start spreading misinformation about the behavior of another node (let us say node A) trying to decrease that node's reputation. If the malicious node is successful, others may stop forwarding node A's packets, resulting in node A losing connectivity. To address this limitation CORE [93] has suggested only allowing a node to spread positive feedback about other nodes. Nodes are not allowed to share negative feedback about a node with other nodes. Another weakness of the reputation systems is when a node cooperates with some nodes and does not cooperate with some other nodes. One such example is when a node drops packets it is supposed to forward from a subset of the nodes. If the percentage of nodes it cooperates with is larger than the number of nodes it does not cooperate with, then it is possible for the node to get away with misbehavior.

Another possible attack on a reputation system is the following. Two nodes may collude and send positive feedback about each other to other nodes. This can help boost the reputation of both nodes. Reputation systems must provide some protection against these types of attacks by giving more weight to direct observations than to indirect reputation. Ya and Mitchell [94] have discussed the challenges with reputation systems that allow nodes to take into account indirect reputation information provided by other nodes. In the absence of a central authority (which is not feasible for MANET anyway), Ya and Mitchell [94] maintain that the risk of denial-of-service attacks in such systems is so high that it does not justify taking into account indirect reputation information. In addition, reputation schemes also suffer from additional bandwidth overhead. They propose [94] a scheme that only relies on local direct evidence for updating the reputation table on each node.

Even though reputation systems are open to a variety of attacks, it still seems possible to use such systems in a cooperative manner rather than having to base all of the decisions exclusively on local information. The reputation systems can be designed so as to require that nodes provide evidence of node misbehavior when they accuse other nodes of misbehavior. For example, certain attacks require that a node send conflicting messages such as hello messages advertising one set of direct links and routing LSU messages advertising a different set of neighbors. A node may be able to use such contradicting messages as evidence of malfeasance. Another possibility is to require multiple nodes to agree through direct evidence that a node is bad. For example, multiple nodes may observe the same malicious behavior through promiscuous monitoring. If several nodes agree that a node is bad, it seems reasonable to take advantage of such information when considering the reputation of a node. It is still possible that multiple nodes may collude to trick other nodes, but such a case might be more difficult. In the case when the network has many colluding nodes, the MANET may not be able to operate at all. This is because a large number of colluding nodes can completely disrupt the basic operations of the network, such as routing, and thereby make it inoperable.

5.4.3.3 Schemes for Stimulating Cooperation An interesting approach for dealing with misbehaving nodes is to provide mechanisms that encourage nodes to be well behaved. These techniques not only apply to malicious nodes that seek to disrupt communication, but also to selfish nodes that do not perform their expected functions as part of the MANET. Since a MANET depends on the cooperation of nodes, selfish behavior is disruptive. The techniques discussed here provide incentives for nodes to cooperate and perform their responsibilities.

Wireless ad hoc networks, as we have seen, depend on every node supporting the various networking functions and other services. Performing these additional functions adds to the node's CPU load and drains its battery because of the increased power consumption. If every node decides to act selfishly and not participate in the routing process or other required services (including intrusion detection), then the whole MANET breaks down and no communication is possible. It is also possible that only a few nodes may decide to act selfishly and stop forwarding packets. This will reduce their CPU utilization and battery consumption. While the network will still be operating, the overall network performance will not be optimal. Buchagge and Boudic [92] have shown that even a small percentage of selfish nodes (<10 percent) can significantly affect the performance of the network. Even though this was shown through simulation for a specific network configuration, intuitively we expect that this will be true in general for a typical network. Marti *et al.* [84] show through simulations that, if 10–40 percent of nodes misbehave (by acting selfishly or because of malicious intent), throughput decreases by about 16–32 percent, although nodes near the misbehaving nodes experience much larger decrease of throughput.

Buttyán and Hubaux [95–97] have proposed schemes using the concept of virtual cash called "nuglets." The schemes assume that each node has a tamper-resistant hardware module. Whenever nodes need to use the network they need to expend some of their virtual cash; the tamper-resistant hardware ensures the integrity of the virtual cash. In the Packet Purse Model variation of the scheme, each packet is loaded with nuglets by the source. Each forwarding node takes some of the cash loaded in the packet for providing the forwarding service. If there are enough nuglets on the packet, the packet will reach the destination. Unfortunately with this scheme, the source of the packet needs to know how many hops the packet needs to travel so that it loads the packet with enough cash to satisfy all the forwarding nodes. Otherwise the packet may not be forwarded to the destination. In a different variation of this scheme called the Packet Trade Model, each packet does not carry cash but is traded by the forwarding nodes for cash. Each forwarding node buys the packet for some cash from the previous node and sells it to the next hop for more cash, gaining some cash in the process. This model overcomes the limitations of the Packet Purse Model because nodes originating a packet do not have to estimate the cash that needs to be loaded on the packet. The disadvantage of the Packet Trade Model is that the source does not need to expend any cash to originate a packet, leaving the scheme open to denial of service attacks from the source.

In the most promising variation of this scheme [97], the tamper-resistant hardware module maintains a "nuglet counter." Whenever the node sends its own packets, the counter is decreased. Whenever the node forwards packets destined for other nodes, the counter is increased. This encourages nodes to participate in the MANET routing process. This is because, if nodes want to originate traffic, their counter needs to remain positive, which requires them to forward packets from other nodes. Under simulations this scheme also seems to provide good throughput for the network, even when there is a significant number of misbehaving nodes. The limitation of the scheme is that it does not propagate knowledge of node misbehavior throughout the network, ostensibly to improve overall performance. It also relies on tamper-resistant hardware, which may not be available (or realistic) in certain applications.

Having considered the types of architectures possible as well as the implications of the choices made, we next look at another important factor. This is the data or evidence that nodes collect. Analysis and decisions are based on the collected evidence.

5.5 EVIDENCE COLLECTION

Detections of intrusions depend very much on the quality of the data on which these decisions are based. Each node has access to data collected using a variety of ways. Potential types of evidence depend on the nature of collection and include:

- local evidence available on the node;
- traffic observed on the broadcast wireless medium through promiscuous monitoring;
- evidence made available by other nodes.

One of the simplifying assumptions we make here is that the data that is required for detecting an attack is not encrypted. This is usually the case for routing messages, IP header information, and possibly content of messages. If a large portion of the network traffic is encrypted, the IDS's ability to analyze intrusion detection information may be significantly limited. Note that the traffic can still be collected. A potential solution to this problem is to make keys available at IDS nodes that can be used for decrypting the relevant data. Distribution of such keys may not always be possible and may potentially introduce additional vulnerabilities into the security scheme. Note also as mentioned earlier that, even when traffic is encrypted at a particular layer, the intrusion detection analysis could operate at a different layer.

We next look at the types of evidence.

5.5.1 Local Evidence

Local evidence is the most reliable information for a node to use when performing intrusion detection. There are several sources of data available, including audit logs of host activity and failed logins as well as network traffic received or relayed by the node. The traffic that can be monitored by a node not only includes traffic that it originates for other nodes and the traffic that is destined for the node but also traffic relayed by this node as part of the routing process.

5.5.2 Promiscuous Monitoring

In wireless networks, radios have the ability to observe all the traffic that is transmitted by every node near them provided the signal is strong enough. The process of promiscuous monitoring itself is not difficult. A wireless node receives all the packets that have strong enough signal for them to be interpreted correctly at the physical and medium access control layer. Under normal conditions the medium access control layer filters out all the packets whose destination layer 2 address (MAC address) does not match the address of the node in question, but when a node is operating in the promiscuous mode, the MAC layer does not filter any packet that it receives but delivers all these packets to the higher layer. This results in the node being able to receive all packets being transmitted in its neighborhood. Note that this includes packets that will not be relayed by the node. The number of neighbors that a node can hear depends on the power levels used by the neighbors. For example, if the power level of the transmitting node is high, then the strength of the signal is high and a single node may be able to hear a large number of its neighbors over a fairly wide area. Operation in promiscuous mode increases the number and types of attack that a node can detect. This is because a node can observe a significant portion of the traffic sent and received by its neighbors and therefore the node can determine whether any of those neighbors are misbehaving. Therefore, several techniques that take advantage of promiscuous monitoring have been proposed. Several detection techniques discussed in this chapter depend on promiscuous monitoring.

Promiscuous monitoring also has significant limitations, though [84]. There are several radios that do not support promiscuous monitoring since the corresponding drivers at the MAC layer might not allow the filtering to be turned off. The promiscuous monitoring mode of operation also increases the power consumption for the radio because it requires radios to monitor transmissions not destined for them. Intrusion detection techniques based on promiscuous mode monitoring may generate a number of false positives (and missed detections) because promiscuous monitoring may not provide an accurate view of successful transmissions of neighboring nodes. A node using promiscuous monitoring may not see all the packets that are received or transmitted by its neighbor. For example, a temporary obstruction or link loss may block some of its neighbor's traffic from being observable. In other cases, a node may move in and out of range from the node monitoring its traffic promiscuously, resulting in an inaccurate view of the traffic.

There is another subtle case that limits the effectiveness of promiscuous monitoring called the hidden terminal effect (see Figure 5.10). Let us assume that node C is monitoring its neighbor, node B. As shown in the figure, node C is within B's transmission range and therefore node C can use promiscuous monitoring to determine whether node B is forwarding packets as expected (and therefore not launching a packet dropping attack). Let us assume that node C has sent a packet to node B that is destined for node A and is now waiting to see whether node B will forward the packet to node A. At the same time node C is within node D's transmission range. If node D transmits some traffic at the same time node B is forwarding node C's packet to node A, then node C will see a collision of node B's and D's transmission and will therefore not be able to see node B's transmission. Node B's transmission to A might actually have been successful

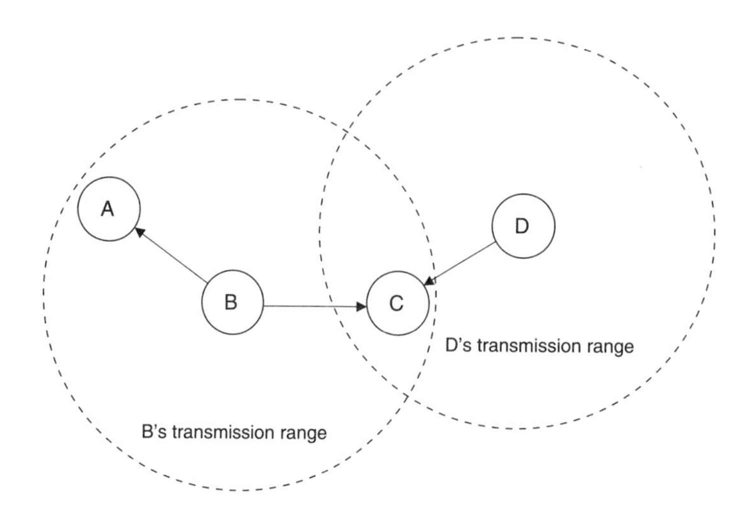

Figure 5.10. Hidden terminal effect.

because node D is out of range of both B and A. The result is that, although B forwarded the packet as expected, node C did not see that. As a result, node C may falsely suspect node B of maliciously dropping the packet.

Another similar collision situation may happen if at the time node B tries to transmit a packet to node A, node A (or even one of its close neighbors) transmits a packet. That will cause a collision that node C may not observe if it is not within A's transmission range. If node B does not retransmit the packet, node A will not receive the packet. Node C actually believes that node B has successfully forwarded the packet and therefore will not be able to detect node B's malicious packet dropping behavior. Node B can actually cause such collisions on purpose to evade node C's promiscuous monitoring. For example, node B may wait until node A begins transmitting a packet to start transmitting node C's packet, causing a collision on purpose.

Node B can also evade node C's monitoring in a different way. If node A is further than node C from node B, then node B may transmit the packet destined for node A without enough power for node A to be able to receive the packet but with enough power for node C to receive the packet. If that happens, node A does not receive the packet while node C believes that node B has actually transmitted the packet.

In spite of the limitations of promiscuous monitoring we believe that, when available, promiscuous monitoring should be used for intrusion detection in MANET. This is because it provides a wealth of data that can be used for detecting attacks without requiring any data exchange by multiple nodes. Promiscuous monitoring, however, cannot be relied on as the single method for collecting observations. Evidence obtained through promiscuous monitoring needs to be checked against other evidence collected through other means (such as data obtained from neighbors). Persistent suspicious behavior observed through promiscuous monitoring is usually a clear indication of intrusion.

5.5.3 Evidence made Available by Other Nodes

In order to detect attacks at a node it is beneficial to leverage observations that other nodes make. Evidence collected by other nodes can help a detector confirm suspicious behavior of other nodes. Unfortunately, sharing a large amount of information across nodes in a MANET environment is problematic. This is due to the bandwidth and power constraints that limit the amount of data that can be communicated across nodes. Even if the hierarchical intrusion detection architecture discussed earlier is used, it may be beneficial to apply further compression in order to reduce the amount of data shared through the hierarchy.

One potential approach is to compress the amount of data that needs to be communicated across nodes and only selectively transmit evidence that is relevant to the detection of attacks of interest [98]. For example, wavelets have been proposed in the past as a way of compressing such evidence, although other compression techniques are applicable.

Even though such compression may be somewhat effective in reducing the level of communication overhead, a different approach is to actually distribute the intrusion detection function itself. In this case nodes share alerts when specific types of events are observed. For example, Little and Ko [99] discuss an approach where nodes look for a specific signature that may be one step in a multistep attack. Once that step is observed, an alert is shared with other nodes. A node can then correlate those alerts and infer when an attack has occurred. This technique is particularly effective for signature-based detection because a signature of an attack may be decomposed into signatures of multiple events that need to be observed across the network. Note, though, that the technique is more general and can be applied to other detection techniques such as anomaly-based detection [100].

5.6 DETECTION OF SPECIFIC ATTACKS

So far we have focused on the concept of intrusion detection in general. Potential general approaches that can be used in a MANET environment have also been discussed. In this section we focus on intrusion detection techniques for detecting attacks of specific types. We focus on two types of attacks:

- packet dropping; and
- attacks against routing.

We focus on these specific attacks because they are representative attacks for MANET. Due to space constraints we cannot discuss all possible attacks.

5.6.1 Detection of Packet Dropping Attacks

One of the attacks of interest in a MANET environment is the packet dropping attack. In a MANET, nodes typically participate in the routing process and are then responsible for forwarding packets as necessary. In the packet dropping attack, a node does not forward packets that it is supposed to. This is potentially a very damaging attack. Routing protocols in MANET rely on each node to forward packets based on the routes that the node has advertised. If the nodes do not forward the packets, the operation of the network can be significantly disrupted. The impact can be exacerbated when misbehaving nodes are in key locations. Subtle variations of such attacks include the following:

- a node dropping packets randomly with some probability smaller than 1;
- a node dropping packets destined for a specific node (for example a key server) while forwarding packets destined for other nodes;

- a node dropping packets of a specific type (e.g. high priority traffic or traffic from some key application or packets of a certain protocol) while forwarding other types of traffic.

Such subtle variations of the attack may be particularly difficult to detect. This is because packets may be dropped as they travel the network not only due to malicious reasons but also for valid reasons. For example, when the network gets overloaded, congestion may result in dropping of packets. In other cases, the lossy nature of wireless links might result in intermittent packet drops. Random dropping of packets by a node can be easily confused for natural losses thus allowing the misbehaving node to evade detection.

Certain techniques have been proposed for detecting packet dropping attacks. The simplest of those approaches relies on the use of promiscuous monitoring. It assumes that some nodes in the network monitor their one-hop neighbors using promiscuous monitoring. Through their monitoring, these nodes can observe whether other nodes are behaving as expected. Statistical detection techniques have to be used to detect malicious packet dropping behavior. This is because some amount of packet dropping may not be due to malicious behavior but rather due to normal (nonmalicious) conditions. Such conditions include the packet losses due to the vagaries of the wireless links, congestion as well as due to node mobility. The limitations of the promiscuous monitoring approach as discussed in Section 5.5.2 can also be addressed by such statistical techniques. The thresholds used in the statistical techniques, however, would be important. If thresholds in the statistical detection scheme are set too low, this technique may result in a large number of false positives. If the thresholds are set too high, malicious users may be able to exploit the detection scheme by dropping packets intermittently while ensuring that the thresholds are not crossed, thereby evading detection.

We now discuss some more elaborate schemes for detecting packet dropping attacks. Rao and Kesidis [101] consider the fact that a node may drop a packet not out of maliciousness but because of congestion. The congestion happens because more packets arrive at that node than can be stored in its buffers before they are forwarded. They define a traffic model that can be used for calculating the expected rate of packet dropping due to congestion. A node monitors a suspicious node through promiscuous monitoring and can observe whether the rate at which a suspicious node is dropping packets is consistent with the expected drop rate due to congestion as calculated. The calculation assumes a Poisson traffic model and a nonoverloaded network. The reason for the latter assumption is to ensure that the statistics of the interpacket arrival times do not change from hop to hop. In order for the monitoring node to accurately assess the behavior of the suspicious node, it also needs to know the total traffic arriving at the suspicious node. The monitoring node achieves that by asking all the neighbors of the suspicious node to provide the current transmission rate towards the suspicious node. This scheme is shown to perform reasonably well. Unfortunately the applicability of the scheme is very limited due to the assumptions of the traffic model (e.g. Poisson arrival times, underloaded network). The scheme also suffers from the limitations of the promiscuous monitoring as discussed earlier.

The limitations of promiscuous monitoring in detecting packet dropping in this and other schemes can be alleviated by requiring multiple nodes to monitor every node. Each node can coordinate with direct neighbors of a node exhibiting suspicious behavior to crosscheck on the behavior of the suspicious node. Such an approach that compares observation from multiple nodes would make it difficult if not impossible for a node to exploit the limitations of promiscuous monitoring for evading detection. Allowing nodes to compare their observations also reduces false alarms because multiple nodes will compare their view of the suspicious node's behavior. Sterne *et al.* [85] describes such a scheme where reports from nodes monitoring their neighbors promiscuously are propagated and consolidated throughout the intrusion detection hierarchy. This results in much more precise intrusion detection with fewer false positives. However, this also causes increased communication overheads.

Anjum and Talpade [102] have proposed a technique that overcomes the limitations of promiscuous monitoring altogether. That technique is called LiPaD (lightweight packet drop detection for ad hoc networks). In this scheme, every node keeps track of each flow that is either originating or terminating, or passing through that node. A flow is defined by the source and destination IP address tuple in the IP header. For each such flow, the node keeps track of the packets the node sends and receives. Periodically, every node reports flow statistics to a coordinator that combines those statistics. The coordinator can then run detection algorithms that take advantage of the exact and actual information about the number of packets each node is forwarding. This is instead of the indirect information based on promiscuous monitoring. The coordinator examines all the data provided by every node and looks for flows where a node is not forwarding as many packets as expected. This can be achieved because the coordinator has knowledge about the number of packets forwarded by each hop.

Figure 5.11 shows one simple such example. In this case, a flow originates on node A and is destined for node D. Packets for this flow are routed through nodes B and C. Each node that is part of the flow (in this case nodes A, B, C, and D), report periodically (let us say every 5 s) to the coordinator (node E) the number of packets they received and transmitted for this flow. Node A will only report the packets originating from it (the flow is assumed to be unidirectional), node B will report the number of packets it received from node A and the number of packets it forwarded to node C, and so on.

The coordinator plays a key role in this architecture because it is the one that correlates the statistics provided by the other nodes. Anjum and Talpade [102] conclude that a single coordinator can handle potentially a few hundred nodes. Larger networks would need multiple coordinators.

This scheme has also been shown to work even if some of the nodes are lying (in their reports to the coordinator). For example, let us assume node C in Figure 5.11 received 20 packets but only forwarded 15. It will either need to report to the coordinator that it

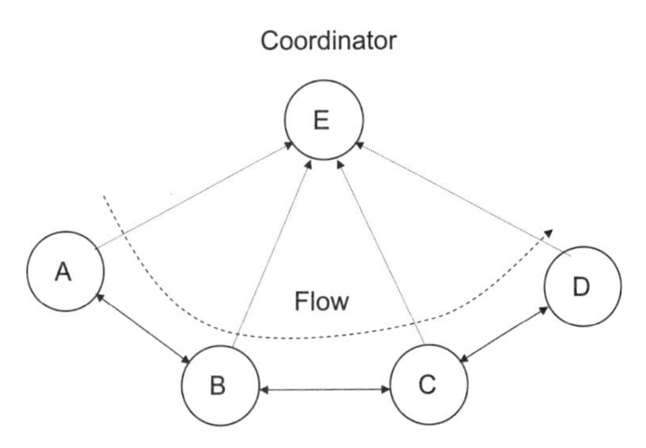

Figure 5.11. Simple packet dropping example.

received 20 packets from node B and forwarded 20 packets to node D (even though it only forwarded 15), or it will need to report that it received 15 packets from node B (even though it actually received 20) and forwarded 15 packets to node D. In the first case Node D's report will contradict node C's report because node D will report that it only received 15 packets. In the second case node B's report will contradict node C's report because node B will report that it sent 20 packets to node C. In either case the coordinator can detect that there is a malicious node and also the pair of nodes in which one is malicious. The coordinator may not necessarily be able to detect which of this pair of nodes is malicious. Assuming that there are enough flows in the network and that the number of well-behaved nodes is sufficiently high, the coordinator can figure out which nodes are malicious because there will be multiple pairs of suspicious nodes. The nodes that are common across multiple pairs will be the malicious nodes. For example, let us assume in our previous example that there is another flow originating in node F, traveling through node C and terminating at node G (see Figure 5.12). If node C starts lying in its reports, contradictions will arise with the reports of multiple nodes. Therefore the coordinator will be able to clearly identify the malicious node. The scheme is not necessarily foolproof because the coordinator may not have sufficient information in all cases to decide about the lying node. Some such cases are when there are not enough flows, when there are no other nodes in the neighborhood, or when a node decides to drop packets only from a subset of nodes. For example, if node C only drops packets destined for node D and tries to contradict only node D's reports, then the coordinator will not be able to decide whether node C or node D is lying.

The advantage of this technique is that the detection algorithm does not rely on promiscuous monitoring. The algorithm has its own significant limitations, however. Reports of nodes to the coordinator may contradict each other on account of the lossy nature of the links rather than malicious behavior. In the example discussed earlier, node A may report the correct number of packet that it forwarded to node B, and node B may report the packets it actually received from node A, but the reports may differ. Node B may not receive all of the packets sent by node A. Some of the packets may be lost due to the lossy link between the two nodes. In that case reports from nodes A and B will be inconsistent and may result in a false positive. The scheme tolerates such

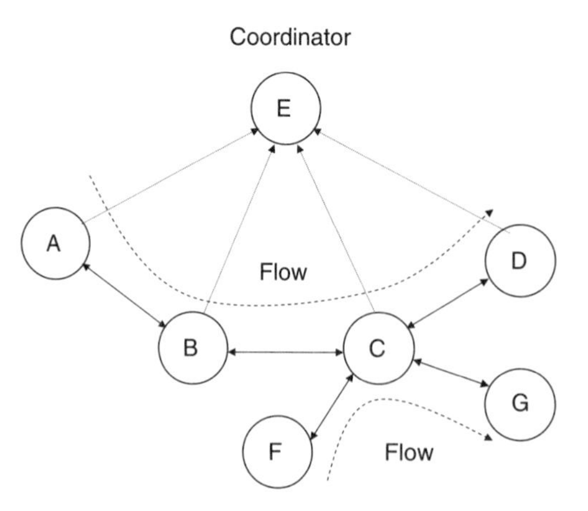

Figure 5.12. Packet dropping example with an additional flow.

losses by allowing for some level of inconsistencies between the reports of nodes based on the expected behavior (i.e. losses) of the wireless link. However, a malicious node may take advantage of such an approach by dropping packets intermittently at a rate below the expected packet loss rate of the link. Of course, this will limit the potential harm that can be caused by that node assuming a reasonably small normal packet loss rate.

Another challenge to the algorithm stems from the fact that nodes in a MANET are not static. The nodes that a flow traverses change constantly. For example node C may temporarily move out of range from node B, forcing the flow to traverse a different path. This can create temporary inconsistencies in the reports of nodes. Another limitation of the algorithm occurs when a coordinator becomes a malicious node itself. The current scheme as explained cannot withstand this potentially dangerous attack. Basically, an adversary can defeat this scheme by taking over a single node namely the coordinator. A potential solution is to assign multiple redundant coordinators and require nodes to report to multiple coordinators so that misbehavior of a single coordinator can be detected by a back-up coordinator. Multiple back-up coordinators can increase the resistance of the scheme against coordinators being taken over, but such an approach will also increase the communication overhead of the scheme as nodes have to provide reports to multiple coordinators.

Another limitation of his algorithm is its communication overhead. Each node in the network is required to report to the coordinator at periodic intervals. This results in a fairly significant amount of data being transmitted by the network. This overhead is generated regardless of whether there is an attack or not. For each flow, the flow id, and the number of received and forwarded packets need to be reported. Anjum and Talpade [102] report that for a 100 node network 1450 bytes of reporting data are generated per reporting period. A long reporting period results in lower overhead but causes increased detection latency. The level of overhead could result in potentially congesting the coordinator since it receives reports from multiple nodes. As a result, short reporting periods not only increase the overhead but potentially require a larger number of coordinators in the network. Assigning subsets of nodes to each coordinator can reduce the effectiveness of the approach because mobility will often result in a node reporting to a different coordinator creating discrepancies in reports. This also points to the need to design schemes for ensuring that the multiple coordinators work in a synchronized fashion.

Some of the limitations of this scheme can be improved by requiring the full reporting mechanism only if there is a suspicion of the presence of malicious nodes. For example only a subset of nodes (let us say only originating and terminating node on the flow) or a small percentage of well-placed nodes may be required to provide such reports. Once discrepancies start to appear, indicating suspicious behavior, the coordinator may request additional nodes to start providing reports on the flows they participate in. This approach can significantly reduce overhead, especially when no malicious nodes are present.

5.6.2 Detection of Attacks Against Routing Protocols

In this section we focus on detection of attacks against routing protocols. Routing is a key component of MANET and therefore it is important to focus on these types of attacks. Routing protocols in general are prone to attacks from malicious nodes, as discussed in Chapter 4. These protocols are usually not designed with security in mind and are often very vulnerable to node misbehavior. This is particularly true for MANET routing protocols because they are dependent on every node participating in the routing process. Making routing protocols efficient often increases the security risk of the protocol and allows a single node to significantly impact the operation of the protocol because of the lack of protocol redundancy.

In Chapter 4 we discussed some potential approaches for modifying routing protocols (or in some cases proposing new routing protocols) to make them more resistant against attacks. In this chapter we discuss approaches for detecting attacks against traditional routing protocols (that may not have the added security features). Some of the detection schemes discussed are also applicable to secure routing protocols. This is because the secure routing protocols do not offer protection against all possible attacks (as discussed in Chapter 4). In this section we focus on how the typical detection techniques, namely anomaly-based detection, signature-based detection, and specification-based detection (as defined in Section 5.1.1.2) can be used for detecting routing attacks. Although our discussion of attacks and detection schemes are fairly general, we provide examples specifically for OLSR and AODV because they represent two different classes of MANET routing protocols, namely proactive and reactive routing protocols.

5.6.2.1 Detection of Routing Attacks A large number of attacks can be detected (or even prevented) by requiring that routing messages be signed, i.e. authenticated (through individual keys) by each node. Attacks in which nodes send routing messages pretending to be a different node (spoofing) can be easily detected this way. Attacks in which nodes modify routing messages in transit with the intention of misleading other nodes can also be detected easily if all routing messages are signed. The identification of the malicious node(s), however, might not be easy in these cases.

Modifying the routing protocols to require node authentication (as discussed in Chapter 4) is a viable approach but has some limitations. Firstly, it increases the overhead since it increases the size of routing messages and the amount of processing needed to process each routing message. This is because every node needs to sign every routing message it transmits. Further, every node needs to verify the authenticity of the incoming routing messages. Another very significant hurdle with requiring that each node authenticate each routing message is due to the requirement of a public key infrastructure when using asymmetric key-based schemes. As seen earlier, this is not easy on account of the features of a MANET. Hence we need to investigate other approaches to detecting routing attacks.

Several techniques can be applied for the detection of attacks against routing protocols. These techniques can be divided into the main categories of anomaly detection, signature detection and specification-based detection. We will discuss next how each of these techniques can be applied to the problem of detecting attacks against the routing protocols. These techniques apply to each of the routing protocols such as OLSR, AODV and DSR and potentially other infrastructure protocols used in MANET (e.g. multicast, session management). Our discussion will be general with some examples of how these techniques can be applied to specific protocols. Each of the techniques is of course most effective when it is modified to suit each protocol since it can then leverage the unique characteristics of the protocol.

5.6.2.2 Anomaly-Based Detection As explained earlier, detectors based on this technique try to detect attacks by looking at activities that vary from the normal expected behavior, usually by utilizing statistical techniques. Usually these techniques utilize thresholds determined during training to detect abnormal activity. In the case of routing there are several measures that can be used for detecting abnormal behaviors. Some examples are as follows:

- *Rate of Routing Updates*—a node may attempt to send routing updates much more often than required by the routing protocol. This may cause instability in the routes, and may drain other nodes' batteries and CPU since the nodes have to process all those updates. Routing updates are usually caused by, for example, node movements, changes in link connectivity, new nodes joining the network, or nodes disappearing from the network. Given a specific application or mission, if there is a reasonable level of understanding of the frequency of these events then it may be possible to determine the expected rate of routing changes for nodes. If a specific node is generating routing updates and changes that significantly exceed that rate, it is often an indication of an attack. It is of course possible for this technique to generate false positives because the routing changes may be generated for valid reasons. For example, in a battlefield environment, sudden movements may be experienced due to sudden unexpected events. As remarked earlier, there is always a trade-off between setting a threshold to such a small level that several false positives are generated and setting a threshold to a large value that may allow attacks to go undetected.
- *Number of Neighbors*—a node may advertise a large number of nodes as its neighbors. As discussed earlier this type of an attack may create a black hole. It may be possible to detect that attack if the number of nodes advertised is abnormally high for the density of the network. This can be especially true if the neighbors of the offending node advertise only a small number of neighbors. Again, it is possible to get false positives because node positions may actually justify significant discrepancies in the number of neighbors. This property is actually an example of the usage of emergent properties [103]. An emergent algorithm leverages the features that no individual node can provide, but that emerge through the collaboration of many nodes. Thus the set of neighbors is one such example of an emergent property.
- *Length of Routes*—this applies to routing protocols that advertise complete routes such as DSR. An attacker may try to insert incorrect routes to cause network instabilities this way. Based on the expected connectivity of the network, changes in routes and their length may help detect attackers when there is a significant variation from expected behavior.

As discussed earlier, determining the right threshold for each metric used for anomaly detection is challenging and may result either in a significant number of false positives or in attacks evading detection. The results of the detection techniques may be improved significantly by taking advantage of other knowledge that may be available. Some examples are as follows:

- *Power Levels*— if the power levels of nodes are known it may be possible to use them when deciding whether a node is malicious based on the number of neighbors it is advertising.
- *Location*—if we know where nodes are located (and this may be possible if location information is available for the nodes), we may be able to determine whether a large

number of neighbors is justified or not in a node's routing advertisement. For example, if we know that a node is really isolated and far away from most other nodes, then the node is definitely malicious if the number of neighbors it advertises significantly exceeds the expected number of neighbors.

- *Speed*—the actual speed of nodes might help in determining if the number of routing updates is justified or not. For example, increased rate of routing updates can be justified if we know that nodes are moving very fast at the time. In such cases a larger threshold value may be used by the anomaly detection schemes.
- *Relative Movements*—If we know the relative movements of nodes, we may be able again to make much better detection decisions. For example, if nodes are moving together and their relative position is not changing then we would not expect large numbers of routing updates. Therefore a much lower threshold may be used to detect an attacker.

Several researchers have proposed a variety of techniques for detecting routing attacks using anomaly detection. For example, Zhang and Lee [104, 105] have discussed the use of anomaly detection for detecting attacks against AODV, dynamic source routing (DSR), and destination-sequenced distance-vector routing (DSDV) and have presented experimental results for the effectiveness of the techniques discussed there.

The big advantage of anomaly-based detection is its simplicity. The various values need to be calculated and compared with the pre-determined thresholds. The big disadvantage is that this technique can only be used for detecting a subset of routing attacks. It is not possible to create appropriate metrics, especially for subtle attacks where a single node is the target of the attack. The other disadvantage of this technique as discussed earlier is the potential generation of false positives.

5.6.2.3 Misuse Detection Vigna *et al.* [5] describe some examples of potential misuse detection schemes for AODV. One potential attack described there is when a node drops AODV RREQ messages or RREP messages. This type of attack significantly disrupts network connectivity by making it more difficult to propagate route information. The detection scheme proposed is based on detectors that monitor the behavior of neighboring nodes. If neighbors do not propagate the routing messages as expected within the expected timeout then an attack is inferred. Several other attacks that can be detected using this technique are also described there.

5.6.2.4 Specification-Based Detection The specification-based intrusion detection technique is usually based on building finite state machines that reflect the expected behavior of a node as discussed earlier. Tseng *et al.* [4] propose a finite state machine for detecting attacks against AODV using such an approach. For example, whenever a node receives a new RREQ it either forwards the RREQ message if it does not know the path to the destination or generates a RREP message. If the node does not do either then it is possibly malicious. Also, if it generates incorrect RREP message advertising incorrect paths, then that is also is an indication of malicious behavior. In order to detect such behavior the detector needs to maintain up-to-date information on the current view of available paths that the monitored node is aware of.

The previous discussion illustrates the challenges associated with the specification-based intrusion detection technique. A detector needs to monitor a node very closely and maintain information about the messages sent and received by the node. The detector then also needs to perform similar calculations as performed by a node executing the routing protocol. Therefore the complexity of the detector is typically similar to the complexity of executing the routing protocol itself. This increases the detector complexity, and the data that needs to be stored on the detector to a level that may not be acceptable. Another problem with specification-based detection is danger from misinterpretation of the protocol when the protocol is modeled in detail. This will lead to false alarms because the alarm may be due to a misinterpretation of the protocol in the detector finite state machine. To avoid this problem it has been proposed to simplify detectors by only modeling key characteristics of the protocol and not necessarily every detail. This decreases complexity and simplifies the detector but leaves open the possibility that an attack exploiting the portions of the protocol behavior that are not modeled by the detector will go undetected.

5.7 SUMMARY

In this chapter, we focused on detection of attacks from nodes that have penetrated the preventive barriers. Nodes that are accepted as part of the MANET can take advantage of their acceptance in the network to launch attacks against other nodes and the network itself. Typically launching such attacks involves taking abnormal actions. As a result such behavior deviates from expected behavior and can be used for identifying the attackers. IDS are typically used for detecting such attacks.

As was discussed throughout this chapter, there are still several questions that need to be researched further. A systematic approach for intrusion detection in MANETs is urgently needed. This would be along the lines of the approach taken in [80, 83], but which would be applicable to general scenarios. It is also important to consider additional MANET characteristics that can be exploited for creating new intrusion detection techniques. For example, it may be possible to exploit information from multiple layers in the stack (e.g. MAC, network layer) to improve detection accuracy through cross layer detection. We expect that additional results in this area will be published in the next few years.

In this chapter we focused mostly on collecting data and then using that data for detecting attacks and the nodes that are responsible for those attacks. In the next chapter we will focus on some approaches that can define the way to recover from the impacts of an attack.

6 Policy Management

6.1 INTRODUCTION

Network management is an important network administrative function for ensuring the efficient operation of a network. Network management typically involves configuring the network when the network is first set up so that it can become operational. The initial configuration is based on the needs of the specific mission and applications that are supported by the network. The ability to configure a network is called configuration management and is just one of the many aspects of successful network management. Network management includes other capabilities such as fault management, performance management, and security management. Fault management involves monitoring the network to detect events that disrupt the operation of the network such as equipment failures and link failures. Performance management is a complementary function to fault management and involves monitoring the network to detect performance degradation in the operations of the network, even though there may not be any actual faults and all of the network elements are still operational. Performance degradations can be detected, for example, when portions of the network are not supporting the needs of the mission and the applications. Such events may be due to congestion in the network links because the corresponding link routes more application traffic than it can handle. The performance degradation can then be addressed by changing the configuration of the network.

Security management is another important component of network management that involves performing all the functions necessary to ensure the secure operation of the network. Key functions of security management include specifying the various parameters needed to support the various security services. Such parameters could include the types of authentication and encryption mechanisms to use, the permissions the various users have, and the identities of users that are allowed to perform specific functions and that can be trusted. Nodes that have been detected misbehaving (e.g. via an IDS) may lose trust and may only be allowed to perform a limited set of functions, if any. Although our focus in this chapter is on the security management aspect of network management, we will start our discussion with network management in general. We do this in order to provide a general context that will frame the security management problem.

Network management has been the focus of research for a long time because it provides for ensuring efficient operation of networks. Several network management systems have been developed for automating the management of networks. These systems, however, have traditionally been developed for networks that are fairly static and do not change frequently, such as enterprise IP networks and Internet Service Provider (ISP) networks. Such networks typically include network equipment (e.g. routers and switches) placed in

Security for Wireless Ad Hoc Networks, by Farooq Anjum and Petros Mouchtaris

specific locations that rarely change. Equipment is connected through wires that have fixed network capacity (such as fibers). Even the management of cellular data networks (which are wireless and support mobility of network elements) has focused mostly on the management of the fixed infrastructure of the cellular network, namely the towers and the networks interconnecting the towers. In the case of static networks, network engineers typically spend a significant effort in designing these networks when they are initially set up. Following this, only incremental changes are typically made to them as the needs of the applications for additional traffic grow. These changes are usually infrequent and incremental, and therefore have not been a significant focus in terms of automation. Changes to configuration of such networks have mostly been a manual process in which experienced network administrators make the changes based on their understanding of the user and application needs and their past experience with such changes.

As networks have grown and become more complex, the management of these networks has become more difficult. The level of change in the network architecture and configuration has also increased. In addition to changes due to addition of new users, introduction of new applications, and increase in the traffic demands, network administrators have to deal with changes intrinsic to the network such as network congestion, where portions of the network become overloaded and require modifications to the architecture, and network attacks, where malicious users attempt to disrupt the operation of the network. Two particular challenging problems associated with existing network management approaches are: (1) dealing with the increasing numbers of controllable parameters in a network; and (2) managing changes in networks through manual processes.

The protocols that a typical router supports have grown in number and complexity and have exposed a variety of new configuration parameters that need to be configured by network administrators. With the introduction of various quality of service (QoS) schemes, MAC layer schemes, access control schemes, packet filtering, and firewall functions, network administrators have many more parameter options to select when configuring the network. Typical network management systems tend to expose all of the possible configuration parameters to network administrators in order to provide flexibility. Selecting the correct values for all such parameters is a difficult and error-prone process. The additional challenge for network administrators is supporting changes to already configured networks. Since networks have become much larger and more complex with an increasing number of configurable parameters, adding new equipment, links, users, and applications to an existing network is much more difficult to do without an automated process. This is because the administrators have a hard time understanding the potential impact of any changes they may make to the network. Manual processes for managing networks today are typically error prone and slow since it takes a significant effort over a long period of time for an administrator to figure out the new configuration. As a result, managing change in today's networks has become a great challenge.

The challenges of today's network management are exacerbated in the wireless ad hoc networking environment. Such networks are usually large and complex because they typically use diverse physical layer, link layer, and network layer techniques. It is very difficult for a human to deal with changes in such networks due to the number of configurable parameters that can potentially impact the operation of the network. Wireless ad hoc networks also normally operate in a very volatile environment. Network conditions change very rapidly as nodes move around. Nodes may lose and regain connectivity as nodes move in and out of coverage. The terrain changes due to the node movements may impact which nodes are connected with each other and the quality of the links connecting them.

It is impossible for a network administrator to deal with all of these changes manually and keep the network in a good state based on the current view. Keeping track of all the changes and reconfiguring the network quickly enough so that the network is configured well for the current state of the network is an impossible task to do manually. This is because it is difficult for a human to select the correct values for each of the configuration parameters. This is a complex optimization problem. Further, by the time a human solves the problem and implements the new configuration throughout the network, it is likely that the state of the network would have changed substantially, thereby making the new configuration suboptimal. Therefore, an approach that automates network management of wireless ad hoc networks is very much needed. We will next discuss a new approach that has been proposed for automating network management, called policy management. We will discuss how it can be applied to the management of a MANET in Section 6.2. We will also discuss how policy management can be used for providing security management in Section 6.3.

6.2 POLICY-BASED NETWORK MANAGEMENT

6.2.1 Overview

A new approach has been proposed in the industry for managing large networks and overcoming the challenges discussed earlier. This approach is called policy-based network management (PBNM). In PBNM, administrators do not have to worry about configuring the large number of parameters in each individual network element. With PBNM, users (or network administrators) specify the high-level requirements of the expected network behavior rather than worrying about specifying each configuration parameter of each network element in detail. The PBNM system is responsible for taking the high-level user requirements and configuring each individual network element automatically without any additional user input in a way that complies with the high-level user requirements, which are typically called policies.

PBNM allows users to specify high-level requirements that typically include:

- Policies that define long-term requirements that need to be enforced throughout the network, such as whether certain types of traffic needs to receive a specific level of QoS or be transmitted at a particular compression level, whether certain types of traffic needs to be encrypted, and so on.
- Policies that define how the network should react when changes are required and specific events occur in the network. PBNM creates an automated feedback loop so that the network is reconfigured without any human in the loop in response to changes in the network state. The reconfiguration is based on policies that have been specified by the administrators beforehand. For example, a policy may specify that, if a user is detected to be attacking the network, the attacker may be disconnected from the network so that the attack is stopped. Another policy may specify that, if the loss rate in the network increases (e.g. due to changes in the terrain due to recent movements), a different transmission scheme may be used that may make network transmissions more tolerant of losses.

At a high level, Figure 6.1 demonstrates the concept of PBNM. PBNM is a general concept that has been discussed in a variety of industry fora and publications. It has been used for

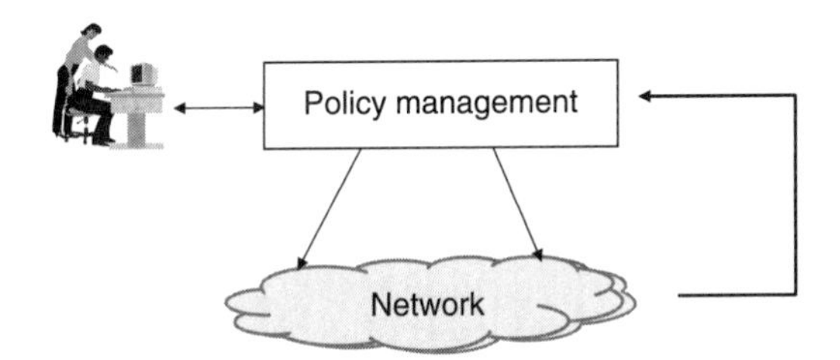

Figure 6.1. Policy-based network management concept.

areas such as QoS management, configuration management, and fault and performance management. In this chapter we will discuss the concept in general and then we will focus primarily on how the PBNM concept can be applied specifically for security management for wireless ad hoc networks. We start by providing a more detailed overview of the architecture used in PBNM. We will then discuss in detail, specific policy languages that have been proposed. Policy languages merit particular attention because expressing policy languages is one of the most critical components of the architecture. We will discuss how the PBNM architecture can be used in a distributed fashion. At the end of this section we will also briefly discuss several standardization activities in this area.

6.2.2 Architecture

We next look at the PBNM architecture in more detail. The PBNM architecture discussed here has been adopted by the IETF and the Distributed Management Task Force (DMTF) (as discussed in Section 6.2.5) and has been a research focus. The PBNM architecture typically incorporates certain components that are shown in Figure 6.2. The components of the PBNM architecture as shown in Figure 6.2 are:

- The policy management tool is used by administrators for defining the high-level policies that need to be implemented throughout the network.

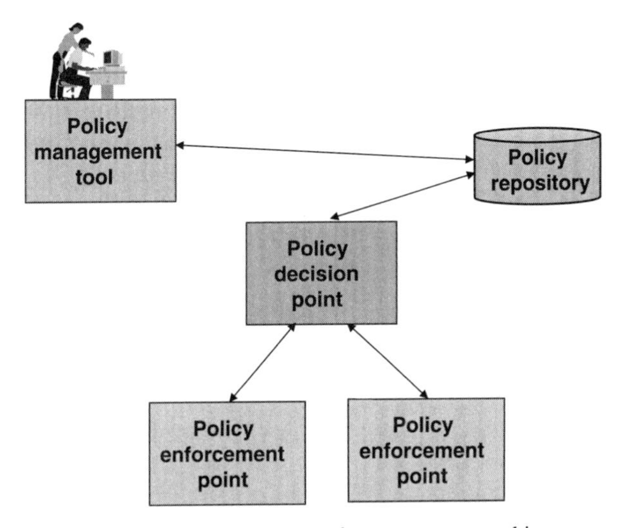

Figure 6.2. Policy-based network management architecture.

- The policy repository is used for storing the policies created by users through the policy management tool.
- The policy decision point (PDP) is responsible for interpreting the various policies defined by the administrators through the policy management tool. The PDP then instructs the network to take appropriate actions based on the current state so that the policies specified by the administrators can be enforced throughout the network. The PDP may receive input from the network regarding the current state of the network and may be notified by the network about specific events that may require reconfiguration of the network. For example the PDP may be notified when network devices fail, when the performance of the network degrades, and when cyber attacks are detected. On receiving such notitifaction, the PDP can reconfigure the network so that the policies can be enforced under the new network conditions.
- The policy enforcement points (PEPs) are responsible for implementing the decisions that the PDP makes for reconfiguring the network. PEPs are usually network devices such as routers, radios, and firewalls, that can implement the actions required for ensuring that the policies are enforced across the network. In some instances, however, it might not be possible to satisfy the policies. This occurs when the PEPs do not have the capabilities required for enforcing the policies defined by the users.

The PBNM architecture addresses both of the challenges with existing network management systems discussed earlier. Configuration of all the individual network parameters is simplified and automated. Administrators do not need to configure each individual parameter. They just define high-level policies and goals through the policy management tool and the PBNM system is then responsible for configuring the detailed individual configuration parameters automatically based on the high-level input. This is accomplished by the PDP, which parses the high-level policies, as discussed earlier, and then configures the PEPs in a way that is consistent with those policies.

The PBNM also deals effectively with changes whether they are introduced by the administrators or due to changes in the network itself. Administrators can introduce new high-level policies or change existing policies through the policy management tool. The PDP is then responsible for parsing the changes introduced by the administrators and reconfiguring the PEPs to ensure that the updated policies are enforced. The PDP also gets notified about changes in the network such as cyber attacks, faults, and performance degradation. The PDP is then responsible for reconfiguring the PEPs based on the new events to ensure that network effectively supports the high-level policies.

The PBNM architecture is a very elegant architecture that can address the challenge of automating network management and in particular dealing with the constant changes that are of particular importance for wireless ad hoc networks. There are two areas of the architecture that have proven to be challenging in implementing the high-level PBNM architecture and have attracted the interest of the research community. The first area is the definition of policy languages that can be used for capturing the high-level goals that need to be enforced by the PBNM architecture. The second area is the problem of extending the PBNM architecture to operate in a fully distributed environment. We will now discuss these two areas in detail.

6.2.3 Policy Languages

The policy language used for allowing users and administrators to define the high-level policies is critical for making the policy management architecture effective. If the policy language is not expressive enough then it cannot capture the user requirements. Therefore it may not be possible to effectively use the policy management architecture for the specific application. If the language is too complex then it may be too difficult for users to understand. As a result, users may not be inclined to use the system.

The best way to define policies is through a natural language that is easy for a user to use and understand. Unfortunately, little progress has been made towards achieving that goal. Three main approaches have been pursued instead by the research community so far for creating policy languages that can be used in the policy management architecture:

- the first approach is based on using structured programs developed specifically for expressing policies;
- the second approach is based on using formal logic languages;
- the third approach is based on using a set of simple well structured rules. These rules are similar to "if ... then ... else" statements.

We will discuss each of these alternatives in more detail next and we will also provide specific examples for each of these approaches that have appeared in the literature.

The first alternative for defining policies through a policy language is to define policies as small structured programs that can be understood by computers. This can be accomplished either by using existing programming languages or by developing new programming languages specifically for defining policies. This approach of specifying policies has been used extensively and successfully because it is fairly general and extensible, and allows the specification of even very complex policies. The Ponder language [106] is one such example and is described at a high-level in Section 6.2.3.1. The generality and flexibility of these approaches is also their main limitation. Reusability of policy management systems that utilize such languages is very limited because standardizing all of the components of the language is a fairly complex task.

The second alternative approach that has appeared in the literature for specifying policies through policy languages is the definition of policies through formal logic. Examples of such languages are described in [107, 108]. These policy languages are based on a formal logic called deontic logic. The Rei policy language from [107] is an example of such language and is described at a high-level in Section 6.2.3.1. Even though these languages are well defined and elegant, their main limitation is that they usually require a strong mathematical background to understand, and have very high complexity for typical environments that limits their applicability in large systems.

The third approach for defining policies through policy languages is the most widely used one. This approach is based on using specific well-structured rules that are usually of one of the following types:

- *condition action rules*, which are rules that are applied to the network when specific conditions are true; an action is implemented when the conditions are satisfied;
- *event condition action rules*, which are triggered when a specific event occurs; when that happens an action is implemented only if the conditions specified in the policy are satisfied;
- *Invariants*, which are rules that express statements that need to be true throughout the operation of the network.

Even though these rules are not as general as the ones that can be defined by programming languages, policies expressed by these simple rules can be used for defining a very wide variety of policies. An example of the use of such rules is the Policy Description Language (PDL) [109] that is described in Section 6.2.3.1. The advantage of using well-structured rules for defining policies is that, since policies are expressed as simple and well-structured rules they can be standardized fairly easily. Hence, this approach supports reusability and interoperability across systems. These types of policies are described in more detail below.

Condition action rules are expressed as "if then" statements. For example, the following syntax can be used:

$$\text{If } (C_1, C_2, \ldots, C_n) \quad \text{then } (A_1, A_2, \ldots, A_m)$$

where C_i is a condition that needs to be satisfied. Conditions are usually statements that are true or false depending on values that specific network or system parameters take. For example, a condition might be true, when a node has been compromised (as detected by the IDS), the CPU utilization goes above 70 percent, or a link is loaded more than 80 percent.

A_i are actions that the policy system needs to implement. These could be revoking the credentials of a compromised node, performing network reconfigurations or changes in system configuration parameters, or implementing new firewall rules.

A simple condition action rule can be the following:

If (node A has been compromised) then (revoke credentials of node A) and (migrate server from node A to node B).

This policy states that, if node A is compromised (based on alerts provided by the IDS), then the credentials of the node need to be revoked and the server running on node A needs to be migrated from node A to node B. A more complex policy may state that:

If (node A has been compromised) or (node A CPU utilization $>$ 70 percent;) or (connectivity to node A is intermittent) then (migrate server from node A to node B).

This policy states that, if the CPU of node A has been compromised, or is overloaded or if the connectivity to node A is intermittent (e.g. due to congestion or link unavailability), then the server needs to be migrated from node A to node B. That would allow other nodes to more easily access the server and receive improved performance.

Event condition action rules are similar to condition action rules that are triggered only when specific events occur. These rules can be described as follows:

$$(E_1, E_2, \ldots, E_k) \text{ causes } (A_1, A_2, \ldots, A_m) \text{ if } (C_1, C_2, \ldots, C_n)$$

where E_i is an event that occurs in the network or in one of the systems. For example, an alert from the IDS system can be an event.

A_i and C_i are events and conditions, respectively, and were explained in the condition action rule description. A simple event condition action rule can be the following:

> (An alert from the IDS indicating node A is compromised) causes (node A's credentials to be revoked) if (IDS system is not compromised).

This policy specifies that, if the IDS sends an alert indicating that node A has been compromised and if the IDS itself has not been compromised (and can be trusted), then the credentials of node A should be revoked.

Invariants are rules that can be used for making assertions and expressing truths that need to be maintained throughout the operation of the network. A simple invariant rule can be the following:

> All ftp traffic from external networks should be blocked

6.2.3.1 Example Policy Languages In this section we describe briefly some example policy languages that have been proposed in the literature and cover all three policy language approaches described earlier.

6.2.3.1.1 Ponder The Ponder language has been developed at the Imperial College over several years [106]. It is a declarative object-oriented language that was developed with a focus on two types of policies: policies that can be used for automating management of networks and security policies. Policies are expressed in a syntax similar to small computer programs. Therefore, Ponder belongs to the first category of policy languages discussed earlier that define policies using structured programs.

Network management policies are called obligation policies and are event-triggered condition action rules. Policies specify actions that need to be applied on *target* objects by a *subject* when a specific event occurs. The following is a simple example of such a policy from [106] which is a loginFailure policy:

```
Inst oblig loginFailure {
        On                  3*loginfail(userid);
        Subject             s=/NRegion/SecAdmin;
        Target   <userT>    t=/NRegion/users^{userid};
        Do                  t.disable( )→ s.log(userid);
}
```

This policy example specifies that, when a user has three login failures, then the administrator will disable the user. The subject in this case is the security administrator for the domain. The /NRegion/SecAdmin *subject* field demonstrates the concept of how objects are grouped in the Ponder language, which is similar to a file system. The grouping may be based on an organizational hierarchy, administrative control, geography, or other attributes. The top level in the hierarchy is called a domain. Domains have subdomains and subdomains may have their own subdomains. The *target* for the policy is the user object that is associated with the three login failures. There are two actions associated with the policy which are separated by the → operator. The first action states that the user that had the three login failures needs to be disabled. The second action states that the security administrator needs to log the id of the user that had the three login failures.

Security policies in Ponder are more complex than network management policies. There are four primitive policy types which can be used for defining more complex security policies. These are: (1) authorization policies that define which member of the domain has access to the various services; (2) refrain policies that define what members of the domain should refrain from doing even if they may be allowed to do that; (3) delegation policies that allow members of the domain to delegate some of their authority to other members; (4) information filtering policies that limit the information that can be disclosed such as for privacy reasons.

We now discuss a simple example of security authorization policy from [106]. The policy language allows the definition of policy types that can be used for instantiating specific policies. Below is an example policy type:

```
Type auth+ PolicyOpsT(subject s, target <PolicyT> t){
        Action load( ), remove( ), enable( ), disable( );}
```

This example specifies that subject *s* is allowed to perform the specific actions (load, remove, enable, and disable) to the target object of type *t*. The following is a specific instantiation of this policy type.

```
Inst auth+ swithPolicyOps=PolicyOpsT
                        (/NetworkAdmins,/Nregion/switches);
```

The security policy is instantiated from the PolicyOpsT type. The policy allows network administrators to load, remove, enable, or disable switches in the domain Nregion.

6.2.3.1.2 Ismene In the Ismene Policy Description Language (IPDL) [110, 111] policies are defined as structured programs and therefore belong to the first category of policy languages discussed earlier. On IPDL a policy is defined as an ordered set of clauses. Each clause is defined by a tuple of *tags*, *conditions*, and *consequences*. *Tags* are used for assigning policies to a specific grouping. Multiple policies that belong to the same grouping (such as key management) are ordered based on which one appears first. *Conditions* test whether a specific measurable aspect of the system is either true or false. *Consequences* are specific actions to be taken if all of the conditions listed in the policy are true. Below are some example policies from [110, 111]:

```
Key_management: GroupIncludes(Manager), GroupSmaller(100) ::
Config(LKHKeyManager(rekeyOnJoin=true,rekeyOnLeave=true));
```

This is a key management policy that specifies that, if the group includes a manager and the group has fewer than 100 members, then LKHKeyManager scheme will be used with rekeying required on new joins and leaves.

```
Join: Credential(Role=Manager, IssuedBy=$Trusted_CA) :: accept
```

This is a join group management policy for authorizing new members to join a group. This policy specifies that, if the role of a node is a manager and the credentials for the node have been issued from a trusted CA, then the node is allowed to join the group.

Another possible policy is as follows:

Intrusion_Response: Alert(CompromisedNode = x,
IssuedBy = $Trusted_IDS):: RevokeCredentials(Node = x)

This policy defines that, if an alert is issued by a trusted IDS stating that node x has been compromised, then the credentials for node x should be revoked.

6.2.3.1.3 Rei Policy Language The Rei language is a policy language based on deontic language and therefore belongs to the second category of policy languages as discussed earlier. The Rei language is described in [107]. Rei defines rights, prohibitions, obligations, and dispensations:

- *rights* are permissions that an entity has and can be expressed as actions that entities are allowed to perform if they meet specific conditions;
- *prohibitions* are statements that do not allow entities to perform specific actions;
- *obligations* are actions that a specific entity must perform usually when specific conditions are satisfied;
- *dispensations* are obligations that are not required to be performed anymore.

Policies are associated with an entity with a construct called "has." Below is an example policy:

Has(administrator, right(LogOnAsRoot, RightsNotRevoked))

This policy states that administrators are allowed to log on as long as their rights have not been revoked. The Rei language has additional constructs for defining actions, delegating rights, making requests, canceling requests, and revoking rights. The details of the Rei language are described in [107].

6.2.3.1.4 Policy Description Language The Policy Description Language (PDL) was developed by Lucent [109] with a focus on network management. PDL was implemented in Lucent's softswitch product for handling voiceover IP calls, but is general and applies to other applications. Policies in PDL are defined as event condition action rules and therefore belong to the third category of policy languages discussed earlier. In PDL, policies are a collection of propositions of two types: policy rules and policy-defined events. These two types of propositions are defined below.

1 Policy rules are statements with the following format:

Event causes action if condition

This type of proposition specifies that, when a specific event occurs, a specific action is triggered by the policy management system if a specific condition is true.

2 Policy defined events (pde) have the following format:

Event triggers pde (m1=t1, . . . , mk=tk) if condition

> This type of proposition specifies that, when a specific event occurs and a specific condition is true, then that triggers a set of events (which are policy-defined events) within a predefined period of time specific to the application with the valuation of *tk* for each attribute *mk*.

The strength of the PDL language is its flexibility in defining events in policies. Events can be either of two types. They can be a basic simple event or they can be complex events that include several events within a prespecified period of time. Complex events include the following:

1 e1 & e2 & . . . & en, which means that all events occurred within the prespecified period of time;
2 e1 | e2| . . . |en, which means that any of the events occurred within the prespecified period of time;
3 !e, which means that there was no occurrence of the event within the prespecified period of time;
4 ^e, which means that there was a sequence of one or more occurrences of the event.

Additional complex events are defined in [109] and are omitted here for brevity.

Some complex functions are also defined and can be used when stating conditions in policies. For example, the function Count counts the number of occurrences of a specific event within the prespecified period of time. Functions Min, Max, Sum, and Avg can also be used for specifying the number of occurrences of a type of event.

Below is an example policy based on the discussion in [109]. The policy shown below defines a policy-defined event as defined earlier called response_mode. The goal of this policy is to trigger a response once the system is receiving a large number of alerts from the IDS. The response_mode event is triggered as follows:

normal_mode, ^(alert)
 triggers response_mode
 if Count(alerts)>t

This policy states that the event response_mode is triggered when the number of alerts is larger than *t* within the prespecified period of time. An example of a policy rule as defined earlier is the following:

response_mode **causes** Raise_Infocon

This example rule specifies that, if the response_mode event is triggered (as defined in the policy-defined event policy discussed earlier), then the action Raise_Infocon is applied. Such action may result for example in higher authentication being required before any actions are allowed by the system. More elaborate policies are discussed in [109].

6.2.4 Distributed Policy Management Architecture

Management of large networks is not usually possible from a single network management system because a single system will need to maintain a large quantity of data. Further, the number of functions that the single system will need to perform may require a very large amount of CPU processing. Traditional network management systems distribute the management functions to several systems and usually multiple layers of management. Unique characteristics of wireless ad hoc networks also make it necessary to distribute the network management functions.

Depending on the specific application that the wireless ad hoc network is used for, additional requirements may exist that mandate a distributed architecture and potentially drive the implementation of a specific distributed architecture. For example, for military applications, wireless ad hoc networks often need to be divided into domains corresponding to specific military units. Brigades usually have their own network and therefore typically require their own network management system. It is also possible that individual companies in a brigade may require their own network management system, especially if the company will often be disconnected from the rest of the brigade. This is to allow the company to manage its own network while it is disconnected from the brigade.

Designing an effective distributed architecture for the specific application that is supported by the network is a challenging problem that involves many trade-offs. There are two general approaches that have been proposed for distributing the network management functions:

- In the *peer-to-peer architecture*, the network is divided into domains and each domain is managed by a single network management system that is fully independent from the network management systems in the other network domains. Coordination among network management systems in different domains may not be possible or may not be allowed because the different domains may belong to different administrative entities. Coordination among the network management systems in each domain may only be allowed if it benefits both domains.
- In the *hierarchical architecture* also, the network is also divided into domains. Each domain has its own management system but each system is not independent of the others. In the hierarchical architecture there is a management system that is at the top of the hierarchy and all other systems are subordinate to that system. Distribution of the network management functions is driven from the top of the hierarchy such that policy enforcement is accomplished across the whole network and is implemented such that it benefits the entire network.

We now describe each of these two approaches in more detail.

6.2.4.1 Peer-to-Peer Architecture In the peer-to-peer architecture, a policy manager is responsible for managing a single network domain. The domain may include just a single network device or several network devices that are connected together. The policy manager includes the policy management tool, the policy decision point, and the policy repository functions of the PBNM architecture described earlier. Policy managers of different domains may need to coordinate for supporting policies that span multiple domains. For example, when a link connecting two domains fails, then policy managers of both domains need to coordinate in order to decide how the traffic should be shared among the available links connecting the two domains. When an intrusion is detected, the policy manager of the domain where the attack is detected may notify Policy Managers in other domains of the attacker in case the attacker tries to launch an attack in other domains.

The policy managers in this architecture have completely independent and often different policies that they are trying to enforce because they may be managing domains that have very different applications, user needs, and networking requirements. The policy managers of the different domains may not need to coordinate and may act completely independent of each other. This could be so particularly in cases where the different domains belong to different organizations that do not wish to coordinate (such as competing commercial organizations) or may not have the need to coordinate if they do not share any common traffic or applications. In some cases, though, it may be beneficial for the organizations to work together and coordinate their network management functions, especially if they share network traffic and applications. In such cases, policy managers of the different domains may share information about specific events such as details about intruders, and specific failures, and may coordinate their policy enforcement actions.

One specific application of the fully distributed peer-to-peer policy management architecture has been explored by DARPA's Dynamic Coalitions program [112, 113]. This program has supported several research efforts in the policy management area and has played a major role in advancing the state of the art in policy management. Several technologies developed under that program are discussed later in this chapter. In DARPA's Dynamic Coalitions program each domain is assumed to correspond to the network of one of the coalition partners (United States, Sweden, Poland, etc.) in a military environment. In such an environment, each domain has its own policies managed by the domain's own policy management system. It might be necessary for policy management systems across coalition partners to cooperate in order to support the communications among coalition partners. Policy coordination occurs only if the coalition partners can negotiate and agree as peers on a set of policies that can be used for supporting the communications among the coalition partners. A hierarchical structure does not fit this environment because none of the partners would be in control of the complete network. A similar situation occurs in commercial applications where there could be multiple competing enterprises that need to communicate and share information. Such communications are typically supported by communication service providers. Usually enterprises have their own policies

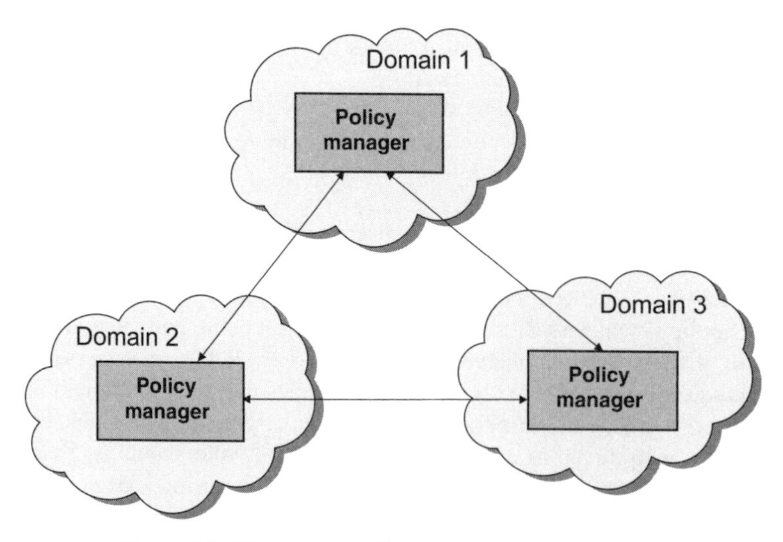

Figure 6.3. Peer-to-peer policy management architecture.

and policy management systems and the service providers that service those enterprises also have their own policies and policy management systems. Supporting communications across enterprises over the service provider networks requires that the policy management systems negotiate as peers in order to agree on how the traffic that crosses the various networks should be managed. Therefore, it is not possible for a single policy management system to be in control of the policy management functions across all of these networks.

One of the challenges in peer-to-peer policy management architectures is negotiating policies across domains that are administered independently. Such negotiations are necessary for coordinating policy enforcement in support of communications across different domains. For example, we can consider two enterprises that need to set up a secure collaborative environment (e.g. a virtual private network) between them. In order to accomplish that, the two organizations need to negotiate a common policy that defines how the secure collaboration will be achieved. Such a policy may define the encryption mechanism to be used when communicating across the domains, the size of keys to be used, the authentication mechanisms to be used, and so on. Each of the enterprises may have its own policies for supporting secure communications within the enterprise. If the two policies in each of the domains do not overlap (for example if they do not share or support a common encryption mechanism that they typically use for secure communication), it may be impossible for the two domains to agree on a common policy to be used for secure communication across the two domains. Assuming that the policies in the two domains overlap in the types of functions and parameters that they allow, then policy negotiation may be possible by selecting the common subset supported by both domains. When more than two parties are involved in the negotiation it is even more complicated to negotiate a common policy that is agreeable across all domains. Such negotiations typically occur today in an offline fashion among the administrators of the different enterprises. This is a slow and error-prone process. The policy management approach provides an environment for automating such a process. Defining the protocol that can be used for negotiating policies across independent domains is still an open research problem.

6.2.4.2 Hierarchical Architecture The alternative distributed policy management architecture is the hierarchical architecture. In the hierarchical architecture (as in the peer-to-peer architecture), each policy manager is responsible for managing a network domain which may contain one or more network devices. The difference in the hierarchical architecture is that the different domains are not independent administratively. There is a relationship among the policy managers of each domain reflected in the hierarchy. This hierarchy affects how the various policy managers coordinate. In the hierarchical architecture there is a policy manager that is at the top of the hierarchy and is responsible for specifying global policies that need to be enforced across the entire network. Each policy manager in the network is responsible for doing its part in support of the global policies that the policy manager at the top of the hierarchy specifies. The topmost policy manager is also responsible for coordinating functions in response to changes that affect the entire network. This is achieved by requiring that each policy manager share with the topmost policy manager all events and data that may be of interest to other domains. The topmost policy manager is then responsible for sharing those events and data as appropriate. Each policy manager typically is also responsible for local policies that are specific to the domain the policy manager is managing. Such policies may not need to be coordinated across multiple domains with other policy managers.

Figure 6.4 shows an example architecture in which there is a policy manager for the top domain in the hierarchy and multiple policy managers in domains lower in the hierarchy. It

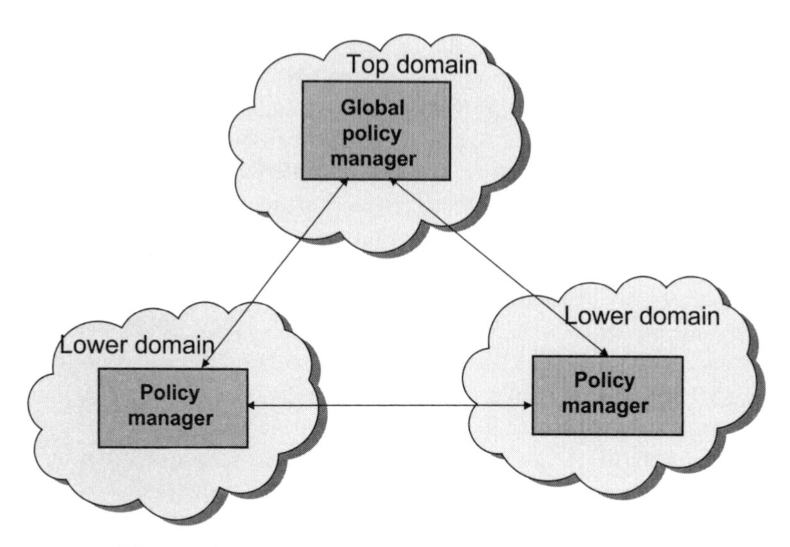

Figure 6.4. Hierarchical policy management architecture.

is possible that there are multiple layers of hierarchy in which a policy manager responsible for a lower domain is itself a top policy manager for multiple layers below it.

In a wireless ad hoc networking environment, connectivity in the network is intermittent and therefore it is possible that policy managers in lower domains will become disconnected from the higher layers in the hierarchy. This is not a problem in the hierarchical architecture. Policy managers that lose connectivity to a policy manager higher in the hierarchy can continue functioning based on the existing policies. A new hierarchy may potentially need to be established (for example through the approaches discussed in Section 6.2.4.3) for coordinating policies across the domains that are still connected, and a policy manager may become the top of the hierarchy temporarily for the layers below it that are still connected to it. Once the original hierarchy is reconnected, the policy managers can reestablish the original hierarchy. At that point, the policy managers would probably have to resynchronize their states. This is because changes that may have occurred during the times that the hierarchy was disconnected could affect policy management across the network. For example, the original top policy manager may need to be notified of any failures, attacks, or performance degradations, that had occurred during the time that it was disconnected from the rest of the network. Policies and states in the higher layers in the hierarchy will probably receive a higher priority when the synchronization happens because they have broader management responsibilities, although more recent events should take priority over past events. Performing such state and policy synchronization is a challenging problem in a very dynamic environment with many policy managers losing and gaining connectivity from the hierarchy. This problem has not been adequately addressed in the literature.

The big advantage of the hierarchical architecture compared to the peer-to-peer architecture is its suitability to environments that are naturally hierarchical. For example, a military environment is typically hierarchical. The top of the hierarchy may be a typical military division headquarters. The policy manager in the division headquarters may define a policy that needs to be enforced across the division. Lower domains in the hierarchy such as brigades, battalions, and companies in the division may have their own policy managers, but will need to comply with the policies specified by the policy manager in the division headquarters. Policy managers lower in the hierarchy still have some flexibility in selecting local policies as long as the local policies are consistent with the global policies that are defined at the top of the hierarchy. Management of networks in commercial enterprises is also naturally hierarchical within enterprises. For example, a top level policy manager (probably located in the headquarters of a corporation) may define a global policy that applies across the entire enterprise while local organizations within the enterprise may define local policies that only apply within that local organization as long as such policies are consistent with the global policy.

Both the peer-to-peer architecture and the hierarchical architecture seem to have important applications in specific environments. It is also possible that the two architectures can work together in support of some applications. For example, in a military environment the hierarchical approach can be used for managing the network of a coalition partner while policy management across coalition partners can be supported by the peer-to-peer architecture. Similarly in a commercial environment the hierarchical architecture can be used for managing secure collaboration within the enterprise while the peer-to-peer architecture is used for managing secure collaboration across multiple enterprises.

6.2.4.3 Use of Clustering Even though policy management has been an area of interest in the past several years, most of the efforts have focused on the application of policy management to enterprise wireline and fairly static networks. Limited research has been conducted on how to apply the policy management concepts to the wireless ad hoc networking environment. The most prominent efforts have been conducted through funding from the Army and have been published in [114–116]. A unique challenge in the wireless ad hoc networking environment is the placement of policy servers within the network. This problem applies both to the peer-to-peer and hierarchical architectures discussed earlier. In a traditional wireline network, the policy manager is placed on specific nodes that are selected by network administrators when the network is first configured. The policy manager is not migrated to a different node unless the node hosting it fails. In the ad hoc networking environment, a static policy management architecture where the policy manager is placed on a predefined node is not viable. This is because nodes could move around and may cross from one domain to another, nodes could get captured and destroyed on a frequent basis, nodes could lose connectivity from other nodes, often forcing the redrawing of domains, and the network could change in unpredictable ways. Hence it might not be appropriate for any predefined nodes to be hosting the policy manager. Therefore, selecting the nodes where to host the policy manager is a challenging research problem in such networks.

Chadha *et al.* [116] have proposed to use clustering for organizing event reporting and dissemination functions as well as for selecting where to place certain policy management functions in a MANET. This concept is similar to the concept used for selecting where to place the intrusion detection correlation capabilities described in Chapter 5. The key idea is the use of clustering algorithms for organizing nodes in groups (called clusters). Such groups are typically groups of nodes that are located close to each other and have direct connectivity to each other. Once clusters have been formed, the clustering algorithms can be used for electing leaders in those groups called clusterheads. The clusterheads can be used for hosting a specific function for the whole group. One such function can be the policy manager or intrusion correlation function as discussed in Chapter 5. Since the clusters are usually organized based on how nodes are connected, clusterheads usually tend to be centrally located within the cluster and have direct connectivity with a large number of nodes belonging to the cluster. Such an approach reduces the overhead usually involved in the communication between the policy manager and other nodes. As networks change due to movements, changes in terrain and connectivity, and node destruction, the structure of the clusters changes dynamically and new clusterheads may be elected. Some clustering algorithms that can be used for organizing nodes in clusters were discussed in Chapter 5. Even though those algorithms were discussed in the context of routing and intrusion detection the same algorithms can be applied here for organizing policy management functions. As was discussed in Chapter 5, creating and maintaining the clusters introduce some level of network overhead. This is a problem in wireless ad hoc network since bandwidth is always limited, but using the clustering approach has the advantage of making the policy management architecture much more dynamic.

6.2.4.4 Conflict Resolution As was discussed in the hierarchical architecture, global policies created at the top policy manager apply throughout the enterprise. Local policies may also be created by local policy managers and are applied locally. It is possible that such policies may conflict with each other (probably due to errors in defining those policies). This is also possible within an enterprise environment if policies are very complex. Dealing with conflicting policies is a challenging research problem that has only been addressed to a limited extent so far. We discuss a simple example of such policy conflicts. A policy may state that no employees should be allowed to perform certain auditing functions on computers. At the same time a different policy may specify that administrators are allowed to perform those auditing functions on computers in order to perform their administrative functions. Since administrators are employees it is clear that there is a conflict between the two policies. Such conflicts also may arise in the hierarchical architecture if local policies are not developed carefully enough and may end up conflicting with the top-level policies, namely the global policies. For example, the policy about administrators discussed earlier may be a local policy while the general policy about employees may be a global policy. Careful design of the policies can help alleviate this risk. Another possible approach for dealing with policy conflicts is prioritizing the various policies. If multiple policies conflict with each other, the policies with the highest priority take precedence. In the example mentioned earlier, by giving the policy that allows the auditing functions of an administrator higher priority than other policies, the policy conflict will be resolved and the system will provide the expected behavior.

6.2.5 IETF and DMTF Standardization Activities

The importance of policy-based management has been well understood within the industry, including the IETF community and other standards organizations. Several working groups within IETF have addressed this topic. For example, The Resource Allocation Protocol (RAP) Working Group has defined the Common Open Policy Service (COPS) protocol [117]. COPS supports communication of policies between PEPs and PDPs associated with resource reservation requests that are part of the Resource Reservation Protocol (RSVP) [118]. When PEPs receive requests for new flows through RSVP request, PEPs communicate with PDPs to request permission to admit the new flows. The PDPs check whether the admission of the new flow is consistent with existing policies and if that is the case they allow the PEPs to accept the new flow. Event though the COPS protocol was originally developed for a specific application (namely admitting RSVP requests) COPS was developed assuming a fairly general architecture that is consistent with the policy-based network management architecture discussed earlier.

COPS is the protocol envisioned between the PDP and the PEP in the PBNM architecture. The same architecture and protocol (i.e. COPS) have been extended by IETF for providing policy management functionality for other applications. For example, even though the original COPS model assumed that COPS would be used by PEPs for outsourcing policy decisions to the PDP as discussed earlier, the COPS protocol was extended to allow PDPs to provision policies in the PEPs proactively by initiating actions themselves [119]. If that model is used, there is no need for PEPs to communicate with the PDP whenever a decision has to be made. Policy enforcement decisions can then be based on pre-provisioned instructions provided by the PDP, making the policy enforcement process more efficient. Even though the COPS specification was developed several years ago, there are only limited deployments of the protocol so far.

In the security management area, IETF has focused specifically on policy management for IP Security Protocol (IPsec) through a working group called the IETF IP Security Policy (IPSP) Working Group. This working group focused on developing an information model for defining IPsec security policies. Parameters included in the information model and controlled by the policies defined by this working group include whether to use encryption or not, the specific algorithms to use for encryption and authentication, and other parameters that are supported by IPsec. This working group also defined a policy specification language that is general, although the group focused on applying this language specifically to IPsec. The working group also created guidelines for provisioning IPsec policies using COPS and SNMP. The IETF work in this area has been fairly narrow in its focus and has seen limited implementation up to this point.

Another industry standardization group that has played a key role in the definition of the policy-based network management architecture and associated standards is the Distributed Management Task Force (DMTF). DMTF has focused on a variety of topics including a common information modeling (CIM) for PBNM and the directory-enabled network (DEN) concept (see www.dmtf.org/spec/cims.html) which focuses on the use of directories. Much of the work created by the DMTF organization has been leveraged by the IETF in its policy management efforts. This has ensured consistency in the standards developed by both organizations. A key component of the DMTF architecture is the use of a directory which has been leveraged by IETF for use in its policy repository using the Lightweight Directory Access Protocol (LDAPv3) [120]. Even though we have not discussed the policy repository concept much, it is an important component of the PBNM architecture. The use of LDAP in the PBNM architecture for storing policies in a standard way allows the policies to become easily accessible from various network and policy management components. This facilitates interoperability and ease of integration of diverse network components in the PBNM architecture.

The policy model that has been adopted by both IETF and DMTF assumes policy rules of the form:

If condition then action

These types of policies were discussed in Section 6.2.3 and, as discussed there, specifying policies in this manner facilitates interoperability and standardization. Some attempts have been made to standardize conditions and general actions to be used when defining these types of policies. For example, conditions defined in terms of times of day and days in week have been standardized. Progress to date is incomplete and further work is needed to fully standardize the policies needed for managing a typical network.

6.3 APPLICATION OF POLICY MANAGEMENT TO SECURITY MANAGEMENT

Policy management applies in general to various aspects of network management including security management. In this section we will be focusing on its application to security management. For example, policy management has been applied to managing access control (e.g. based on the level of trust of each node), which is a key security management area. One of the approaches for applying policy management to managing access control has been the use of policies for defining access control for users based on their roles rather than defining access control manually for each user which is called role-based access control. We discuss this briefly in Section 6.3.1. Security policies have also been utilized for structuring security authorizations in a concept called trust management. Trust management is described in Section 6.3.2. Firewalls are a key capability used widely for providing security in networks. In Section 6.3.3 we will apply the concept of policy management to the management of firewalls. The last topic discussed in this section is the application of policy management to policy enforcement (Section 6.3.4). We next discuss each of these topics.

6.3.1 Role-Based Access Control (RBAC)

Role-based access control (RBAC) is a widely used security scheme for providing access control. Although it is not really a policy management system, it can be considered as a vehicle for implementing access control policies, which is one of the most important aspect of security policies.

RBAC assigns access privileges not to specific users but to specific roles. For example, a security administrator may have access privileges to significant services that other users may not. Tying access privileges to roles rather than individual users simplifies the management of access control because permissions only need to be entered once for the role and then each user with the same role receives the same permission, avoiding a repetitious and error-prone task. Also, changes in the security policies of the organization can be easily implemented by modifying permissions to roles and not to every individual user.

It is also possible in RBAC systems to allow users to inherit permission from another role and then expand on the permissions for that role. For example, a security administrator may inherit all of the privileges that a regular user has and then obtain additional privileges that are unique to the administrator role. This further simplifies managing changes to security policies and provides the ability to consistently manage access control.

6.3.2 Trust Management and the KeyNote System

Trust management is a common framework for representing and managing security policies, security credentials, and trust relationships. The concept was introduced by Blaze *et al.* [121]. A key advantage of this concept is that it allows applications to treat these three key security functions consistently and in a unified manner reducing overall complexity and risk of security flaws.

Blaze *et al.* [122] discuss the KeyNote trust management system that has been developed for implementing the trust management concept and is representative of the concept. In the KeyNote system security policies are defined as small programming units that authorize specific users to perform specific actions under certain conditions.

The KeyNote system has the following concepts:

- Principals are entities that can be authorized to perform certain functions. They may be users, objects, programs, etc.
- Policies and credentials are specified using a concept called assertion. Assertions are the basic programming units that define the conditions under which a principal authorizes actions requested by other principals. Policies are authorized by a principal called "Policy" who is the root of the trust hierarchy.
- Actions are a collection of attribute-value pairs. Applications make queries to the KeyNote system requesting whether a particular set of actions is authorized or not. The KeyNote system determines compliance based on the security policies and returns a policy compliance value (e.g. authorized, unauthorized).

An example policy based on discussion in [Bl99] is shown below:

```
KeyNote-Version: 2
Authorizer: "POLICY"
Licensees: 1-of("RSA:abc1234", IDS 1,
               "RSA:abc1234", IDS 2,
               ...,"RSA:abc1234", IDS N)
Conditions: (app_domain == "RevokeCredentials")
node state==Compromised);
```

This policy specifies that any of the N IDS systems can provide authorization to the "RevokeCredentials" application to revoke the credentials of a node if the state of the node has been deemed to be compromised by the IDS. In this example, the RSA symbol specifies the type of algorithms to be used for the authorization.

6.3.2.1 STRONGMAN The KeyNote architecture has been extended in what is called the STRONGMAN. The STRONGMAN (scalable trust of next generation management) approach [123] (see Figure 6.5) assumes that high-level policies are specific to applications and therefore policy languages specific to the application are used for specifying high-level policies. All the high-level policies are then translated into a common low-level policy language. The low-level policy language hides the implementation complexities from the high-level policy engines. The low-level policies are defined using the KeyNote system.

The STRONGMAN architecture relies on the KeyNote system. KeyNote is a simple trust management system that provides compliance checking. In other words, it supports checking whether a proposed action complies with the local policies. Policies can be broken into smaller pieces which are signed assertions called credentials. Credentials can be distributed over the network and then local nodes can make decisions based on those credentials. Credentials signed by multiple parties can be considered when making a specific decision. Each service that needs to determine whether to permit or decline specific requests can utilize the local compliance checker to make that determination.

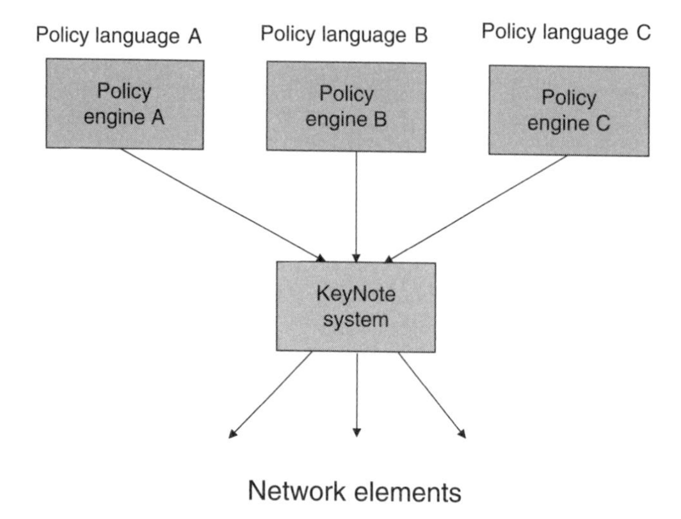

Figure 6.5. STRONGMAN architecture.

Since policy is expressed as credentials issued to users, it is not necessary to distribute policies throughout the network to all policy enforcement points. Users are required to provide credentials to prove that they are allowed to perform specific functions. With time, policy enforcement points learn the various policies that need to be enforced. Credentials may age with time to allow revocation of credentials.

6.3.3 Firewall Management

In traditional enterprises, firewalls are probably the key devices for securing enterprises from outside threats. Firewalls are usually positioned at the boundaries of enterprises with external networks and limit the traffic that is allowed to enter the enterprise. The main idea behind firewalls is the belief that the fewer the traffic types that are allowed to enter the enterprise network the smaller the risk from a malicious outside user. Firewalls are often even placed inside the enterprise to provide additional layers of protection in case the external perimeter is penetrated. This also helps limit access to portions of the network that host critical servers with even stricter traffic filters. Thus, it results in enhanced protection against outsiders as well as against potential malicious insiders (i.e. users operating within the enterprise network). Ensuring that firewalls in an enterprise are configured correctly is a challenging problem. To a large extent configuration of firewalls today is done manually. Firewalls as they exist in enterprise networks today are not directly applicable to wireless ad hoc networks. This is because in the ad hoc environment, it is not possible to identify traffic concentration points where a firewall can be placed to filter most of the traffic. The concept of a firewall can be extended though. Section 6.3.4 discusses the concept of a distributed firewall where every node assumes some of the firewall responsibilities and thereby traffic filtering can be achieved.

Burns *et al.* [124] propose policy management as a way of automating management of network security, including management of firewalls. The key idea is to define enterprise security policies using a simple descriptive language that captures the enterprise security requirements. A tool is then responsible for configuring the network devices automatically so that the security policies that were defined via the policy language are enforced at all times. This is accomplished by automating the configuration of the enterprise firewalls, and also potentially other devices in the network that have the ability to restrict the flow of traffic such as routers. The policy language that is described in [124] allows for the creation of rules that either allow specific types of traffic between nodes in the network or deny specific types of traffic. For example, a policy rule may state that external nodes are not allowed to telnet to any internal nodes.

Even though this concept was originally introduced for providing security management in enterprises, it can also be applied to the ad hoc networking environment. A key application of the security policy management concept described in [124] is the use of the approach described there to enable rapid changes in the security policies enforced in the network with limited or no manual intervention. This feature is critical for wireless ad hoc networks where changes in the network are frequent. The policy management can ensure that the security policies are enforced throughout the network as nodes move around and network connectivity changes.

The policy management system can also allow administrators to specify different policies for different conditions which are general and can be defined before the network becomes operational. For example, in a military environment different policies may be defined for different Infocon levels (these levels are dependent on whether the system is under attack and the severity of the attack). These policies may allow or prohibit different types of services and actions depending on the current Infocon level. For example, when the Infocon level is high (i.e. the network is under attack), stricter rules limiting access to critical servers from users that have low authorization may be enforced even though such access may be acceptable when the Infocon level is low. Another possibility is that actions allowed for a node under normal conditions may not be allowed if the node is considered to exhibit suspicious behavior. A policy may declare that all nodes with valid credentials can access a critical server while all suspiciously behaving nodes may not be allowed access to critical servers because they may be attempting to penetrate the server. Access to less critical servers may be allowed so that at least some basic functionality is allowed on suspicious nodes until it can be determined whether they are really malicious or not.

The policy management approach basically provides a lot of flexibility for changing what traffic is allowed to flow through the network under different conditions without requiring any manual intervention from the administrators. The automation provided by the policy management system also allows for changes in the security policies to be immediately enforced. This is particularly important for ad hoc networks because the conditions change greatly in such networks. Further, making changes to each individual device manually to enforce a security policy is highly ineffective, or impossible in many cases.

6.3.4 Policy Enforcement in a Wireless Ad Hoc Network

Policy enforcement is always dependent on the ability of the various devices in the network to implement the policies, which may include permitting or blocking traffic from specific hosts or applications, permitting or blocking traffic of specific type, and so on. Network management policies in traditional enterprise wireline networks are typically enforced by the routers in the network or other network devices. In wireless ad hoc networks every node is typically involved in the networking functions such as routing. It is therefore advantageous for each node to act as a policy enforcement point (PEP) and have the ability to accept requests from the policy management system and enforce the various policies based on those requests.

In wireless ad hoc networks there are usually no traffic concentration points where most traffic from the outside can be inspected and filtered. In these networks, as nodes move around, the boundaries of the network change and therefore every node or at least most nodes may become the boundary of the network with other external networks. Since every node may become part of the network boundary in order to provide effective protection from the outside every node in a wireless ad hoc network should have some policy enforcement capabilities.

In wireless ad hoc networks there is also a significant threat from malicious or misbehaving insiders (as was discussed in Chapter 1). Furthermore, in many applications envisioned for such networks, nodes are allowed to join the network dynamically. Such nodes can provide networking services, thereby becoming part of the critical infrastructure of the network. As discussed in Chapter 1, few malicious or misbehaving nodes can affect the network significantly and disrupt communications in the network. It is therefore important in such an environment for the network to be able to limit the access of insiders (i.e. all the nodes) to the network and services provided over the network. This also leads to the requirement that policy enforcement points in wireless ad hoc networks must have the ability to protect each node from all other nodes (even other insider nodes). Therefore, a fully distributed solution where each node has policy enforcement capabilities is imperative for securing a wireless ad hoc networking environment.

One potential approach for providing distributed policy enforcement capability is utilizing a fully distributed implementation of a firewall. Such an approach has been proposed in [125]. The distributed firewall in [125] was originally proposed for protecting the hosts and the network from insiders in a large enterprise environment. Since insiders may be anywhere in the enterprise, the traditional approach of placing a few firewalls at the network boundaries is not a viable solution for protecting the network from malicious insiders. Protecting the network from insiders in an enterprise environment also requires a fully distributed solution. This leads us to believe that a solution such as the one described in [125] can be adapted to the wireless ad hoc networking environment.

The key concept proposed in [125] is the use of a Network Interface Card (NIC) at each host, which is a hardened tamper-resistant device that incorporates firewall capabilities. The NIC is a nonbypassable interface to the network that has its own processor and memory that is not accessible from the host operating system or the applications running on the host the NIC is protecting. Therefore, the NIC cannot be easily compromised by malicious users. The NIC is controlled only by a policy server that distributes new packet filtering firewall rules (i.e. access control policies) during start up and whenever new or updated policies need to be enforced. In [125] the policy server is a centralized well-protected entity in the enterprise environment that can be used to define the security policies to be enforced by the distributed firewall implementation. The policy server needs to be well protected, because a compromised policy server can be used to open up the defenses of all nodes by implementing policies that would allow any traffic to go through the NIC. The concept of the distributed firewall is shown in detail in Figure 6.6.

Although it seems straightforward to adapt this architecture to the wireless ad hoc networking environment, there are some challenges. One of the challenges is that the interface from a host to the network in a typical enterprise environment is based on Ethernet. In a wireless ad hoc network the network interface is not usually Ethernet-based but most

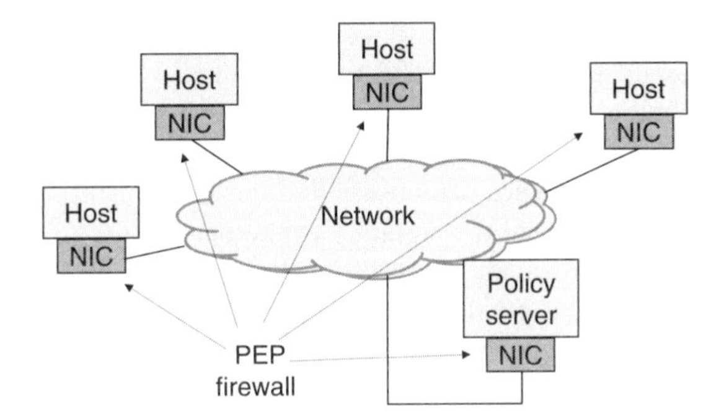

Figure 6.6. Distributed firewall architecture.

likely some type of a radio interface. NIC cards that implement a distributed firewall function in a radio environment are somewhat more complex, although it seems feasible to develop tamper-proof nonbypassable NIC cards with independent processor and memory that can operate in a radio environment. There will probably be some impact on the power consumption of a node if it uses a wireless NIC card with the capabilities discussed here. This may make the proposed architecture inapplicable to devices that have significant power constraints. However, for applications that do not have such constraints, it should be possible to implement the concept of distributed firewall on a nonbypassable NIC for a radio environment.

The other challenge with the architecture in Figure 6.6 is the applicability of the centralized policy server in a wireless ad hoc networking environment. In general, it is very difficult to ensure that a centralized policy server will be able to communicate with all NICs at all times. In a wireless ad hoc network environment, connectivity to nodes is intermittent and it is impossible to ensure connectivity between the policy server and all NICs. Making the policy server distributed may alleviate this concern because then one could locate policy server functions in multiple locations only a few hops from each node, limiting the possibility that connectivity between a policy server and an NIC will be lost. Such an approach increases the vulnerability of the solution, however. If a malicious user succeeds in penetrating a policy server it will gain control of all NIC cards controlled by that policy server and therefore will be able to disable all of those nodes. The more policy servers there are, the more likely it is that one may be compromised by a malicious user. In a military environment for example, policy servers available at the lower echelons of the force (so that connectivity can be ensured at all times) can possibly be physically captured, which will make it easier for a malicious user to compromise them. There is definitely a trade-off between making the policy server available at all times to the NICs and the threat of a compromised policy server. The right architecture and level of distribution is really dependent on the specifics of the applications.

6.4 SUMMARY

Managing access of users to the network and other systems is typically a slow and manual process. This is a challenging problem for MANET in particular because of the increased volatility due to node movements, connectivity changes, networks joining, and splitting, which necessitates frequent changes. Policy-based network management allows administrators to define high-level security requirements, therefore allowing tools to automate the security management process. This scheme can also be leveraged for automating response to attacks. In the previous chapter we discussed how IDS can be used for detecting attacks and the source of those attacks. Policies can define what actions can be taken in response to attacks. These policies can ensure that the level of response depends on the potential impact of the attack and the certainty of the detection.

Even though policy management is a well-developed technology for network management, there are many areas that need further investigation. Such areas include approaches for negotiating policies in a distributed policy management implementation, synchronizing policies across multiple distributed policy management systems, and resolving policy conflicts in such an environment.

So far in this book we have focused on fundamental technologies for making wireless ad hoc networks more secure. We focused on preventive technologies that make it more difficult for attackers to launch attacks, technologies for detecting attacks from nodes that have managed to penetrate the preventive barriers, and technologies for responding to attacks once the attacks have been detected. We will next focus on a topic of special interest, namely secure localization.

7 Secure Localization

7.1 INTRODUCTION

A basic feature of many wireless networks is the ability of the entities in the network to be mobile. As a result, the location of these entities keeps on changing with time. The availability of location information of these network entities is expected to radically alter the way in which wireless networks function. It is widely expected that knowledge of the physical location of a network entity will enable radically new applications and services. Such information will also help find critical resources faster and will improve the security of wireless networks. In addition knowledge of the location of entities in a wireless network makes it possible to perform existing functions more efficiently.

To illustrate this, let us consider the example of routing, which is a very important function in ad hoc networks. There have been several proposals to incorporate location information into routing. The resulting geographic routing protocols, such as GPSR (greedy perimeter stateless routing) and DREAM (distance routing effect algorithm for mobility) [126–129], have been shown to provide remarkable performance improvement over routing protocols that do not consider location information [130–132]. These protocols take packet-forwarding decisions based on the location information of the neighboring nodes as well as the location information of the destination node. As a result the geographic routing protocols incur low route discovery overhead which leads to conservation of resources. Further, these protocols also do not require that nodes maintain information on a per-destination basis. The information about the location of the neighbors combined with location information about the destination is typically sufficient to route the packets. Even traditional routing protocols benefit tremendously when provided with location information. For example, the authors in [133] have shown that using location information with the routing protocol DSR improves the data packet delivery by about 40 percent while end-to-end delay is reduced by approximately 20 percent. These percentages increase at higher speeds, achieving data packet delivery improvements of about 130 percent in some cases.

Use of geographical location information has also been proposed for multicast routing in ad hoc networks [134]. Location information is proposed to be used here in order to achieve energy-efficient multicast communication. Another interesting application that makes use of location information is Geographic hash tables (GHT) [135]. GHT is a data centric storage system for sensor networks. The idea behind GHT is to associate data with a key. The keys are hashed into geographic locations. The data associated with the key is then stored at the node that has the smallest distance to the location indicated by the hash of the key. This makes it easy to locate and update the data. The issue of

Security for Wireless Ad Hoc Networks, by Farooq Anjum and Petros Mouchtaris

efficient data aggregation in sensor networks is also proposed to be addressed by making use of location information [136]. The basic idea behind data aggregation is to combine data generated by different sensor nodes. If the location of the sensor node that generated the data is known, then data generated by sensor nodes positioned in the same locality can be naturally aggregated. This works very well when the objective of the sensor network is to obtain information such as temperature and humidity of an area.

Other applications such as target tracking, environment monitoring, and several security mechanisms [137] also depend on location information. In the case of examples such as target tracking and environment monitoring, the data might have to be accompanied by information about the location where the data was generated. On the other hand security mechanisms to protect against wormhole attacks (see Chapter 4) require location information for their basic functioning. Mechanisms to protect against attacks such as Sybil can also be buttressed using the location information to cross-check the other information. In addition, there have recently also been proposals for location-based keying [67]. The idea behind location-based keys is to tie the private key of a node to both its identity and geographic location. This is opposed to tying the key just to the identity as done in conventional schemes. The use of location information for authentication in addition to the use of identity has also been proposed [138]. Thus, users who need access to the network need not only have the proper identity but also be in the proper location.

It is important to point out here that these location-based applications, services, and security mechanisms are sensitive to the accuracy of the location information. For example, the authors in [139, 140] have investigated the effect of localization errors (due to accidental and not malicious reasons) on the performance of both GHT and a geographic MANET routing protocol GPSR. They found that realistic localization errors can lead to incorrect behavior and significant performance degradation. While it may be possible to modify the applications so that their performance is robust when dealing with statistically inaccurate information, it is a different matter when considering location information that is inaccurate due to the presence of an adversary. This is due to the fact that it is not possible to factor the inaccuracy resulting from maliciousness. In such a case, the gains afforded by the use of location information under normal conditions can very easily be erased. It is also trivial to conclude that other applications such as data aggregation and target tracking will also suffer in the presence of maliciously modified location information.

Therefore it is necessary to investigate location determination techniques that have been designed with a focus on security. This is more so considering that the type of networks that we are focusing on, namely ad hoc and sensor networks, naturally attract adversaries given their typical deployment scenarios. *The secure location determination techniques should prevent both malicious insider nodes from misrepresenting their location and outside entities from interfering with the location determination process being followed by the system components.*

Thus the security mechanisms for the location determination schemes must be designed to be robust against two types of attackers. These are the internal and the external attackers. Internal attackers are those insider nodes that can authenticate themselves to the system and are hence part of the system. Such nodes can report false location information or interfere with the localization process with the objective of cheating the location determination system. External attackers are the outside entities that are not part of the system and hence do not know any system secrets. The aim of such attackers is to convince an honest node or the location determination infrastructure or both that the node is present at a location different than its true position. They attempt to do this by modifying the measured positions and distances of nodes.

In this chapter our focus will be on location determination techniques in ad hoc (multi-hop) networks and the mechanisms proposed to secure these. The rest of the chapter is organized as follows. We start off by explaining the various schemes for location determination and pointing out the vulnerabilities in these schemes in Section 7.2. We study the schemes by dividing them into two phases which we explain later; computation phase in Section 7.2.2 and ranging phase in Section 7.2.1. We consider secure localization schemes in Section 7.3. We present a summary in Section 7.4.

7.2 LOCALIZATION

Given the importance of knowing the location of a node in the network, a natural question is whether it would be feasible to determine the location before deployment. If so, the location of the node can be preprogrammed into the node before deployment. However, this will be unlikely in a majority of the cases. Due to random deployment of the nodes in an ad hoc network, the location of each node cannot typically be predetermined. Further, in case of sensor networks, sensors are typically quite light and hence could be easily displaced by various environmental factors such as wind or water. Mobility of these nodes is also another important factor that has to be considered. Thus, in many cases the location of the nodes would have to be determined after their deployment.

The objective of location determination, also called localization in the following, is to determine the location of one or more wireless devices based on measurements or on observations. There have been many localization schemes proposed for wireless networks. These schemes typically require methods adapted to the underlying technology allowable or being used in the network. These schemes typically consist of two phases namely *ranging* and *computation*. During the ranging phase, a node in the network gathers information related to its distances from one or more reference points in the network. The location of the reference point is assumed to be known. The computation phase then consists of the node taking the information gathered during the ranging phase and combining this information with other data such as the known locations of the reference points to compute its location. This phase typically involves mathematical operations.

We next illustrate the above using an example. Consider a network that uses radio frequency (RF) signals. Consider two nodes A and B in such a network such that node B is the reference point. Hence, the location of node B is known. During the ranging phase, we find the distance between these two nodes by determining the time taken by the RF signal (also called the time of flight) to traverse between them and multiplying this by the speed of the RF waves (speed of light). During the computation phase, we then try to estimate the location of node A using the information gathered during the ranging phase in addition to the locations of the reference point(s) such as node B above. Unfortunately, in this case the information gathered during the ranging phase is not sufficient for us to achieve the objective of estimating the location of node A. As a result the computation phase will be incomplete unless we have some additional information. This additional information might be the angle between the two points or the distance of the point in question with two additional points with known locations. We show this in Figure 7.1 by considering just the range information between nodes A and B as well as assuming that the angle of arrival of the signal from node B is known at node A.

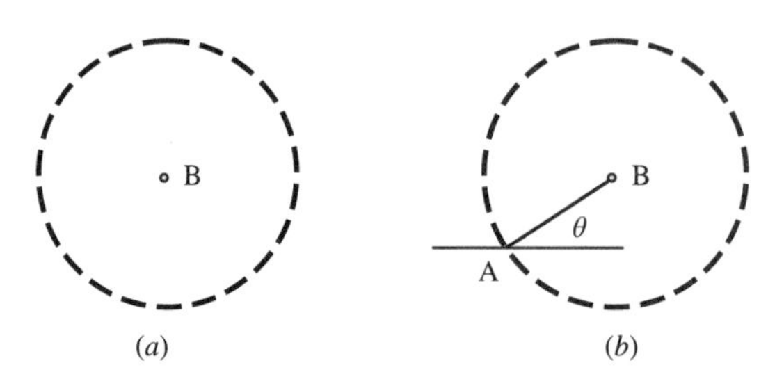

Figure 7.1. Localization example.

Schemes based on ultrasound, infrared, Bluetooth, and 802.11 RF networks using the above approach have also been proposed. Schemes that estimate the location based on measuring a different parameter during the ranging phase have also been proposed. Some such parameters include time difference of arrival, angle of arrival, received signal strength indicator (RSSI) with RF networks, and so on. As we will see later, there are drawbacks with these measurement-based location estimation approaches due to which other schemes based only on observations have also been suggested. Such observation-based location estimation approaches obviate the need for precise measurements. In this case the ranging phase consists of range estimates based on observations.

The information gathering required during the ranging phase of the operation of a location-determination system typically consists of two main components that make for the two-tier network architecture. One of the two tiers forms the reference system and is made from a small number of nodes called the *locators or anchor points*. These are the reference points such as node B that we saw in our earlier example. These anchors are capable of determining their own location either through GPS (Section 7.2.1.1.1) or through manual configuration during deployment. It need not always be true[12] that locators make up the reference system, although in this chapter we do not consider other types of reference systems.

The second of the two tiers is made up of a large number of resource constrained regular devices, also called *system nodes* in this chapter. Locators as well as the more ordinary and numerous system nodes are randomly deployed and communicate in the ad hoc mode. The system nodes infer their location using the spatial relationship (distance between themselves and the anchors) between themselves and the reference points while leveraging the fact that the locations of the reference points are known.

In order to allow the system nodes to infer the spatial relationship, the reference points transmit beacons. The beacons are then used by the system nodes to estimate the spatial relationships (distance between themselves and the anchors) using measurements of basic properties such as time of arrival, time difference of arrival, angle of arrival, received signal strength, or hop based distance. Each system node can then estimate its own location when it has enough measurements from different beacons. The measurements obtained from the beacons corresponding to different locators can be considered as the

[12]For example, the position and orientation of a mobile robot can be obtained from the visual cues obtained from the color cylinders placed in the field of view of the robot [141]. In this case the reference system can be assumed to be that given by the color cylinders.

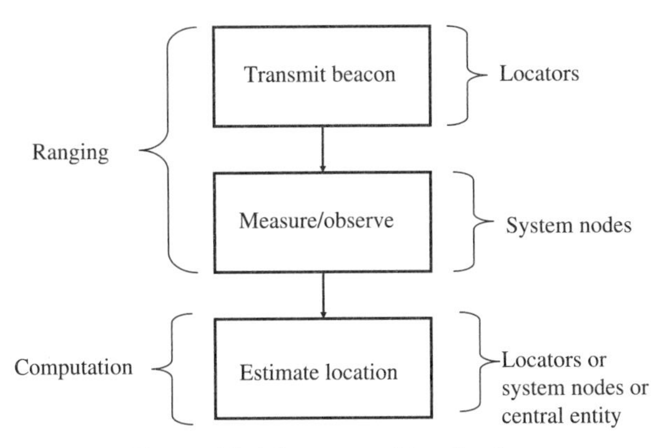

Figure 7.2. The stages of localization.

constraints that the location of the system node must satisfy. Using this approach the estimation of the location would involve obtaining a mathematical solution that satisfies these constraints while minimizing the estimation error.

The information contained in the beacons would depend on the measurements as well as on the method to infer the location from the measurements. Typically the location of the anchor that is the source of the beacon is carried in the beacon. Additional information might also be contained in the beacon, depending on the scheme. For example, information about the signal strength at the transmitter might be contained in the beacons in systems based on the received signal strength method.[13]

An implicit assumption above is that the system nodes would be computing their own locations, but this need not always be so. A complementary system could be designed where the locators or some other special nodes in the system have the responsibility to compute the location of the system nodes. Of course in this case the system nodes would have to either transmit the information that they collected from the beacons or transmit other signals to the locators. We summarize this discussion in Figure 7.2, where we show the various stages involved in determining the location of a system node in a network.

We next briefly explain several different localization schemes. We consider the ranging and computation phases separately. We first consider the ranging phase in Section 7.2.1 and explain the different type of ranging techniques. Following that, we focus on the computation phase in Section 7.2.2 and point out the several algorithms that are typically used in this phase.

We would like to remark here that there has been much of work related to location determination, but the emphasis in a vast majority of these schemes is on determining the location while assuming that none of the users in the system is malicious. Many of these localization schemes can be easily hijacked by an adversary, resulting in incorrect location estimation by the nodes in the network. On account of this as well as the need for location accuracy, secure localization is receiving a lot of attention from researchers. We will discuss secure localization in the next section.

7.2.1 Ranging

We consider two main families of ranging schemes. The first type, the range-dependent scheme, involves measuring some quantities from the beacons. The other type, the

[13]Note that proposals for beaconless schemes also exist [142], but we do not consider those in this chapter.

range-independent scheme, only needs the existence of beacon signals and does not require any measurements to be taken. We look at each of these schemes in Sections 7.2.1.1 and 7.2.1.2, respectively. In these sections we briefly explain several techniques that belong to these families and also point out the vulnerabilities in each of these techniques.

7.2.1.1 Range-Dependent Localization We consider the range-dependent schemes in this section. While protocols based on these techniques produce more accurate location estimations, they also generally result in more expensive systems due to the cost of the hardware needed for the measurements. Range-dependent localization techniques involve measuring the physical properties such as time of arrival (TOA), time difference of arrival (TDOA), received signal strength (RSS), or angle of arrival (AOA). These measurements can then be used to compute the distance to the locators, taking advantage of physical properties. We next consider these schemes in more detail, starting with the widely known global positioning system (GPS).

7.2.1.1.1 GPS The first thing that comes to mind when we talk of location determination is the concept of global positioning systems [143]. This is an attestation to the widespread use of GPS these days. Thus a natural question to ask is the reason for studying the problem of location determination given the concept of GPS. Before addressing this question we look at the concept of GPS briefly and then point out the shortcomings of GPS.

GPS is a constellation of 27 Earth orbiting satellites of which 24 are in operation and three are redundant. The satellite network was developed and deployed by the U.S. military as a navigation system in 1993, but is now freely available. Each of these 3000–4000 lb solar-powered satellites makes two rotations of the Earth every day at about 12,000 miles. The orbits of the satellites are arranged such that at any time four satellites are visible anywhere on Earth. Signals from at least four satellites are collected by a GPS receiver, which determines the distance to each satellite as we explain next and then uses this information to deduce its own location using trilateration (http://electronics.howstuffworks.com). A typical GPS radio signal has a strength of about (1×10^{-16}) W at the Earth's surface. This is roughly equivalent to seeing a 25 W light bulb present in Tokyo from Los Angeles [144]. The GPS system provides a location resolution of approximately 5–6 m on average [144]. GPS also provides devices with accurate time references. LORAN [145] is a similar system but is based on the use of ground-based beacons instead of satellites. We do not consider LORAN here.

We show the operation of GPS in Figure 7.3. Here we show the GPS receiver marked as R, receiving signals from four satellites. From these signals the receiver can determine the range from each of the satellites given as d1, d2, d3, and d4. Additionally, the receiver also needs information about the positions of the four satellites, shown as X1, X2, X3, and X4 in the figure. Thus, a GPS receiver needs to know two things in order to determine the location using trilateration:

- it needs the location of at least three satellites; and
- it needs to know the distance between itself and each of those satellites.

Based on this information, a receiver can draw spheres centered on each of the satellites with the radius of each sphere being equal to the distance between the GPS receiver and the satellite on which the sphere is centered. The spheres will hopefully intersect at only one point on the Earth's surface which corresponds to the location of the GPS receiver.

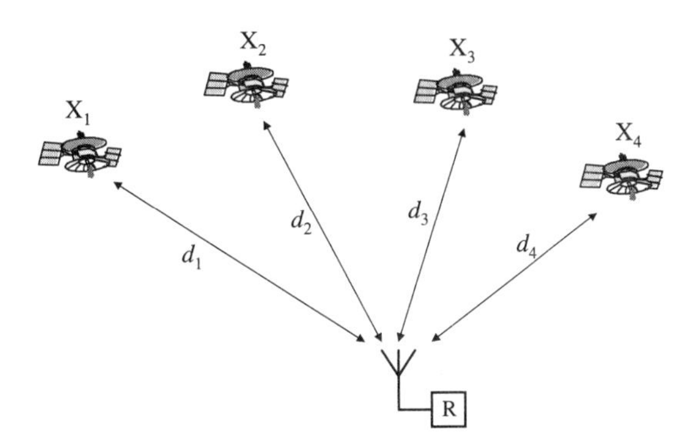

Figure 7.3. GPS operation.

Determining the location of each of the satellites is particularly easy since the satellites travel in precise and predictable orbits. The GPS receiver contains an almanac that contains information about the exact position of each of the satellites. Any change to the orbit of the satellites due to factors such as gravitational pull of the moon is negated by the military, which monitors the exact positions of the satellites and makes any adjustments as necessary. The arriving signals from satellites contain information about the identity of the satellite that originated the signal and then, using the almanac, the GPS receiver can calculate the location from which the signal was created.

Determining the distance between the GPS receiver and a satellite is also theoretically easy, but has some practical problems. The basic idea is to determine the time it takes for the signal to travel between the satellite and the GPS receiver. This is done by ensuring that a satellite begins transmitting a pseudorandom code at a given time. The GPS receiver is also expected to generate the same sequence of bits starting at the exact same given time but without transmitting these bits. Thus, when a satellite's signal reaches the receiver, the sequence of bits received will lag behind the sequence of bits being generated but not transmitted by the receiver. This delay, which can be measured, corresponds to the time of flight of the signal. The distance between the GPS receiver and the satellite can be obtained by multiplying the time of flight by the speed of light. Practical problems arise due to the need for nanosecond precision synchronization needed between the satellites and the GPS receiver. In addition, problems can also arise when the radio signals bounce off large objects such as skyscrapers, or when the speed of the radio signals varies in Earth's atmosphere. Techniques to overcome these have been designed (http://electronics.howstuffworks.com), but we do not consider those here since that is not our focus.

Each GPS satellite broadcasts two signals: a civilian signal and a military signal. The civilian signal, which is intended for normal civilian usage, is neither encrypted nor authenticated. As a result, the civilian GPS signals are not secure. Note that authentication is more important here as compared with encryption. This is because the GPS receiver needs to have guarantees of the source of the signal. The signal itself does not need to be hidden (encrypted) as any GPS receiver is expected to have access to the signal. Relatively unsophisticated spoofing attacks against most civilian GPS receivers, however, can be detected by simple software changes to the receivers [144]. In addition, minor modifications to the satellites can also protect against attacks such as signal synthesis without resorting to the use of keys to authenticate the signal [146]. The idea in [146] is to use a TESLA-like principle of publishing the key used after certain duration. More sophisticated attacks would be difficult to detect though. The military signal, on the other hand, is meant for use by the U.S. Department of Defense and is encrypted using symmetric keys. While plans are underway to upgrade the existing GPS system, they do not include securing the civilian signals by authenticating them.

The GPS system suffers from several problems due to which it cannot always be used for localization. We start off by looking at the problems relating to security. Firstly, the GPS system provides a node only with its own location; a node cannot determine the position of other nodes. In order for a node to determine the position of the other node, it will have to trust the information provided by the other node. This implies that GPS would be useful mainly for cooperative scenarios. The second problem that is more serious is that the current commercial system does not have any built-in security. As a result, GPS receivers can be fooled with fake GPS signals using a GPS satellite simulator, which leads the receivers drawing wrong conclusions about their location. The GPS satellite simulators are legitimately used to test new products and cost about U.S. $10,000–50,000. These satellite simulators produce fake GPS signals that are stronger than the real signals. A GPS receiver accepts these stronger signals while ignoring the weaker, authentic signals. This problem of spoofing only affects the civilian GPS signals. The third problem is jamming of GPS signals, which leads to failure to determine the location. Jamming is less surreptitious than spoofing, since GPS receivers are generally aware of jamming due to the loss of GPS signals [144]. Note, though, that GPS jamming will affect both civilian and military GPS signals.

The next set of problems impacts the working and resource utilization. GPS signals do not work indoors on account of the difficulty of receiving satellite signals indoors. Therefore GPS receivers cannot be used for indoor operation or for determining the location in dense urban environments. In addition, the power consumption of GPS receivers will reduce the battery life. Finally, the GPS receivers also cost tens if not hundreds of dollars and hence cannot be justified in situations where cost is an important factor, such as in sensor networks. Since sensors are intended to be low-cost disposable devices, solutions based on GPS will be impractical given the hardware and power constraints.

In order to address some of these weaknesses in GPS we would need to secure the civilian GPS signals. If this is done following the symmetric key approach, then key management will become a difficult problem. A better solution is to use the asymmetric key approach to authenticate the GPS signals (with each GPS satellite digitally signing its signal while also using mechanisms to provide replay protection). This is the approach expected from the newer version of the military GPS signals. We are not aware of any current effort towards securing the civilian GPS signals.

Note, though, that these solutions to secure the GPS signals are impractical when dealing with localization in cost-sensitive networks, as mentioned earlier. Further scenarios such as indoor locationing would also not benefit from this approach. As explained earlier, this approach would also not allow other nodes in the network or an authority to verify the correctness of a node's position unless a trust relationship exists between these entities. For all these reasons, we need to investigate alternative localization mechanisms.

7.2.1.1.2 Time-of-Flight Techniques The basic concept behind the time-of-flight technique is that there is a direct relationship linking the distance between two points and the time needed for a signal to propagate between these two points. Note that the time needed depends on the propagation speed of the signal. The distance between the points can then be estimated by measuring the time needed and multiplying this by the speed of propagation. This idea is not a new one and dates back to the birth of radar systems.

This is even found in nature, where bats use ultrasonic time-of-flight ranging. The GPS concept that we saw previously also makes use of the time-of-flight technique.

Time-of-flight technique can be used with several underlying technologies such as visible light, infrared (IR), radio frequency (RF), and ultrasound (US). In each of these cases the time needed for the RF or sound signal to propagate between the sender and the receiver is measured. This time is called the time of flight. The distance between the sender and the receiver is then estimated based on the time of flight multiplied by the speed of the corresponding waves.

A potential drawback of these systems is that, since these systems (barring ultrasound based systems) operate at the speed of light, the devices require fast-processing hardware (typically nanosecond precision) for measuring the time. On the other hand, ultrasound-only systems do not need nanosecond precision hardware. However, the US systems can be mainly used indoors. In addition, such signals are harmful to animals.

This technique can be used in two modes, one-way and two-way. In the one-way mode the sender transmits a single bit (signal) while the receiver measures the time at which the bit was received. The receiver can then compute the distance between the two. Note that in this case the sender will not be able to compute the distance unless the sender gets the time-of-flight information from the receiver. Further, this mode also requires that both the sender and the receiver be synchronized in time.

In case of the two-way mode, the sender transmits a single bit (signal) and the receiver is expected to reflect back the bit towards the sender as soon as the receiver receives the bit. The sender thereby measures the time taken for the round trip and uses this to compute the distance between the two. In this case the receiver cannot compute the distance. While the two-way mode does not require that the sender and receiver be synchronized in time, it does require that the receiver be able to reflect back the bit without any delay. In reality, a sequence of bits might be used in both these modes instead of a single bit.

We next analyze this technique for possible ways in which an adversary can misuse such a localization system. We assume that the signals transmitted by the source are neither encrypted nor authenticated. We initially look at systems based on signals that propagate at the speed of light such as RF, infrared, and visible light. We then focus on the ultrasound-based systems. Consider the one-way mode. An important property of the time-of-flight technique is that, if the underlying technology is IR or visible light or RF, then an attacker of either type (internal or external) cannot speed up the signal. As a result an external attacker can only increase the estimated distance between two nodes by jamming and replaying the signal later. Such an attacker will not be able to decrease the estimated distance. An internal attacker other than the sender and the receiver can also achieve the same effect. This will result in an increased time of flight estimate, thereby leading to an enlarged estimated distance. A compromised sender or a receiver on the other hand can lie about the time at which the signal was transmitted or received. Note that a compromised receiver might not always make much sense since the receiver would only be fooling itself by modifying the time at which the signal was received from the sender. In those scenarios,[14] where a compromised sender or receiver exists, the estimated distance can be either larger or smaller than the actual distance.

Now consider the two-way mode. In this case an external attacker can cause both distance reduction and distance enlargement attacks. Distance reduction is possible by having the adversary position a node close to the source, which can respond back faster than the

[14]For example A can modify a GPS receiver and give it to B so as to fool B.

intended destination. Note that this attack is possible due to lack of source authentication of the reflected signal. Distance bounding technique, which we consider in Section 7.3.1, addresses this vulnerability. Distance enlargement can be done as earlier by jamming and replaying the signal later. A compromised receiver without any other colluding node, however, can only increase the distance by sending the response long after receiving the signal. A compromised sender acting alone though can cause both enlargement as well as reduction of the estimated distance.

In addition, when the underlying technology is IR or visible light, then a problem is the need for a clear path between the sender and the receiver. Both these modes are hence sensitive to the topology. An attacker can try to take advantage of multipath in order to bias the estimation of distance. This can be done by placing a barrier sufficiently close to the transmitter, thereby removing the line of sight signal.

We next consider the security vulnerabilities of US systems. US-based systems are susceptible to the attacks described earlier for both the one-way mode and the two-way mode. Additionally, US-based systems also have the following vulnerability. This is based on leveraging the property that US-based systems operate with sound signals which are several orders of magnitude slower than RF signals. As a result, it would be possible for an attacker to hear the US signal at the sender and send it faster towards a colluding node near the destination by using RF signals. The colluding node could then recreate the US signal such that the destination hears this recreated US signal. This would result in the nodes appearing closer (distance reduction) as the estimated distance will be less than the actual distance.

Localization systems can also be designed using the time-of-flight technique and more than one technology. For example, if both US and RF signals are used together, then it would be possible to determine the distance between two devices (sender and receiver) without the need for any time synchronization. This is because of the large difference in the speed of light (3×10^8 m/s) and speed of sound (350 m/s). As a result the RF signal is used for synchronization between the transmitter and the receiver by ignoring the time of flight of this signal.

Thus, in this case the sender sends RF and US signals at nearly the same time and the time at which the RF signal is received is used by the receiver as the reference time. We show this in Figure 7.4. Then the receiver simply measures the difference between the time at which it received the RF and the time at which it received the US signal. This time difference multiplied by the speed of sound gives the distance between the two devices. Thus, we see that the ultrasound signal is used for ranging. This is the approach used in [147]. Using more than one technology though makes it possible for the attacker to launch attacks that exploit the difference in propagation speeds to skew the distance estimates.

7.2.1.1.3 Received Signal Strength Techniques Location systems based on inferring the distance from the signal strength have also been designed. The underlying principle here is that signal strength changes as a function of the distance. This technique can be used to design systems that are hardware-constrained (for example unable to deploy nanosecond precision hardware). In this case either theoretical or empirical models are used to translate signal strength measurements into distance estimates.

Using one approach the distance can be estimated based on the transmitted and received signal strengths at each of the nodes. This could be achieved by having either of nodes report on the transmitted or received power levels to the other node. The distance can

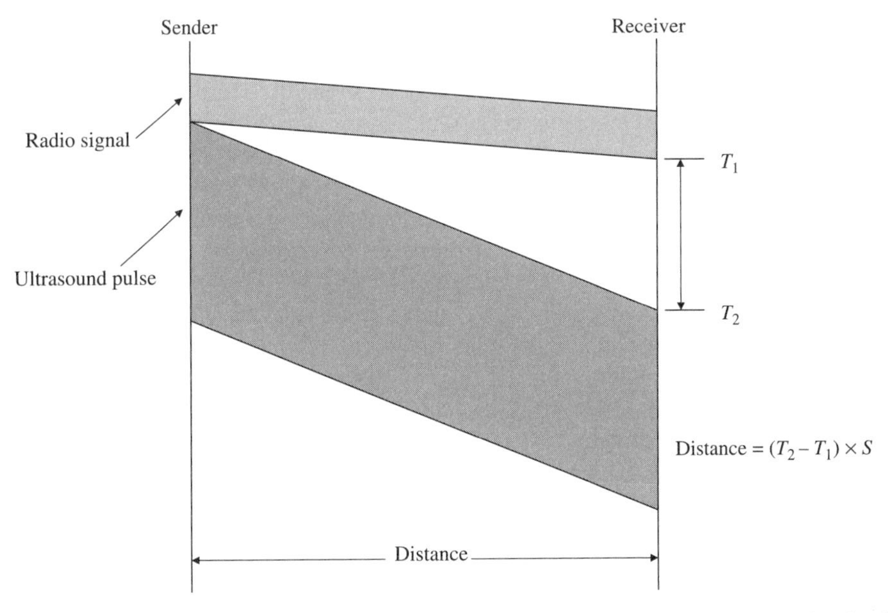

Figure 7.4. Distance estimation using time-of-flight measurements with both ultrasound and radio signals.

then be estimated using the values of these power levels along with a theoretical model. For example, the free space model assumes that signal strength decays inversely to the square of the distance.

The other approach involves the usage of empirical models. In this case the signal strength at various locations is measured in an offline phase and a signal strength (SS) map prepared. During the operation of the system, which corresponds to the online phase, the node that has access to such an SS map can measure the signal strength of the other node. Using the SS map, the node can then estimate the actual location (and not just the distance between them) of the other node by determining the value of the location that has the closest SS measurements. We see from here that this approach requires considerable preplanning. Additionally, both these approaches suffer from inaccuracies due to multipath fading, interference, irregular signal propagation characteristics and so on. An adversary can also make use of these properties to easily fool the system.

Techniques based on received signal strength are cooperative techniques. For example, the first approach above based on the use of theoretical models requires a trust relationship between the two nodes. It will be very easy for either of those nodes to cheat on the measured distance by reporting falsely on the power level. The second approach based on the use of empirical models requires that the nodes do not modify their power levels during transmission. In addition, this technique can also be attacked by an attacker, either external[15] or internal, by walking around in order to create the SS map. Such a map can then be used to fool either the theoretical models or the empirical models by determining the power level at which to transmit.

[15]Assuming that the external attacker has access to the area.

Additionally it would be possible for an attacker (either external or internal) to modify the measured distance by jamming the mutual communication of the nodes and then replaying these signals with higher or lower signal strengths. The adversary can attempt to change the underlying propagation physics, say by introducing an absorbing barrier between the transmitter and the receiver. This would result in signal attenuation, leading to significantly lower received signal strength (RSS). As a result the receiver might conclude that it is further from the transmitter than it actually is. Ambient noise introduced by the adversary as well as changes in the environment would also affect the signal strength measurements. Given the several easy ways in which an adversary can attack a SS based locationing system, the above approaches have not been used so far in the design of secure locationing systems for ad hoc networks.

Another novel localization approach that leverages signal strength measurement has been recently proposed [148]. This is called the radio interferometric positioning system (RIPS). RIPS proposes to use two transmitters which are transmitting at frequencies that are nearly equal. The interference caused by these two signals results in a composite signal that has a low frequency and whose envelope signal can be measured using the RSSI indicator on low precision RF chips. RIPS proposes to then use two receivers to calculate the relative phase offset of the composite signal at the two receivers. This relative phase offset between the two receivers depends only on the four distances between the two transmitters and two receivers. By using relative phase offsets calculated at different frequencies, RIPS provides a method to calculate the relative positions of the transmitters and the receivers. The security aspects of RIPS need to be investigated, however.

7.2.1.1.4 Other Techniques Other techniques can also be designed such as the angle-of-arrival (AoA) techniques or the time-difference-of-arrival (TDoA) techniques. AoA systems estimate the location based on measurements of the angle of arrival. This requires hardware that is capable of measuring the angle at which a signal arrives. APS [149] is a system where the AoA technique is used to measure relative angle between two nodes, which is then used to calculate the distance between them using triangulation.

The TDoA technique consists of determining the times taken by the same signal to reach different points. This is typically made use of in cellular networks where the signal from the mobile device is received at three different stationary base stations, each of which notes the time at which they observe the signal. The time difference in arrival of the signal at the three synchronized base stations is then used to estimate the location of the mobile device. In TDoA systems an internal attacker can send the signals at different times to the three base stations while fooling these base stations into assuming that they are the same signal thereby resulting in false location estimation.

7.2.1.2 Range-Independent Localization Techniques The other category of ranging schemes is the range-independent technique, which is a less accurate but more cost-effective alternative to the range-dependent approach. The characteristic of range-independent localization schemes is that, when using these schemes the system nodes can determine their location without measuring any property related to physical distance such as time, angle, or power. Since the nodes do not have to perform any measurements there is typically no need for any special hardware.[16] We next look at some schemes that belong to this family.

[16]Sometimes directional antennae might be needed to receive/send the beacons.

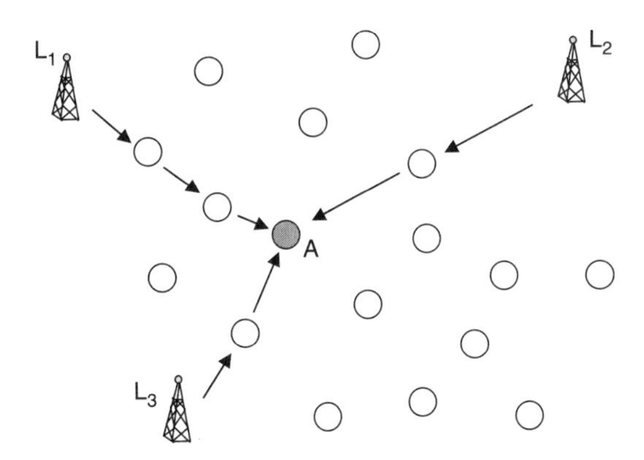

Figure 7.5. Hop count-based scheme. The shaded node determines the hop count from the three anchors and uses this to determine its location.

7.2.1.2.1 Hop Count-Based Schemes Hop count-based localization techniques use a mechanism similar to classical distance vector routing. An illustrative example of this technique is the DV-hop localization scheme [150]. Several anchor nodes are assumed to exist in the network, as shown in Figure 7.5. An anchor node broadcasts a beacon, which will be flooded in the entire network. The beacon contains the location of the anchor node and also has a parameter called the hop count, which is initialized to one. Each node that receives this beacon will copy the value of the hop count from the source of the beacon into its own database while also incrementing the hop count value. The beacon is then further transmitted. Beacons from the same source that are received with a higher hop count value than that maintained by the node are ignored. As a result all the nodes in the network will receive the shortest distance to multiple anchor nodes in terms of number of hops.

In order to convert the hop count into physical distance, the system estimates the average distance per hop. The anchors in the network estimate the average distance per hop by making use of the location and the hop count information for all the anchors inside the network. Using the average distance per hop and the number of hops to a locator, a node can calculate the estimated distance to an anchor. Once a node has the estimated distance from three or more anchors, it uses trilateration (see next section) to estimate its own location. It is obvious here that obtaining the correct hop counts between the system nodes and every anchor node is critical. The more anchors that a node can hear from, the more precise the localization can be. A similar approach is also taken in [151]. The latter proposes a scheme called amorphous localization. In this case each system node uses a mechanism similar to DV-hop to obtain the hop distance. A different approach using mathematical formulae is proposed for estimating the average distance of a single hop.

An adversary (who can be either an external attacker or an internal attacker) can try to either manipulate the hop count measurement or the translation from hop count to physical distance. Note that manipulating the hop count measurement will also result in an incorrect translation from hop count to physical distance. The hop count measurements can be manipulated using physical layer attacks such as increasing/decreasing the transmission power or using network layer attacks such as jamming the area between two nodes (this might result in beacons taking a longer route), forming a wormhole (this will result in

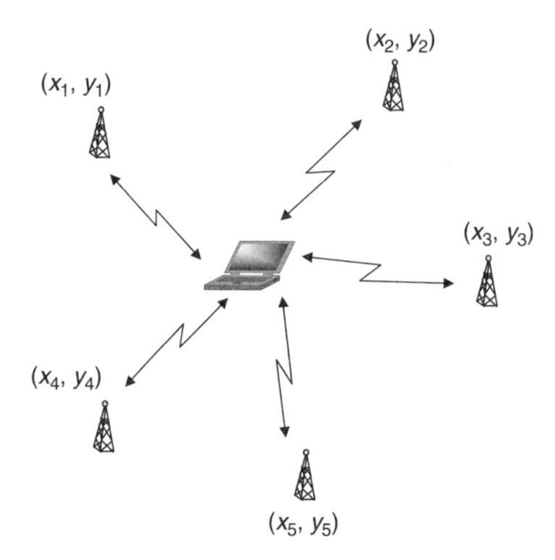

Figure 7.6. Neighbor location where the node estimates its location as the centroid of the locations of anchors that it hears.

shortening the path and hence a much smaller hop count), removing or displacing some nodes (this will modify the path), and so on.

7.2.1.2.2 Neighbor Location This is a range-independent, proximity-based technique based on leveraging the location of neighbors [152]. In this case, given a set of locator nodes, a wireless device can localize by calculating the centroid of the locations of anchors that they hear. The implicit assumption here is that the various nodes are uniformly distributed. The main advantages of this scheme are its simplicity, low overhead and ease of implementation. The drawback is that it can lead to very coarse-grained location determination. The accuracy of localization can be improved using different power levels, thereby changing the transmission range of the locators. This will change the set of locators in the neighborhood of a system node. This approach has been proposed in [153]. Note that an adversary can modify the effective radio region. In addition, the adversary can make use of jammers placed strategically to bias the location estimate.

7.2.1.2.3 Region Inclusion A representative example here is APIT (approximate point in triangle). This is a range-independent scheme [154]. In this case the area is divided into several triangular regions with anchor nodes forming the vertices of the triangles. The anchor nodes transmit beacons. A system node that seeks to determine its location determines the triangles in which the node resides using the transmitted beacons. The test used to determine this is called the APIT test. In this test, a node chooses three anchors from the list of anchors whose beacons it could hear and tests whether it is inside the triangle formed by connecting these three anchors. The test consists of the node checking the signal strengths at all its neighbors of the three locators that form the vertices of the triangle. A node is determined to be further away if its signal strength is smaller. Now in APIT, if no neighbor of the node is further from/closer to all the three anchors simultaneously, then the node assumes that it is inside the triangle. Otherwise, the node assumes that it is outside the triangle. We illustrate the working of APIT in Figure 7.7. In (*a*) node M assumes that it is inside the triangle since none of its neighbors

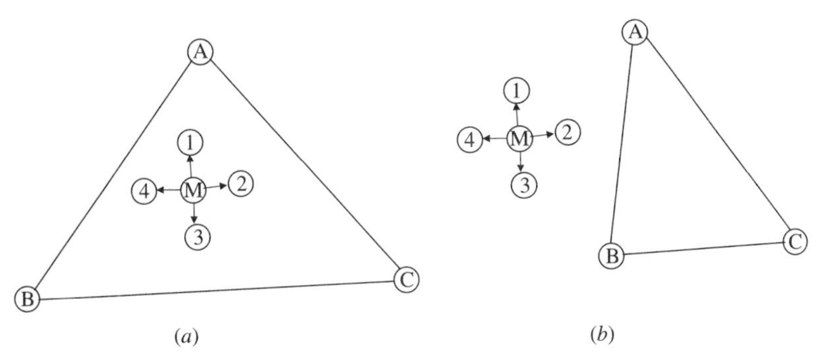

Figure 7.7. Illustration of APIT test.

is closer to or further from all three anchors. In (*b*) node M assumes that it is outside the triangle since its neighbor 2 is closer to all three anchors. Note that this is a range-independent scheme since no correlation is done between the absolute distance and the signal strength.

Note that it is possible for APIT to make an incorrect decision. Some examples are given in Figure 7.8. In (*a*) node M incorrectly assumes that it outside the triangle since it has a neighbor (node 3) that is furthest from all three anchors. In (*b*) node M incorrectly assumes that it is inside the triangle since none of its neighbors is closer or further from all three anchors. The authors in [154] have investigated and report that percentage of such errors is small when considering nonmalicious environments.

This APIT test is repeated with different combinations of anchors until either all the combinations are exhausted or the required accuracy is achieved. The center of gravity of the intersection of all the triangles that a node resides in is assumed to be the estimated location of the node.

7.2.2 Computation

Having described the various ranging techniques we next discuss the different strategies for computation. The first question that arises is about the entity responsible for the computation operation. Either the system nodes or the anchor nodes can perform the computation depending on who has the distance information obtained during the ranging phase. A centralized entity can also be responsible for this operation. There are several ways in

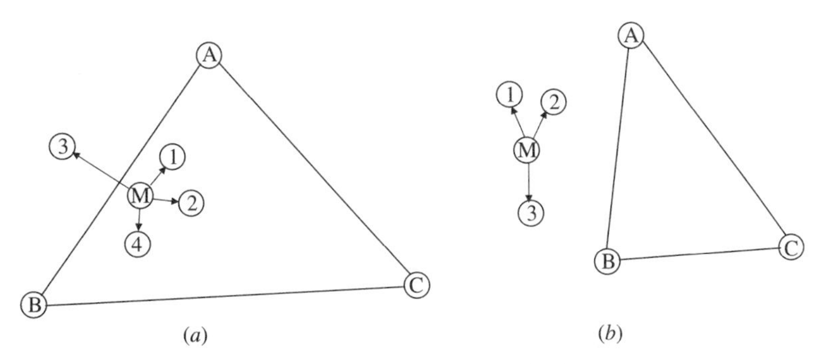

Figure 7.8. Error scenarios for APIT test.

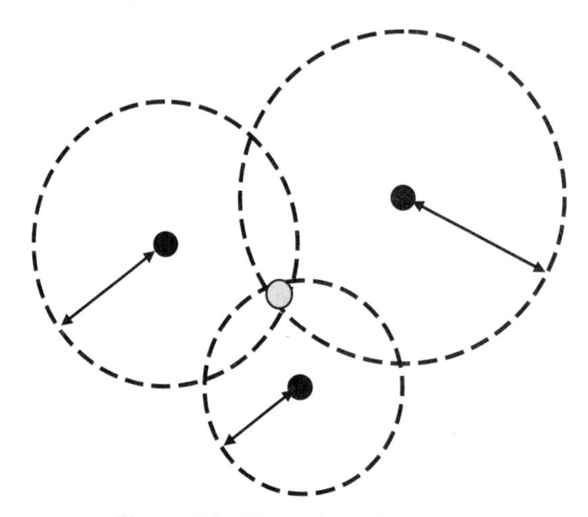

Figure 7.9. Illustrating trilateration.

which the information gathered during the ranging phase along with the location of the anchor points can be used to estimate the location of system nodes in the network. We explain some of these algorithms here.

7.2.2.1 ***Trilateration*** This is the most basic and intuitive method. In this case we assume that the ranging phase has resulted in distances from three noncollinear anchor points. A circle is then drawn with each anchor point as center and the corresponding range as the radius as shown in Figure 7.9. The point of intersection of these circles corresponds to the location of the point in question. It is obvious that in order to determine the position of a device in two dimensions at least three reference points are required. A location in three dimensions requires at least four reference points.

7.2.2.2 ***Triangulation*** In this case we assume that, during the ranging phase, the angles to at least two anchor points have been determined. Following this, the location of the node is calculated using the trigonometric laws of sines and cosines as we show in Figure 7.10. We would also like to remark here that in many cases trilateration is loosely referred to as triangulation.

7.2.2.3 ***Maximum Likelihood*** In the earlier examples, the basic assumption was that the ranging phase produced the minimal number of observations needed to compute the location. Implicitly it was also assumed that these observations are accurate. In reality, this will not be so. We might have more than the minimum number of observations from the ranging phase. For example, this could be because a node obtains range measurements from more than three different anchor points. In such a case we might have an error between the distances measured during the ranging phase and the distances computed during the computation phase.

This set of techniques estimates the location of the node by minimizing the differences between the measured distances and the computed distances. For example the widely used technique here is the minimum mean squared estimate technique (MMSE), which provides a maximum likelihood estimate. The objective of MMSE is to minimize the square of the differences between the measured distances and the computed distances.

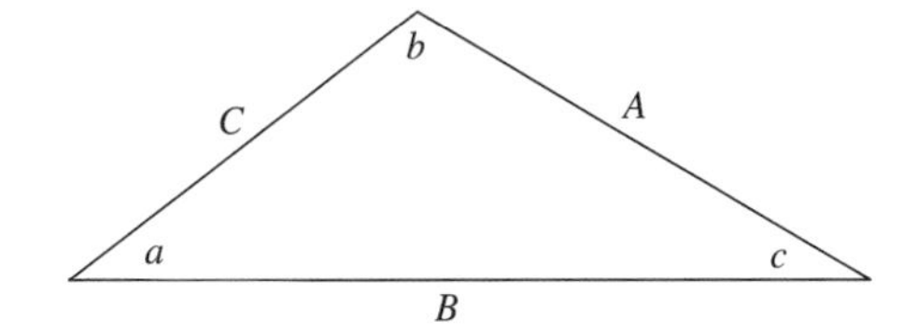

Sines rule $A/\sin(a) = B/\sin(b) = C/\sin(c)$

Cosines rule

$$C^2 = A^2 + \mathrm{B}^2 + 2AB\cos(c)$$
$$B^2 = A^2 + C^2 - 2BC\cos(b)$$
$$A^2 = B^2 + C^2 - 2BC\cos(a)$$

Figure 7.10. Illustrating triangulation.

This case when range information from more than three locations is used for the MMSE algorithm is typically referred to as multilateration. The technique is illustrated in Figure 7.11. In this example, the system node performs ranging with four anchor points. The circles drawn from each anchor point do not intersect on a single point. The location is estimated by minimizing the distance from all four circles. That location is shown in the middle of the four points on each circle. We would also like to remark here that MMSE might also be combined with trilateration or triangulation. This could be needed as the three measurements in either case might not be consistent. This could result in a region estimate for the location instead of a point estimate. MMSE technique could be used to get a point estimate in the estimated region.

7.2.2.4 ***Mapping (RSSI)*** In this approach, the quantities used for ranging, such as time of flight and signal strength, are determined at various locations during an offline phase and a map prepared. During the ranging operation of the real time phase, the system

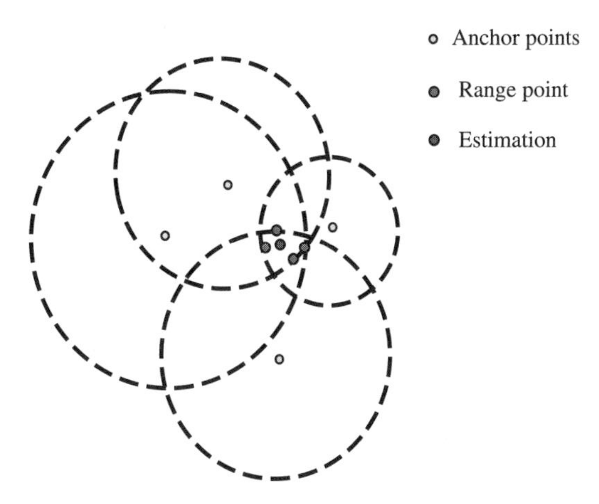

Figure 7.11. Illustration of MMSE technique.

nodes measure the quantities. The measured quantities are then compared with the map prepared during the offline phase. The location that offers the best match is the estimated location. This approach is typically used when using signal strengths to estimate the location.

7.2.2.5 ***Others*** In addition, there are other approaches, chief among them being the probabilistic approaches. In this case, probability distributions are computed and the location is estimated based on the computed distributions. This approach is also typically used in combination with signal strength measurement approaches.

7.2.3 Attacks

We have seen several different techniques for the ranging as well as the computation phases of localization. It is clear at this point that the various ranging schemes have several vulnerabilities associated with them. We summarize the various properties of the different ranging schemes and their associated vulnerabilities in Table 7.1. It is possible to address some of these vulnerabilities through cryptography while the other vulnerabilities would need additional tools. We see from this table that it is possible to modify

TABLE 7.1. Properties of Various Algorithms and Vulnerabilities of These Properties.

Property	Example Algoritms	Attack Threats
One-way RF time-of-flight	GPS	Distance enlargement by delaying transmission of response message, by removing direct path and using multipath, false position reports and position spoofing with GPS. Distance reduction possible in restricted conditions
Two-way RF time-of-flight		Distance enlargement and reduction by removing direct path, by having a different node reply
RF-US ToF	Cricket	Distance enlargement and reduction by exploiting difference in propagation speeds
RSS	RADAR	Distance enlargement and reduction by removing direct path, by changing propagation loss model, by changing ambient channel noise, by changing transmission powers
AoA	APS	Distance enlargement and reduction by removing direct path and using multipath, by changing signal arrival angle
TDoA	AHLos	Distance enlargement and reduction by modifying time synchronization
Hop count-based	DV-Hop	Distance enlargement and reduction by jamming, by forming wormholes, by manipulating the radio range, by physically removing/displacing nodes
Neighbor location	Centroid method	Distance enlargement and reduction by jamming, enlarging radio region, replaying, moving locators
Region inclusion	APIT	Distance enlargement and reduction by forming wormholes, by jamming

(both increase and decrease) the estimated distances in nearly all these techniques. The computation schemes on the other hand can be influenced indirectly by corrupting the ranging schemes, and hence we do not consider them explicitly here.

7.3 SECURE LOCALIZATION

As seen from the previous section, location is an important aspect for several scenarios involving ad hoc or sensor networks. Given that these infrastructureless wireless networks may be deployed in hostile environments, they would be susceptible to a variety of attacks such as wormhole or Sybil in addition to the traditional attacks such as replay and jamming that could significantly impact the accuracy of the localization process. Hence, it is necessary to devise localization techniques that are robust even in the face of such attacks. In this section we investigate how to secure the localization process. We are not concerned with attacks against any other network protocol.

Secure localization is focused on enabling nodes in a sensor or ad hoc network to determine their location securely. The secure localization problem has been solved using one of the following two approaches:

- no specific adversarial model;
- specific adversarial model.

The first approach does not consider any specific adversarial models. Instead the assumption here is that some fraction of localization-related information generated during the ranging phase such as signal strengths or time of flight is corrupted. The corruption can be either due to network faults or to malicious attacks. Typically this approach proposes the use of statistical methods to filter out the corrupted data during the computation phase. The location is then estimated by considering only a subset of data that is statistically expected to be trustworthy. Thus, this approach generally focuses on ensuring the robustness of the computing phase of localization against corrupt data.

The second approach to secure localization is to propose techniques while considering specific adversarial models. This approach involves examining the effects of potential attacks that an adversary can launch against the proposed localization techniques and illustrating the strengths of the proposed techniques to withstand particular attacks. We would like to remark here, though, that for some schemes the distinction between these two approaches is blurry. Such schemes try to ensure that corrupt data generated during the ranging phase is detected and filtered out. Thus, the data that the computation phase makes use of is guaranteed to be authentic with a high degree of probability. Note that use of robust algorithms during the computation phase will strengthen the security of the entire system.

Secure localization consists of several subproblems other than secure location determination. Such problems include location privacy and location reporting. Location privacy aims to maintain the privacy of the location so that the granularity of the location information of a node made available to the various entities in the network can be controlled. The objective of location reporting is to ensure that the nodes report their location information securely. A secure location service may consist of one or more of these components. Note that enabling a node to compute its location securely requires a different set

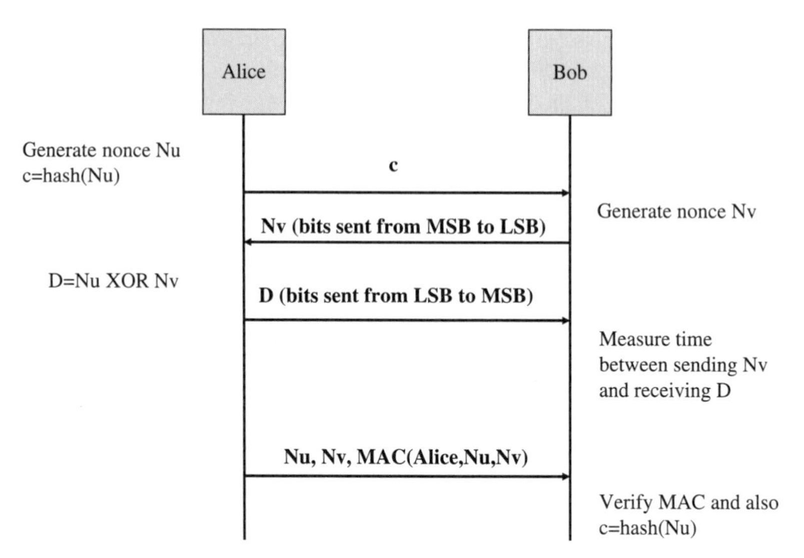

Figure 7.12. Distance bounding protocol.

of techniques as compared with the problem of having the node securely report its location to some other entity inside the network or to the problem of guaranteeing the privacy of the location information of a node.

In the rest of this section we will consider secure location determination. The secure localization techniques are based either on the principle of range dependent localization or on the principle of range independent localization or on a hybrid of the two. We start off by looking at a technique that addresses some (though not all) of the security vulnerabilities present in the time-of-flight technique. Following these techniques we will also study a scheme proposed for location verification. We will not look at location privacy in this chapter.

7.3.1 Distance Bounding Techniques

The distance bounding protocol was first introduced in [155]. This protocol addresses a vulnerability in the two-way mode of the time-of-flight technique. We have seen in Section 7.2.1.1.2 that the two-way mode of the time-of-flight technique is vulnerable to distance reduction attack when an adversary reflects back the sequence of bits transmitted by the first party. This is possible due to the lack of source authentication.

To prevent the above problem, the time-of-flight technique will have to be enhanced to provide for authentication resulting in a protocol exchange such as given in Figure 7.12. In order to explain this protocol we consider two parties, Alice and Bob, who need to determine their mutual distance. Alice is the claimant and Bob is the verifier since Bob wishes to verify or determine the distance (not location) of Alice. In the first step of this protocol, Alice commits to a random value by hashing it.[17] Note that other modes of commitment are also possible. The hashed value is then transmitted by Alice to the verifier, Bob. Bob replies with a challenge nonce transmitted in reverse bit order. As soon as the last bit of the

[17]See Chapter 2 for an explanation of the hashing concept.

challenge has been sent, Bob will start a timer. Alice is expected to XOR the nonce received from Bob with the nonce that Alice committed to in the first step and respond immediately with the result. Bob will stop the timer once he receives this response from Alice and converts the time elapsed to a distance measurement. Finally, in the last step Alice authenticates herself to Bob and reveals the original random number. This message contains a MAC value encrypted using a key shared (using out of band methods previously) between Alice and Bob. The presence of this MAC value provides guarantees to Bob about the identity of Alice. As a result, Bob is sure that the distance just determined or verified indeed corresponds to Alice.[18] Note that Alice is not given any guarantee about the verifier, Bob, since the protocol does not provide for mutual authentication. Thus Alice can be subjected to attack by fake verifiers, giving rise to the topic of privacy, which we ignore in this chapter.

The above protocol is also robust to a malicious third party Carol responding earlier than Bob, as mentioned above. This is achieved by Bob using the value of the original random number as chosen by Alice to check on the veracity of the other steps to conclude that it was indeed Alice that sent the responses. An implicit assumption here is that the key shared between Alice and Bob is not known to Carol. Mutually authenticated distance bounding is provided in [27]. This protocol can be used to prevent attacks by fake verifiers, as mentioned above. Distance bounding is susceptible to attacks such as the jam and replay by a third party, however. Hence, distance enlargement might happen when using distance bounding, although distance reduction attacks are not possible. Furthermore, distance bounding is also susceptible to maliciousness exhibited by either Alice or Bob.

The described protocol assumes that the devices concerned are capable of nanosecond precision message processing (for the initiator or claimant such as Alice above) and time measurement (for the verifier such as Bob above). This dictates the need for dedicated hardware. If the nodes do not possess the capability of nanosecond message processing but are capable of nanosecond precision time measurement, then it might be possible to modify the above protocol to allow the nodes to report on the transmission and reception times. Such a protocol would also be insensitive to the delays associated with channel access. The mutual distance can then be calculated using these time measurements (two on the initiator and two on the verifier). Such an approach called authenticated ranging is proposed in [156]. Authenticated ranging, though, is vulnerable to both distance reduction and distance enlargement.

We would like to remark here that nanosecond processing and time measurements are currently possible only with dedicated hardware. RF time-of-flight systems based on ultra wide band (UWB) technology can achieve nanosecond precision. This can result in determining location of objects to a resolution of up to 15 cm [157]. The system proposed in [157] has a range of 100 m indoors and 2 km in open spaces.

A distance verification scheme that leverages the distance bounding technique has been proposed in [158]. The authors in [158] propose a protocol called the ECHO protocol. The ECHO protocol exploits the distance bounding technique using a combination of RF and US signals. The system is assumed to consist of a verifier and a claimant. The claimant claims to be within a region of interest and the verifier is supposed to verify this claim.

[18] In order to determine the location of Alice, we would need at least three verifiers followed by trilateration or multilateration if the number of verifiers was more than three. Distance bounding combined with multilateration is the focus of verifiable multilateration, which we look at next.

The scheme is simple. The claimant node who claims to be in the region of interest sends this claim to a verifier node. The verifier node rejects the claims if the claimant node does not claim to be in the proper region. Therefore, assuming that the claimant claims to be in the proper region, the verifier transmits a nonce to the claimant using RF signals in order to verify the claim. The claimant is expected to immediately echo the transmitted nonce back to the verifier using ultrasound signals. The verifier node can then calculate from the time taken to receive the nonce back whether the claimant is in the proper region assuming that the region corresponds to a circle. Note that the time taken consists of the time taken for the claimant to receive the RF signal from the verifier and for the verifier to receive the US signal from the claimant. This assumes that the processing time at the claimant in order to receive the RF packet and return the US packet can be ignored.

The verifier can draw a circular region around itself that corresponds to the region of interest and thereby determine the permissible time within which the nonce must be received back. There would be two reasons why the nonce would not be received back in the tolerable time. It could be due to the claimant being present outside this allowable circular region. Such a claimant will not be able to transmit the nonce back such that it is received by the verifier within the acceptable time. Another reason why the nonce might not be received within the allowable time could be due to a processing delay at the claimant. This delay could result in the claimant being falsely inferred to be outside the region of interest. This is a serious problem.

To address this, the authors propose the idea of a buffer region. Let us assume that the allowable response time translates to the fact that the claimant can be in a circle of radius R around the verifier. Now given a zero processing delay, the allowable region could correspond to the circle of radius R centered at the verifier's location, but when the processing delay is nonzero then the verifier node is expected to shrink the allowable region in which claims are accepted. So given a processing delay of d, the allowable region would be the region that remains when an annulus of width ds is removed from a circle of radius R; s here indicates the speed of sound. This smaller allowable region is called the region of acceptance. This is shown in Figure 7.13.

Note that if the claimant tries to cheat by delaying the response then the claimant is only hurting itself. In fact, the longer it takes for the verifier to receive the nonce back, the farther away the claimant is assumed to be. A claimant can also not cheat by starting

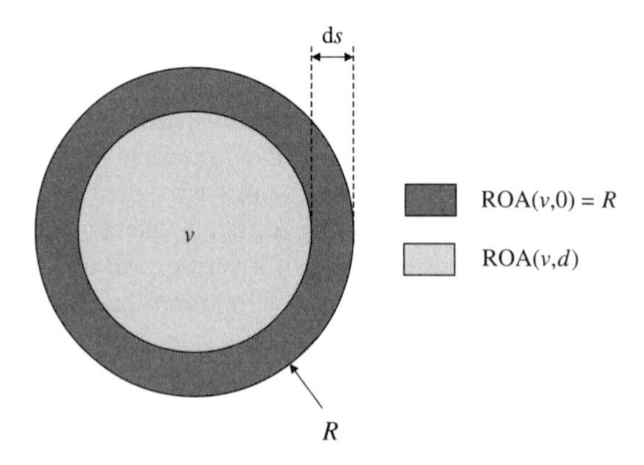

Figure 7.13. Single verifier.

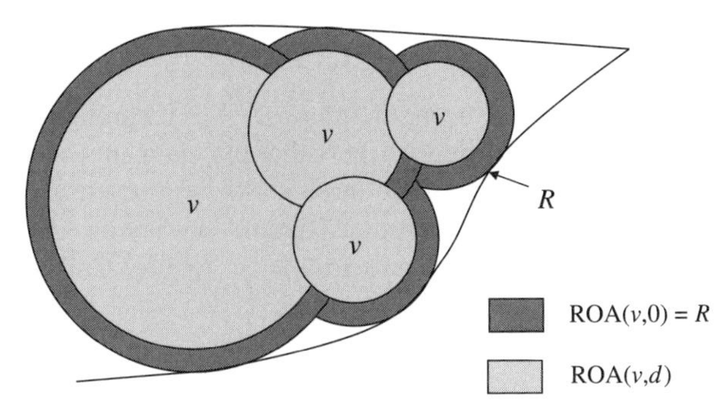

Figure 7.14. Multiple verifiers.

the transmission of the US response early on account of the fact that he has to use the nonce present in the RF signal. This combination of the speed of light and sound ensure the security of the scheme while using low-resolution hardware.

The description so far assumes a region of acceptance that corresponds to a circle of radius *R* around a single verifier. In reality the region of acceptance might be more complex in shape. Therefore, multiple verifiers might be needed in this case. In this case, a verifier is first selected based on the location of the claimant. If no such verifier exists then the protocol is aborted since the claim cannot be verified, but if one or more verifiers exist, then one of them is chosen and the protocol explained earlier is executed to verify if the claimant is in the region of interest. This is shown in Figure 7.14. In order to ensure that at least one verifier exists to verify the claim, it might be necessary to decide on the placement of the verifiers.

It might be possible to consider other variants of the protocol. For example, it might be possible to use RF signals in both directions. This is the approach assumed by [159], but this would require nanosecond precision hardware, and even with such hardware the size of the annulus allowed to address the processing delays would become very large, thereby making it impractical to use such a scheme. Otherwise, one will also have to limit the time it takes to process the signal, thereby requiring nanosecond precision processing hardware. To avoid these problems, it would be necessary to use sound in one direction.

Another scheme based on the verifier using US to transmit the nonce and RF to receive the response could suffer from a remote bugging problem. The authors claim that if sound is used in the outgoing direction then an attacker might be able to bounce a laser off a window with the window being within a range *R* from the verifier. Analyzing the return signal to detect the vibration of the window would allow a sophisticated attacker to determine the nonce transmitted as a US signal. This can be considered to be a form of remote "bugging." Of course, one can argue that a similar attack can also be launched when an RF signal is used to transmit the nonce and a US signal is used to transmit the response. In this case an adversary will have to perform what the authors call "remote actuation" as opposed to the remote "bugging" above. In this case an attacker present far outside the circle of radius *R* around a verifier would have to intercept the nonce transmitted by the verifier from afar, say using a large high-gain antenna, call up some person in the circle of radius *R* around the verifier and convince this person to put the call on speakerphone and transmit the appropriate ultrasonic signal over the telephone. Note that the callee in the circle of radius *R* cannot be considered to be a colluder because in that case all this would not be necessary. The authors argue that this is more difficult as compared with remote "bugging."

Note that the location resolution using the proposed ECHO protocol is coarse. A fine-grained approach is to use the multiple verifiers that the claimants can hear and use the intersection of the regions corresponding to these verifiers to obtain a finer resolution. It would be necessary, though, to ensure that the claimant does not use the lack of time synchronization, if any, between the verifiers to fool the system. Collusion-based attacks might also be easier in such a case.

7.3.2 Verifiable Multilateration

In [156, 160], the authors propose a mechanism called verifiable multilateration (VM) to perform secure location computation and verification. VM can be considered as a range-dependent mechanism. It is based on using multilateration with distance bounding. Distance bounding is used to estimate the range securely while multilateration is used to compute the location from the estimated ranges. Since the two-way mode of the ToF technique is used, there is no need for the base stations to be tightly synchronized with respect to their clocks. Verifiable multilateration can be used for secure location determination in a variety of systems although in [160] the authors have mainly considered the use of VM in sensor networks.

The main idea behind VM is as follows. VM proposes the use of the distance bounding property (Section 7.3.1), due to which the location can only be subjected to a distance enlargement attack (assuming RF signals). Thus, the attacker cannot reduce the measured distance of the claimant to the verifier but can only increase it. VM leverages this property as follows. Consider a claimant and many verifiers. VM tries to ensure that the claimant is inside a triangle formed by three verifiers. Assuming that such a triangle can be found, the claimant performs distance bounding with the three verifiers that form the vertices of the triangle to determine the range from each of the verifiers in the ranging phase. These

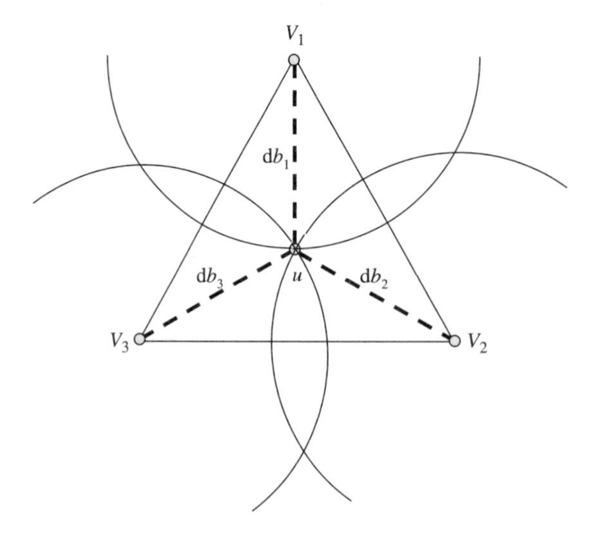

Figure 7.15. Basic VM with three verifiers.

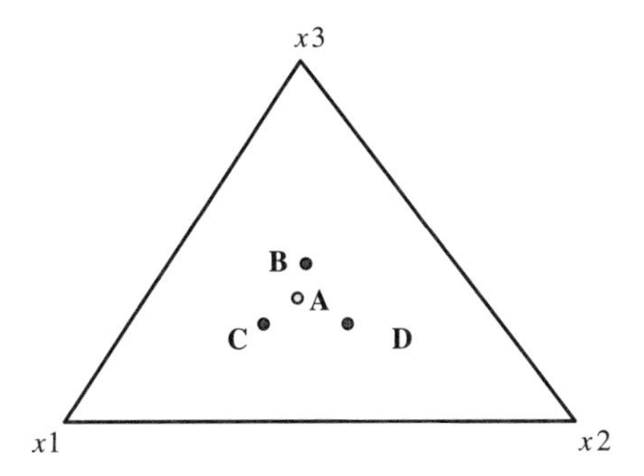

Figure 7.16. Basic concept of VM.

ranges are then used to compute the location of the claimant. This is illustrated in Figure 7.15.

Now consider that the claimant's location has to be faked to another location inside the same triangle. This could be either because the claimant is malicious or because an external attacker desires it. In such a case the claimant's distance to one of the verifiers has to be increased while the range to at least one of the other verifiers has to be reduced in order to keep the position consistent. This is due to the fact that, when an object located within a triangle moves to a different position within the triangle, the object will certainly reduce its distance to at least one of the three vertices of the triangle while possibly increasing the distance to at least one of the other two vertices in order to ensure that the resulting location is consistent. We show this in Figure 7.16. Here a node present at location A desires to appear at one of locations B or C or D, but this requires distance reduction to at least one of the other two verifiers, which is impossible given the basic property of distance bounding. Thus, such an attack can be detected since distance reduction is not possible in this scheme.

The authors in [160] provide an algorithm based on VM. In the first step of the algorithm (which corresponds to the ranging phase), all the verifiers within the RF range of the claimant perform distance bounding to the claimant and obtain the distance bounds. These distance bounds as well as the identities (more precisely the locations) of the verifiers are then reported to a central authority for use in the computation phase in order to determine the location of the verifier. The central authority then computes an appropriate estimate of the claimant's position using this data and a computing technique such as MMSE (that ensures robust position computation).

Finally, in step 3 of the algorithm the central authority executes two tests. The first test checks if the distance between each of the verifiers and the claimant using the computed position differs from the measured distance bound by less than a threshold. This test verifies whether any of the distance bounds were enlarged as compared with the computed location. The second test checks if the calculated location of the claimant falls within at least one physical triangle formed by a triplet of verifiers. If both these tests are true then the authority accepts the computed location as correct. If either of these tests fails, then the computed location is not accepted. If any of the distance bounds are incompatible using the estimated position, then it implies that there might be a possible enlargement attack. The authority can then try to detect the enlarged distances. This might be possible if a large number of verification triangles can be formed around the claimant. When

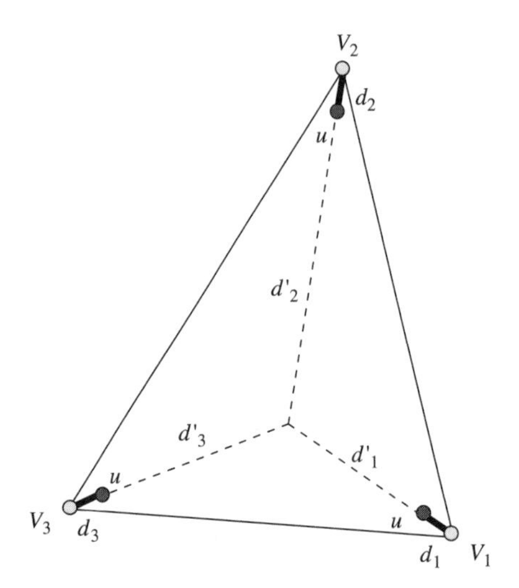

Figure 7.17. Attacks against the VM.

identified, such enlarged distances can be filtered out and the location recomputed using the remaining distance bounds.

It can be seen that VM prevents attackers from spoofing positions of honest system nodes. The VM-based system is also resilient to attacks where the attacker can selectively jam any transmission. Jamming results in the claimants not being able to receive signals used for distance bounding from the verifiers. Such a DOS attack is detectable. Of course, this could result in not being able to determine the location,[19] but would not result in a false location computation.

The VM technique is not completely secure, however. Consider the case when the adversary has control over several devices. The adversary can then place these devices (clones) within the triangle such that each device is close to one of the verifiers, as shown in Figure 7.17. Each of the clones can prove to its corresponding base station that it is at any distance larger than the actual distance, and if the three clones can appear as a single device to the verifiers then the adversary can prove to be at any distance to the verifiers.

Some approaches to solving this problem would be to either make use of tamper-proof claimant devices or to ensure that the adversary would not be able to figure out the actual position of the verifiers. Note, though, that tamper proofing is not a safe solution [90]. Making it difficult for the adversary to figure out the position of the verifiers can be done either by letting the verifiers be mobile or by making use of covert verifiers [161]. Another possibility would be for the verifiers to perform device fingerprinting [162] (to characterize the signal transmission), thereby detecting whenever multiple devices are using the same identity. In addition, this approach is also vulnerable to compromised verifiers.

Given the basic VM idea, the authors then use it to propose a secure range-based location determination system for sensor networks called secure positioning for sensor

[19]This assumes that jamming affects the signals from all the verifiers in the region. If jamming still allows signals from some verifiers, it might be possible to determine the location.

networks (SPINE). SPINE estimates the location of a sensor by verifying the distances of the sensor to at least three reference points. The location estimation is performed at a central location. Once a sensor's location is computed and relayed back to the sensor successfully, it can also become a reference point. SPINE requires deployment of a high number of reference points to achieve localization.

7.3.3 Directional Antennae-Based Schemes

We first look at two related schemes, secure range-independent localization (SeRLoc) [163] and high-resolution range-independent localization (HiRLoc) [164]. Both these schemes require special hardware in the form of directional antennae and yet these are range-independent schemes. We start with SeRLoc. SeRLoc is a decentralized, range-independent localization scheme geared for wireless sensor networks. Localization is achieved passively in SeRLoc since there are no message exchanges required between any pair of nodes. A system using SeRLoc is able to detect attacks such as the wormhole attack, Sybil attack (see Chapter 4), and so on, with a very high probability. The system is not resistant to jamming, however, so an attacker capable of jamming can interfere with the location estimation of any sensor.

The anchor points in a system using SeRLoc possess additional capabilities in the form of several directional antennas. These anchors know their location as well as the orientation of each of their antennas. System nodes are equipped only with omnidirectional antennas. The transmission power of a system node is also assumed to be smaller than the transmission power of an anchor node.

Each anchor transmits information securely in beacons. Each beacon contains information about (a) the coordinates of the anchor and (b) the angles of the antenna boundary lines with respect to a common global axis. Thus the information transmitted by the same anchor point at different antenna sectors varies on account of the different angular information. Each beacon is encrypted using a globally shared symmetric key.[20] Every sensor shares a symmetric pairwise key with every locator in the system. Note that the number of locators in the network is assumed to be quite small.

A system node can possibly receive multiple beacons from several anchors. A node that receives a beacon transmitted at a specific antenna sector of an anchor has to be included in that sector. A node can determine the sector given the range of the anchor, the coordinates of the transmitting anchor as well as the sector boundary information in the form of angles. A node that receives multiple beacons from several anchors determines the sectors associated with each of the beacons. The location of this system node is then determined to be the center of gravity (CoG) of the overlapping region of the different sectors. Note that the CoG is the least square error solution given that a node can lie with equal probability in any point in the overlapping region. See Figure 7.18 for details. Here we see that the system node hears beacons from the anchors L1–L4 and determines its position as the CoG of the overlapping region between the four antenna sectors. The system node estimates its location as the center of gravity of the overlapping region when considering the sectors from the four anchor nodes that it can receive beacons from.

It can be argued that it is computationally expensive for each system node to analytically determine the overlapping region based on intersection of the various sectors. To

[20]Thus, attacks can be launched either by compromised system nodes or by nodes that have access to the shared key. In addition, replay attacks are also possible.

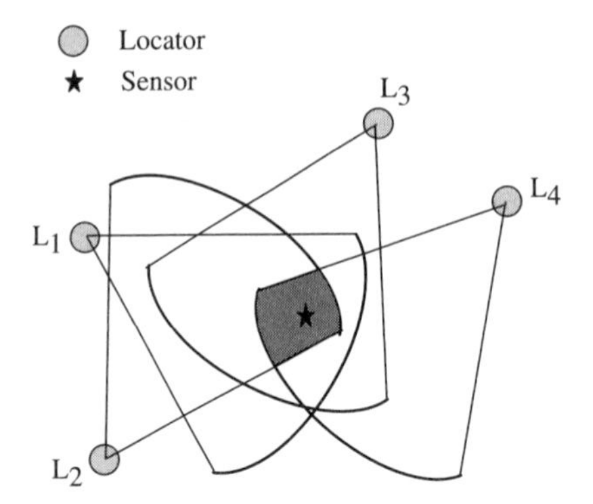

Figure 7.18. Operation of SeRLoc.

address this, the authors have proposed to use a grid scoring system. Here the entire network area is divided into several equally spaced points. Each of the points has an initial score of zero. A system node on receiving a beacon, increments the score of those points that correspond to the sector of the received beacon. This is repeated for each beacon that the sensor receives. The overlapping region at the end corresponds to those points that have the highest score. The location of the node is then determined as earlier to be the centroid of all the grid points that constitute the overlapping region.

SeRLoc can also be used in scenarios where some of the nodes are mobile. In this case both the anchors and the nodes need to update their current location estimation. In order to achieve this the anchors are assumed to acquire their updated position using external means such as GPS signals or GPS-enabled fly-over nodes.[21] The nodes then re-estimate their location based on the new beacons broadcast by the anchors with the new coordinates and sector information.

We next look at some of the threats and explain how SeRLoc addresses these threats. The use of a globally shared key to transmit beacons might result in a malicious node injecting bogus beacons into the network (since the malicious node has knowledge of the global key). To prevent nodes from broadcasting spurious beacons, sensors authenticate the source of the beacons using the one-way property of hash functions (discussed in Chapter 2). In this case, each anchor has a unique password that is hashed n times. The anchor contains the value of all the n hash values. On the other hand, each sensor contains a table with the id of each anchor and the corresponding final (nth) hash value. Now, a jth beacon broadcast from the anchor includes the jth hash value. The system node can verify the authenticity of this beacons by hashing this received value an additional $n - j$ times and verifying with the stored value. A drawback here is that the system designer has to decide beforehand about the length of the hash chain.

The wormhole attack (discussed in Chapter 4) is expected to be detected using two properties. The first property is called the single message per sector per locator property while the second property is called the communication range property. The first property states that reception of multiple messages authenticated with the same hash value is due to replay, multipath effects or imperfect sectorization. This assumes that the beacons will not have to be retransmitted. This also requires that beacons be transmitted simultaneously from all the anchors, which might be difficult to ensure in practice. Due to this property, an adversary

[21]Note, though, the problems associated with GPS as discussed earlier.

cannot replay beacons originating from anchors directly heard by a node. We would like to remark here that this property is needed on account of the format of the beacon messages. These beacon messages contain information about the location, angles, and identity of the anchor and are encrypted using the globally shared symmetric key. In addition authenticity of the messages is guaranteed due to the use of the values from the one-way hash chain that the anchor has, as explained previously. Thus, an adversary who has access to the globally shared symmetric key can retransmit the beacon with modified location and angular information. Such an adversary, though, has to use the same hash value due to the one-way property of the hash values and also due to the fact that the adversary does not know the hash chain. It is such attacks that are intended to be detected by using this property.

The second property states that a node cannot hear two anchors that are apart by more than twice their maximum range. In other words, all anchors heard by a node should lie within a circle of radius R (where R is the maximum transmission range of the anchor). Note that the first property is not sufficient to detect wormhole attacks in those cases where the intruder replays the beacons from an anchor that cannot be heard directly by the node due to range limitations. The second property is intended to address those cases. Due to the second property an adversary cannot replay beacons originating from an anchor that is at a distance of more than twice the range from any other anchor that is heard by the node in question.

The two properties are then combined and it is shown that in such a case a wormhole attack can be detected with a probability close to one in nearly all cases. This is because these properties prevent replay of signals from anchors in the system node's neighborhood and also prevent transmission of signals from anchors not in the neighborhood. Thus, this minimizes the main cause of wormholes. A problem, though, when the node detects that it is being subjected to a wormhole attack is to distinguish between the honest and corrupt beacons. This differentiation is necessary to allow the sensor to locate itself correctly by eliminating the malicious beacons. The detection is expected to be addressed by using an algorithm called *attach to closer locator algorithm* (ACLA). The basic idea of ACLA is for the node to broadcast a randomly generated nonce and its own id. Every anchor that hears this broadcast replies with a beacon that includes the localization information and the nonce encrypted using the pairwise key shared between the node and the anchor transmitting the beacon. The node then identifies the first anchor whose reply it receives back and determines the set of anchors whose sectors overlap with this first anchor. The location of the node is then determined based on this set of anchors. The assumption here is that the closest anchor to the node always replies first while replies that use the wormhole will be delayed.

Sybil attack (discussed in Chapter 4) is also addressed by SeRLoc. In this case the adversary might be more interested in assuming identities of anchors rather than the identities of system nodes. Since system nodes do not rely on other system nodes to compute their location, an attacker will have no incentive to assume node identities. To assume the identities of the anchors, an adversary will have to compromise the globally shared key. The mechanism to protect against the Sybil attack relies on the fact that, when nodes are randomly deployed, every node will be within range of the same average number of anchors. This implies that a node under the Sybil attack will hear an unusually high number of anchors. Hence, there exists a threshold value which is the maximum allowable number of anchors that can be heard by each node. This threshold value can be determined based on the knowledge of the anchor distribution. When a node hears more than this threshold number of anchors, it assumes that it is under the Sybil attack and then executes the ACLA in order to determine the valid set of anchors from which to infer its location.

The authors have also investigated the performance of SeRLoc and have shown that it has higher accuracy than other range-independent localization schemes proposed in [151, 152, 154, 165]. We would like to remark here that the authors in [163] basically provide assurances about the robustness of SeRLoc against various attacks. The guarantees provided are probabilistic in nature and this would typically be the approach that needs to be taken for several problems in this area. Providing absolute security guarantees might be very costly and hence impractical in many cases.

SeRLoc mainly uses the communication range constraint property of the physical medium to allow nodes to determine their location even in the presence of adversaries. Secure localization is achieved by sensors relying on localization information transmitted from anchors (functioning as reference points) with known location and orientation, but in order to increase the localization accuracy, one would need to deploy more locators or use more directional antennas on each locator. High-resolution range-independent localization (HiRLoc) [164] addresses these weaknesses in SeRLoc.

HiRLoc is very similar to SeRLoc. The differences are that the anchors are assumed to be capable of varying their transmission range from zero to the maximum value R using power control. The anchors are also assumed to have the capability to change their antenna direction either by changing their own orientation or by rotating the directional antennas. HiRLoc achieves improved location resolution compared with SeRLoc at the cost of increased computational complexity and communication.

As in the case of SeRLoc, in order to determine their location system nodes rely on beacon information transmitted from the anchors. The beacons are transmitted securely and contain information about the anchor's coordinates, angles of the sector boundary lines and in addition the communication range of the anchor. The anchors are assumed to have the capability to change their orientation over time. On changing the orientation the anchors retransmit beacons in order to improve the accuracy of the location estimate.

A system node determines the sector associated with each received beacon. Note that a system node can hear beacons from multiple anchors as also multiple beacons from the same anchor. The latter is not the case with SeRLoc. The location of the system node is then the region of intersection (ROI) of all the sectors. The ROI can be calculated after every round of beacon transmissions. In order to increase the localization accuracy, it would be necessary to reduce the size of the ROI. This can be achieved either by reducing the size of the sector areas or by increasing the number of intersecting sectors.

Sensors using SeRLoc compute their location by using only one beacon transmission from each anchor. Subsequent rounds of transmissions in SeRLoc contain identical sector information as the first round of transmission. Hence reduction of ROI in the case of SeRLoc can be achieved by either increasing the anchor density or by using narrower antenna sectors to reduce the size of the sectors, but this involves either increasing the number of devices with special capabilities (anchors) or using more complex hardware at each anchor (which allows for more antenna sectors); both these choices increase the costs involved. HiRLoc uses a different approach which takes advantage of the temporal dimension in order to reduce the ROI. The anchors, either by varying the antenna direction or by varying the communication range, provide different localization information at consecutive beacon transmissions. Variation of communication range can be done by decreasing the transmission power. Note that the beacons need to contain information about the range being used. Figure 7.19 illustrates how either of these mechanisms is effective at reducing the ROI. Figure 7.19(a) and (b) show how the ROI can be reduced

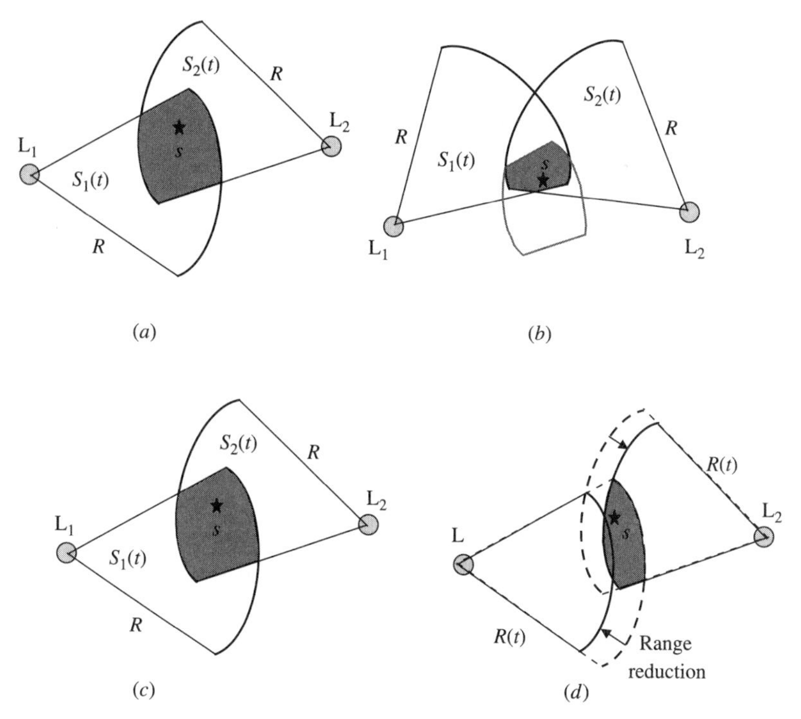

Figure 7.19. Operation of HiRLoc.

by the rotation of the antennas by some angle. Figure 7.19(c) and (d) show how the ROI can be reduced by reduction of the transmission range of the locators.

Once information has been collected in the ranging phase, it then has to be leveraged during the computing phase in order to estimate the location of the node. In [164] the authors also consider two different ways of computing the ROI using the information from the beacons received. The first method, called HiRLoc-I, involves collecting the beacons over several transmission rounds and computing the intersection of the corresponding sectors. The second method, called HiRLoc-II, involves estimating the ROI after each round of transmission. The ROI in the successor round is obtained as the intersection of the ROI of the previous round with the sectors corresponding to the beacons received in the successor round. Both these methods, although resulting in the same estimate of ROI once all transmission rounds have been completed, exhibit different properties. HiRLoc-I makes use of a smaller number of sectors compared with HiRLoc-II. This is because use of the communication range variation results in several beacons being discarded. Note that only the beacon corresponding to the smallest range that the node can receive is included. HiRLoc-II has an extra computational step on account of the need to determine the intersection of the ROI with the previous estimate at each transmission round. At the same time HiRLoc-II provides the node with an estimate of its location at any time instead of having to wait for several transmission rounds, as in the case of HiRLoc-I. As in the case of SeRLoc, each node uses a majority vote-based scheme in order to reduce the computational complexity associated with calculating the ROI.

The defenses against attacks, such as wormhole, Sybil, and impersonation, are all based on similar mechanisms as in the case of SeRLoc, and hence we do not explain those here.

In case of both SeRLoc as well as HiRLoc, an attacker can try to compromise a threshold number of anchors. As a result an attacker can successfully cause faulty estimation of the locations of the system nodes. In addition an attacker can also interfere with the localization process in both SeRLoc and HiRLoc by selectively jamming transmissions of locators. Jamming resistance has been added to HiRLoc and the resulting proposal is explained in Section 7.3.5. Jamming resistance is achieved by using more complex hardware, which is capable of nanosecond precision.

7.3.4 Transmission Range Variation-Based Schemes

We next discuss a range-independent scheme proposed in [166]. This is a centralized scheme where a central entity in the network determines the location of the various system nodes using the information in the beacons transmitted by the anchors. The most distinguishing characteristic of this scheme as compared with VM, SeRLoc, and HiRLoc is the absence of requirement for any special hardware.

This scheme exploits the property due to which the anchors in an ad hoc network are able to transmit at different power levels. For example, the output power range of mica2 sensor nodes varies from -20 to 10 dBm and can be controlled via software. Similarly the WINS sensor nodes [167] can transmit at 15 distinct power levels ranging from -9.3 to 15.6 dBm (0.12–36.31 mW). Nodes in an ad hoc network based on the 802.11 technology can also transmit at different power levels. Use of a different power level will result in a different transmission range.

The proposed scheme assumes that each system node is within the maximum transmission range of multiple anchors. Each anchor transmits a beacon at each different power level. Each beacon contains information about the power level, identity of the anchor (source of the beacon), a timestamp, and a nonce (random number). The last two items are used to preserve freshness and prevent replay attacks. Each beacon is encrypted using the key shared between the anchors and the centralized entity.

As a result, every system node will receive multiple beacons from several anchors at any given point in time. Thus, these beacons correspond to a unique set of nonces. This set depends on the power levels that each anchor utilizes and on the distance of the system node to the various anchors. The system node is expected to transmit the encrypted messages in each beacon that it receives to the centralized entity. The location of the node can then be determined by the centralized entity based on the set of messages (which correspond to a unique set of nonces) transmitted back by the node. Note that the presence of multiple anchors (at least three) makes it possible to securely bind the location of a sensor node to a small area.

We illustrate this using an example. Consider an anchor which has three different transmission power levels, as shown in Figure 7.20. At different power levels the transmission range of the corresponding anchor will be limited by the radii R_1–R_3. The corresponding regions are given by circles C_1–C_3 respectively centered on the anchor location. We represent each encrypted beacon by the corresponding nonce in the beacon for ease of explanation. Then each anchor would be broadcasting the nonces N_1–N_3 at the respective power levels. Thus N_1 will be heard only within the circle C_1, while N_3 will be heard at all points within C_3 as seen from the figure. Thus a node close to the anchor would be able to hear all the transmitted nonces from that anchor.

Now consider three anchors denoted as AN_1, AN_2, and AN_3, as shown in Figure 7.21. Let N_{ij} represent the jth nonce from the ith anchor. For example, N_{12} represents the nonce

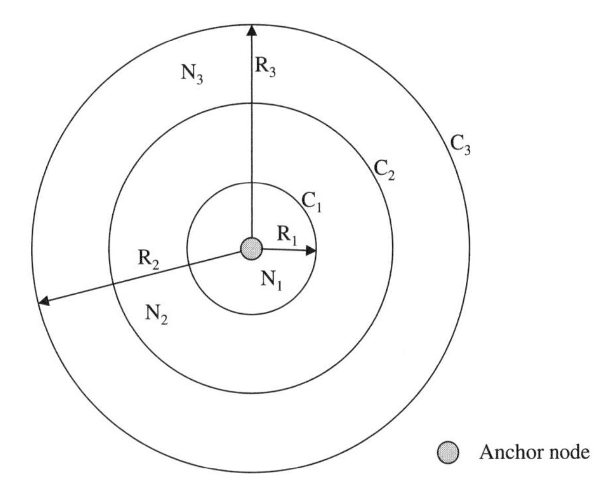

Figure 7.20. An anchor with three power levels.

corresponding to circle C_2 of the first AN. The system node is assumed to be present in the shaded area. Such a node will receive all beacons with nonces corresponding to the set $\{N_{12}, N_{13}, N_{22}, N_{23}, N_{33}\}$. The system node would then respond by retransmitting the encrypted beacons corresponding to this set of nonces to the centralized entity. It can be easily verified from the figure that the set of beacons (and hence nonces) heard at the shaded region is unique to the region. The centralized entity would then be able to determine the location of the system node by decrypting the received beacon messages

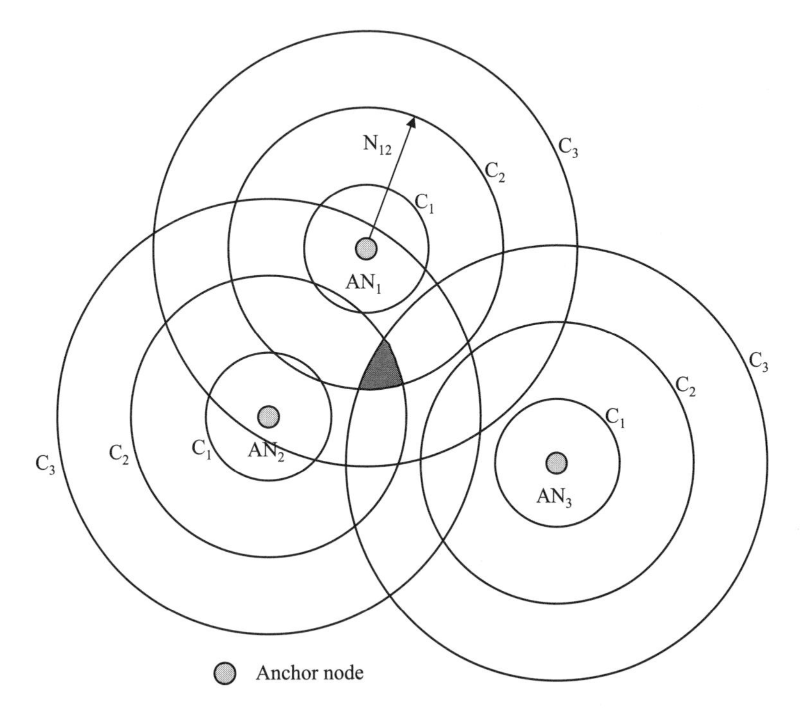

Figure 7.21. Travarsel operation.

and using the decrypted information provided the system node is present in the "triangular" region formed by the three anchors as vertices.

A question at this point though is related to the shape of the transmission range boundary for a given power level transmission by an anchor. We assumed this to be circular in Figure 7.21. To determine this, the authors in [166] carried out an experiment in an open area using a Cisco AP1100 access point as an anchor and a Toshiba PDA as a system node. As the experiment was done using Wi-Fi hardware, the results should be similar when using sensor nodes. The transmitting power of the access point was set to a low power of 1 mW (10 dBm). Figure 7.22 shows the contour of the −90 dBm signal at different angles around the access point. The measurements were obtained for only one quadrant and extrapolated for the rest. The actual measured range is plotted in solid lines in this figure. Using the measured values, the range for the other angles is extrapolated and is shown by dotted lines. We can see from the figure that the transmission range boundary for a 10 dBm power level can be approximated as a circle for open space deployments. We also observe from this figure that for realistic localization, especially in open spaces, power levels far smaller than 10 dBm will have to be used. This can be obtained by controlling the transmission power from sensor nodes such as mica2.

The authors in [166] consider a threat model that consists of a single node having been compromised, thereby ruling out collaborative attacks between nodes. The intention of the

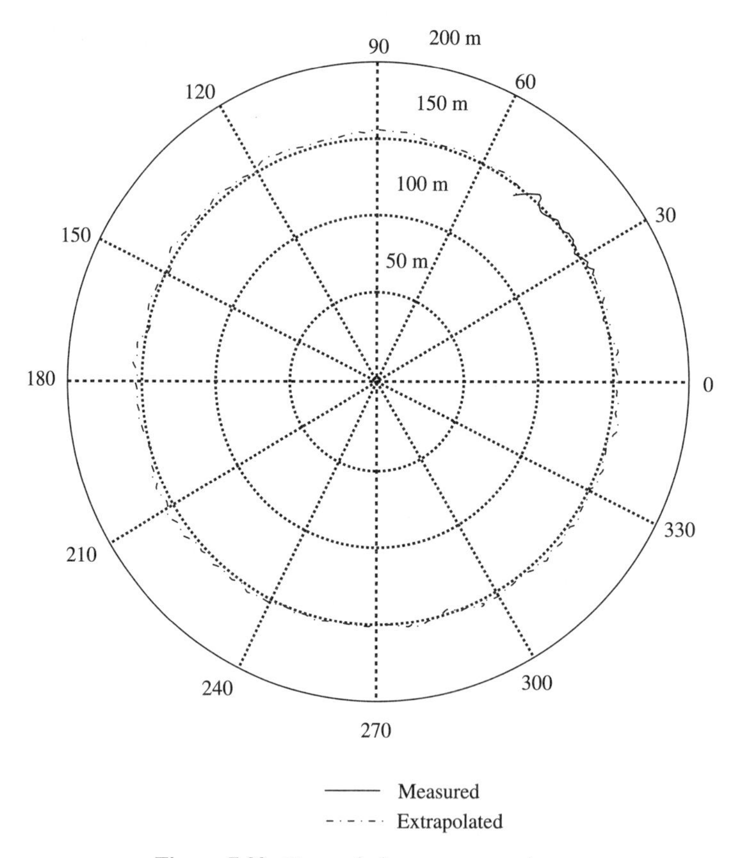

Figure 7.22. Transmission range experiment.

malicious user in this case is to convince the location determination system that the node is present in a region where it physically is not. The intruder is assumed to be limited in its capabilities in that he can only deploy a single node and does not have specialized hardware such as high gain or directional antennae. The system components such as the various anchors and the sink are all assumed to be trusted.

Given such a threat model, it is easy to see that a system based on this idea of varying the transmission powers will be secure when the number of power levels is large enough. In this case any sensor node acting alone cannot falsify its location. When the number of power levels is small (the more realistic case), then a malicious node can appear to be in a neighboring region. For example, consider Figure 7.23. Here a node present in location marked as 1 hears the set of nonces $\{N_{13}, N_{14}, N_{15}, N_{23}, N_{24}, N_{25}, N_{33}, N_{34}, N_{35}\}$ where, as earlier, N_{ij} represent the jth nonce from the ith anchor. Such a node can drop nonce N_{23} from the set of nonces it hears and thereby appear to be in location 2, which corresponds to the nonce set $\{N_{13}, N_{14}, N_{15}, N_{24}, N_{25}, N_{33}, N_{34}, N_{35}\}$. If the node is present in location 2, however, it cannot pretend to be in location 1, although it can pretend to be in location 3 by skipping nonce N_{24}. On the other hand, a node present in location 4 can hear a set of nonces $\{N_{13}, N_{14}, N_{15}, N_{24}, N_{25}, N_{34}, N_{35}\}$ and such a node cannot spoof its location at all. Similarly, a node in location 3 also cannot spoof its location. A localization scheme results in secure localization when applied to such locations, which are the nonspoofable locations. When the number of power levels increases, such nonspoofable locations dominate.

Of course, it would also be possible to increase the number of anchor points, which could also result in increase in the number of nonspoofable areas. Another way of looking at this is to assume that the resolution of localization is larger than each subregion, where a subregion is an area such as 1 in Figure 7.23, which has a unique set of beacons associated with it. Another way to make it difficult for the malicious node to spoof an area by using the above procedure (of ignoring some nonces) is to make it difficult for this node to relate the nonce transmitted to the power level. This could be done by transmitting the nonces randomly and not in the sequence of increasing or decreasing power levels.

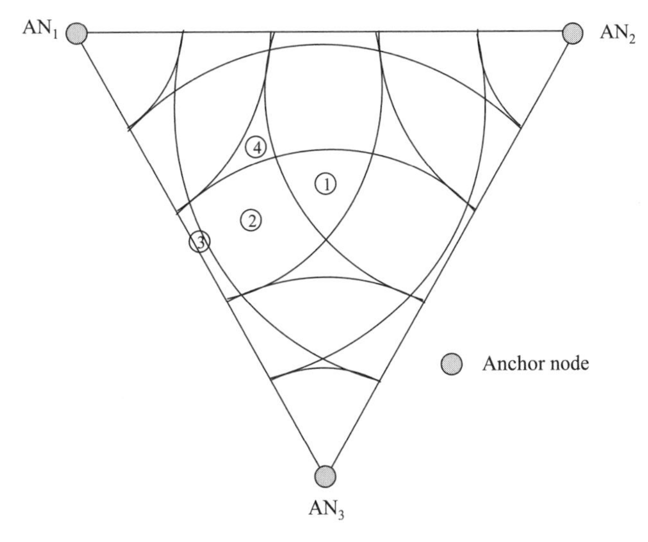

Figure 7.23. Possible spoofing.

Note that, if the malicious node studies and tries to infer the power levels of all the transmissions at different regions, then the probability for the node being able to spoof its location increases.

It should be noted that such a localization scheme based on varying the transmission powers can be used in conjunction with the other conventional distance estimation schemes. For example, the SS of the sensor node transmission may be recorded by the anchors while the collected nonce set is transmitted back to sink by the sensor node. An SS-based localization scheme may then be used in addition to the proposed scheme in order to enhance the distance estimation.

7.3.5 Hybrid Schemes

We next look at a scheme that uses a combination of both range-dependent and range-independent techniques. This is called robust position estimation (ROPE) [168]. ROPE is a hybrid of the ideas used in SeRLoc and VM, as explained in Sections 7.3.2 and 7.3.3. ROPE provides secure location determination without centralized management and is also not vulnerable to jamming (as opposed to SeRLoc). Further ROPE requires deployment of significantly smaller numbers of anchors. The basic idea behind ROPE is to combat location spoofing by combining the geometric properties of SeRLoc with the distance bounding of VM.

The operation of ROPE is as follows. A system node is assumed to have the capability of nanosecond precision time measurements and processing. As for the other schemes seen earlier, the system consists of two tiers of nodes, nodes and anchors. In ROPE nodes initialize the localization process. A node that wishes to determine its location broadcasts its identity along with a random nonce. Any anchor that can communicate with the node performs distance bounding with it. If the node obtains a list of three or more anchors, it checks to see if it can perform VM with these anchors. VM can only be performed if the node lies inside a triangle formed by the three anchors. If VM is possible then the node computes its location and informs all the anchors that responded to its query earlier that the location has been estimated using a "location computed" message. This communication of the node with the anchors is encrypted using a pairwise key. Following this the algorithm is terminated, but if VM is not possible then the node does not send any "location computed" messages to the anchors. In this case the anchors transmit beacons with information similar to that proposed in SeRLoc. As earlier, computationally efficient cryptographic primitives are used to secure the beacon transmissions. The system node then collects all valid beacons transmitted by all locators within range and calculates the ROI from the information in these beacons in a manner similar to that done by SeRLoc or HiRLoc. We show the operation of ROPE in Figure 7.24. In (*a*) the system node performs distance bounding with the two locators it can hear. VM cannot be performed in this case. In (*b*) the system node computes the ROI of the beacons from the four locators it can hear.

The feature of ROPE whereby the nodes initiate the localization process allows for mobility to be easily incorporated. This is because the nodes might request new localization when they believe that their position is outdated. As a result, the anchors would not have to send out the beacons periodically. Note, though, that because the anchors transmit individual beacons for each sensor node, this approach would not be scalable when many nodes have to be localized.

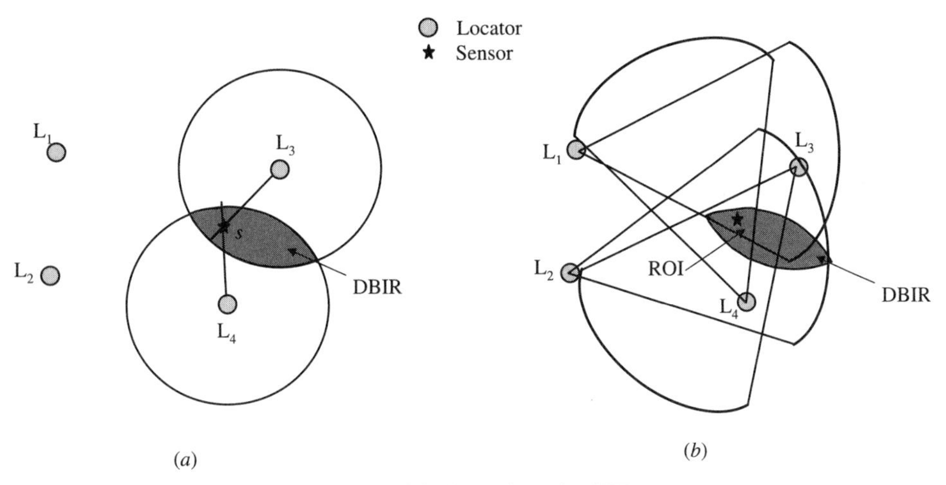

Figure 7.24. Operation of ROPE.

7.3.6 Malicious Beacons

We have so far considered the case whereby the anchors are assumed to be valid, but it is possible for an adversary to compromise the anchors, thereby resulting in the beacons carrying false information. Note that replay protection mechanisms will be of no use in this case. It is also possible for an adversary to replay the beacon signals intercepted in other locations. This latter attack has been considered in the form of wormhole attacks earlier. Thus, this results in an incorrect estimation of their location by the system nodes.

An approach to detecting and removing compromised anchor nodes is given in [169]. The basic idea here is to take advantage of the fact that the locations of the anchors are known. These known locations are then used to check violations of any constraints that these locations and measurements derived from the corresponding beacon signals must satisfy. A beacon signal that violates any constraints is then marked as malicious.

As earlier, the system consists of both anchors and system nodes, but anchor nodes in this proposal perform two functions. The first function, as earlier, is to transmit beacon signals that will be used by the system nodes to infer their location. The second function is to determine the veracity of the information transmitted in the beacons by other anchor nodes. An anchor node that wishes to determine the genuineness of the information propagated by another anchor node (called the target) sends a request message to the latter. The target then responds back with a beacon signal that includes the location of the target. The first anchor can then use the property of the beacon signal such as time of flight to estimate the distance between the two. In addition, since the first anchor also knows its own location, it can also calculate the distance between itself and the target. The first anchor then compares the estimated distance and the calculated distance and concludes that the target is malicious if the difference between these two quantities is larger than a threshold.

A problem arises if the target can infer that it is being queried by another anchor node for the purpose of verifying the target's information. In such a case a malicious target might transmit the correct information in a normal beacon signal to the first anchor. To prevent this, the authors propose that every anchor use an identity which corresponds to a system node when performing the second function. Of course, in this case it is possible for a malicious anchor to determine the location of the request coming in (by using techniques such as time-of-flight) and send out the normal beacon if the location corresponds to an anchor node. This also assumes that a malicious anchor node is aware of the locations of other anchor nodes. The authors in this work ignore this, though, by assuming that a malicious anchor node would not be able to differentiate between the requests coming from an anchor node or a nonanchor node.

Another approach for dealing with malicious anchor nodes is given in [170]. They present two different ideas to deal with this problem. The first is based on the observation that every node typically uses the mean square error calculation to estimate its location. The outliers can then be filtered using the calculated error. To exploit this observation a node first determines its location using the MMSE-based method and information from all the anchors it hears from. The node then assesses if the estimated location is consistent with the information transmitted by all the anchors. If the estimated distance is consistent then the process terminates, else the node attempts to identify and remove the most "inconsistent" reference and repeats the process. The process continues until either a set of consistent location references is obtained or a conclusion is reached that such a set is impossible to find (since the number of location references is less than three).

Given a set L of anchors and a threshold to be used to determine inconsistency, a naive approach to computing the largest set of consistent location references is to check all subsets of L starting with the entire set. This could be done until a subset of L is found that is consistent or it can be concluded that it is not possible to find such a subset. As is obvious, this will be computationally inefficient, which is a big disadvantage given the resource constraints on nodes in ad hoc networks. To address this, the authors in [170] also propose a greedy algorithm which works in rounds. In the first round, the set of all location references is used to verify that they are consistent. If these are consistent then the algorithm outputs the estimated location and terminates. Otherwise, the algorithm considers all subsets of location references with one fewer location reference than in the previous round and chooses the subset with the least mean square error as the input to the next round. The algorithm continues until a subset is found that is consistent or it can be concluded that it is not possible to find such a subset.

The second idea [170] is to use an iteratively refined voting scheme in order to tolerate the malicious location information. The target field is divided into a grid of cells. Each node determines the likelihood of being present in each cell using the information in the beacons received by it. Each cell in which the sensor node can be present is given a "vote." Finally, the cell(s) with the highest votes is selected and the center of this cell(s) is the estimated location. Note that both these ideas assume that the number of benign beacons is more than that of the malicious beacons. Thus, in order to defeat these schemes, an attacker will have to ensure that the number of malicious beacon signals is more than the number of benign beacon signals. Another way for the adversary to defeat these schemes is by ensuring that he is not too aggressive. Note that this might result in some falsification of the location determination, which might not be significant.

Another approach that does not consider any specific attacks but rather focuses on ensuring that statistical robustness is introduced in the computation phase of the localization process is taken in [171]. The advantage of this approach is that it achieves robustness against novel and unforeseen attacks. The authors illustrate here that the impact of outliers can be limited by employing a least median squares (LMS) technique. Note that the authors here exploit the basic property that MMSE estimates are less robust in the face of malicious behavior as compared with minimum median square estimates. As for the earlier case, an implicit assumption here is that the majority of observations collected by the sensor nodes from the beacon signals are benign.

7.4 SUMMARY

In this chapter, we have considered secure location determination in ad hoc networks. We started the chapter by discussing some basic techniques proposed for location determination and pointed out the security vulnerabilities associated with these techniques. Following that we explained several proposals for secure location determination. Several open problems exist though. The issue of cost and efficiency requires more investigation. Approaches that combine several techniques to result in robust secure localization also need to be examined in more detail. Experimental results are also required in order to understand the actual working of several of these proposals.

8 Conclusions and Future Research

Our focus in this book has been on security in ad hoc networks. As discussed earlier, security of a system can be addressed in a multipronged fashion, the three prongs being prevention, detection, and response. We started out by considering the approach of prevention in ad hoc networks. This was in the form of the problem of key management. Following that we explained the problem of secure routing in ad hoc networks. In both these cases, we explained the reasons for adopting a different approach in ad hoc networks for solving these problems. This is as compared with similar problems in fixed networks. We then focused on the detection and recovery aspects in the chapters on intrusion detection and policy management. The important problem of location discovery in a secure manner was also addressed.

We would like to point out here that we have only addressed the tip of the iceberg in this book by considering the basic problems. Several advanced problems such as secure group communication, secure time synchronization, secure MAC protocols, interplay of security and performance, trust establishment, management, privacy and so on have not been addressed in this book. There are many open problems still in all these areas. In addition, we have also not considered in detail mechanisms based on the use of advanced cryptographic primitives such as secret sharing, threshold cryptography, and identity-based systems.

The basic security problems in MANETs have also been explained in the context of ad hoc and sensor networks. Another important application of MANETs which will impact the commercial world is expected to be in the area of vehicular networks. We expect this to be a very important future research area. Such networks have been garnering a lot of interest recently and are expected to be the subject of increasing attention in the future. These networks that allow communication between vehicles as well as between vehicles and the infrastructure are mainly intended to improve traffic safety. It is believed that nearly 50 percent of the nearly 43,000 deaths that occur annually on U.S. highways are caused by roadway departure and intersection incidents. Similar numbers are also reported for the number of people killed in Europe (http://europa.eu.int/comm/transport/care/). The annual costs associated with traffic accidents in the United States, including hospital bills and property damage, have been estimated to total about U.S.$1 trillion [172]. The number of automobiles in the world is also increasing very rapidly and is disproportionate, especially in some of the developed countries. For example, there are about 240 million cars in the United States for a population of about 300 million.[22] The numbers for Europe are 230 and 740 million, respectively, while for Japan they are 73

[22]Organisation internationale des constructeurs of Automobiles (OICA) statistics 2003.

and 130 million, respectively. All this leads to frequent traffic jams, pollution, scarce parking spaces, and so on.

Communication between vehicles can be expected to lead to a significant improvement in the safety-related statistics.[23] Hence, there have been several initiatives to make this possible. The Federal Communications Commission (FCC) of the United States in 1999 allocated a block of spectrum in the 5.850–5.925 GHz spectrum that is expected to be used to enhance the safety and efficiency of the highway system. The most prominent industrial effort in this domain is carried out by the Dedicated Short Range Communication (DSRC) consortium, especially the IEEE P1556 Working Group (Security and Privacy of Vehicle and Roadside Communications including Smart Card Communications). In December 2003, the FCC approved 75 MHz of spectrum for DSRC, and the resulting DSRC system is expected to be the first wide-scale vehicular ad hoc network (VANET) in North America.

In Japan, two DSRC standards have been adopted (the ARIB STD-T75 in 2001 and the ARIB STD-T88 in 2004), and Japanese auto manufacturers are working with the Ministry of Land, Infrastructure, and Transportation in the third phase of an ambitious Advanced Safety Vehicle project. The German Ministry of Education and Research has sponsored the Fleetnet and Network on Wheels projects. Throughout the world, there are many national/international projects in government, industry, and academia devoted to vehicular networks.

One of the challenges facing vehicular networks is related to security. Security in such networks also has many challenges relating to privacy, key distribution, attacks, and so on. This is a crucial problem in such networks, especially due to the effects on life and property. For example, an attacker might try to insert or modify life-critical information. In such instances it is vital that the system tries to prevent such an event from occurring and, if this is not successful, then the detection and recovery mechanisms should be strong enough. Little attention has been devoted to this problem yet. We next briefly look at the problems as well as proposed solutions in this area. We expect that much attention will be given to this area by researchers in the future.

8.1 VEHICULAR NETWORKS

A vehicular network can be looked at as a special form of an ad hoc network. The objective of such networks could be manifold. Examples include enhancing the driving experience, increasing safety as well as reducing resource wastage. The characteristics of these networks, however, imply that the research done for ad hoc networks cannot be directly adapted to these special type of networks. For example, the speeds in vehicular networks are expected to be an order of magnitude higher as compared with the speeds studied so far in ad hoc networks. On the other hand the resource constraints such as power, memory, and computation can be considered not to exist in vehicles. Further, other factors such as the size of the network, relevance of the geographic position as well as sporadic connectivity add to the challenges. The proposed solutions will also have to address the deployment problem, where the number of vehicles that can leverage vehicular communications will be minuscule initially. This number will increase over time. All this implies that much research needs to be done either to adapt the present solutions for general ad hoc

[23]This would be in addition to the safety measures mandated currently, such as seat belts and airbags.

networks to vehicular networks or to design completely new solutions tailored towards vehicular networks.

There have recently been many initiatives both in Europe as well as in the United States focused on various aspects of vehicular networks. Some of these are

- Vehicular Infrastructure Integration (VII);
- Global System for Telematics (GST);
- Cooperative Vehicle Infrastructure System (CVIS);
- Car2Car Communication Consortium;
- Network on Wheels (NOW);
- DSRC.

VII is an initiative of the U.S. Department of Transportation (DOT) to build a nationwide vehicle-to-vehicle and vehicle-to-infrastructure communications system. This system is expected to support public and private telematics applications targeted towards saving lives and improving the quality of life. The target deployment of this infrastructure is around 2011. The catalyst for this initiative was the fact that more than half of the deaths occurring annually on U.S. highways are caused by roadway departure and intersection incidents. It is expected that this can be reduced effectively using communications.

Another initiative in this area is GST. This is a project sponsored by the European Union focused on creating an open system for service delivery and execution in various vehicles such as cars, trucks, and trains. The objective of GST is to create an environment in which innovative telematics services can be developed and delivered economically. This is thereby expected to spur the variety of telematics services available to manufacturers and consumers.

CVIS is another initiative of the European Union. As opposed to GST whose main focus is on the environment within the vehicle, CVIS targets the communication between the vehicles and the roadside infrastructure. It seeks to make use of open standards-based communication platforms both in the vehicle and for the fixed roadside infrastructure.

The Car2Car Communication Consortium was founded by six car manufacturers, Audi, BMW, DaimlerChrysler, Fiat, Renault, and Volkswagen. The objective of this group is to create a European standard for car-to-car communication. The consortium is targeting to have completed prototypes participating in field trials by March 2006.

NOW was founded by several automobile manufacturers in combination with other companies. The founding members of this initiative are DaimlerChrysler, BMW, Volkswagen, Fraunhofer Institute for Open Communication Systems, NEC Deutschland GmbH, and Siemens AG. It was started in 2004. In addition, this initiative also partners with several universities. This project is mainly a German research project and supported by the federal government there. The objectives of this group are to focus on the communications protocols and data security for car-to-car communications. NOW is focusing on 802.11 technology and IPv6 to develop "inter-vehicle communication based on ad hoc networking principles".

Another standard in this area being developed primarily by ASTM and IEEE committees is the 5.9 GHz DSRC. DSRC is a short to medium range communications service based on an extension to 802.11 technology that supports both public safety and private operations in roadside-to-vehicle and vehicle-to-vehicle communication (www.leearmstrong.com/DSRC/DSRCHomeset.htm). DSRC provides very high data rates and thereby can be considered as a complement to cellular communications.

8.1.1 Differences with MANET

Vehicular networks typically make use of multihop wireless links as in typical MANETs or sensor networks, but as mentioned earlier they have significant differences with these other types of networks. These differences include [173]

- network dynamics;
- scale;
- resource constraints;
- deployment;
- applications;
- error tolerance;
- security issues.

The network dynamics that characterize vehicular networks (VANETs) are very different from the network dynamics that characterize MANETs and sensor networks. VANETs exhibit speeds which are much higher, typically in the range of tens of miles per hour. This can result in very short connection times between neighboring vehicles. The trajectories of the vehicles are also determined by the roads and are not random. This could have an impact on both message dissemination and privacy.

Another important feature of VANETs would be the number of nodes that constitute the network. Given the hundreds of millions of vehicles, each of which could be a potential member of the network, the size of the network would be huge. A saving grace, though, is the fact that communication in this network would be mainly local with a natural partitioning based on the geographical location of the members. This feature could be made use of in order to design scalable solutions.

Resource constraints are a major factor in sensor networks and also in case of mobile ad hoc networks, but this is not the case in vehicular networks. Vehicles are expected to have sufficient computational and power resources. In fact, each vehicle is expected to contain hundreds of microprocessors. Hence, solutions based on resource-inefficient operations such as asymmetric primitives can also be used in such networks.

Another important feature of VANETs is related to deployment. Considerable penetration of VANETs is expected only around 2014 [174], but the initial deployment of vehicles with the basic capabilities to function as nodes in VANETs is expected in the next couple of years and hence the network should be operational in stages. This implies that the solutions proposed to various problems in such networks must function even with a low number of network nodes. Such solutions can also not assume the presence of a network infrastructure since such an infrastructure might not be deployed in the initial stages.

The types of applications used in such networks are also expected to be different. While the final word on the applications that will become popular in such networks can only be given after such networks are widespread, it is anticipated that safety-related applications will dominate. Other applications related to traffic information, for example, may also become widespread. At this point, however, it seems as though applications intended for human consumption such as streaming audio or video might not be deployed on VANETs. This is not only due to the different characteristics of VANETs, but also due to the fact that alternative networks such as cellular and WiMAX would be available to carry application traffic intended for human consumption. Note that various security attributes might be mandatory in safety-related applications.

Several applications in VANETs would require real-time responses. This applies especially to safety-related applications where a difference of a few milliseconds can make a difference between having a ten car pileup vs a fender bender. Furthermore, the tolerance for errors in such applications might also be very low given the effects on lives that such applications can have. Thus, we can see that the requirements in such networks are very stringent. This implies that these networks would be very sensitive to denial of service attacks.

These networks will have several security problems that would be different than the corresponding problems in ad hoc or sensor networks. Routing in these networks is expected to be based on the location. Hence, secure geographic routing protocols that can leverage the trajectory knowledge would be needed. This also implies that the secure location determination problem would be very important here. The resources devoted to this problem would be much more than are possible in sensor or ad hoc networks. Key management in these networks can also take advantage of factors such as the ability to use asymmetric operations and the existence of motor vehicle administration that can hand out private keys.

8.1.2 Open Problems and Solutions

Vehicular networks will also see similar types of attacks as in ad hoc or sensor networks. Thus we will have the normal classification of insider vs outsider attacks, active vs passive attacks, malicious vs rational attacks, and so on. What is different, though, could be the impact of the attacks. Ad hoc networks have not been adopted widely by the general population, but this will not be the case with vehicular networks, and this is the main reason for the difference in the degree of impact of the attacks. Attacks that can bypass the inbuilt privacy mechanisms can be used to track many people successfully. Combining this with the regulatory and legal mechanisms as well as the desire for privacy by many people can lead to an explosive situation. Delaying the inbuilt mechanisms to prevent collisions by a few seconds can make all the difference between life and death. The resulting lawsuits hold the potential to bankrupt a company.

There are several challenges related to vehicular networks. The spectrum of applications varies from safety to infotainment (information plus entertainment) applications. Design of such applications will indeed be a challenge. The architecture and design of platforms to be implemented within the vehicles to support such a diversity of applications is also a challenge. The security mechanisms to be used in both the applications and the platforms have to be designed keeping in mind the special constraints in these networks.

Another area of investigation is related to the communication protocols as well as the data security algorithms for inter-vehicle as well as vehicle to infrastructure communication. The radio systems to be used also require investigation. Currently many of the efforts assume radio systems based on 802.11 technologies.

Given this, there are several challenges with a direct impact on security that have to be mastered. In the rest of this section we look at some of these problems.

8.1.2.1 Identification Vehicles typically have a permanent identity that corresponds to the vehicular identity number (VIN) and an ephemeral identity that corresponds to the license plate. The latter is normally provided by the registration authorities. In addition, vehicles might be provided with an electronic license plate. The certified identity that a vehicle provides via a wireless link is called an electronic license plate [175]. Electronic license plates can be used by a vehicle to become aware of its surroundings. This can be possible if every vehicle periodically broadcasts a beacon containing information which includes the electronic license plate. Other items included in the beacons could include clock, current speed, and direction. Information provided by such beacons could also be used by an infrastructure to suggest the most efficient routes to a vehicle. Such information could also aid in finding drivers who flee the scene of an accident. Recovery of stolen cars can also be made easier.

However, such capabilities also bring with them vulnerabilities that can be leveraged by the adversary if left unprotected. A person who desires to steal a car can choose to disable the transmission of beacons if this is possible. Impersonation of another vehicle could lead to problems for the genuine owner. Jamming of signals that vehicles exchange could also lead to problems if the system is completely dependent on such information. Another problem is also related to privacy. The information in beacons could also be used to track the location of vehicles. This might be undesirable in situations where the vehicle is not involved in any incident.

An approach to enable privacy in vehicles is to broadcast pseudonyms instead of the actual identities in the beacons. Further, the pseudonyms must change over time. Further, to make it possible to trace the pseudonyms to the actual owner in exceptional situations (such as accident liabilities), the pseudonyms must have been provided by regional or national authorities. This implies that an important research problem is to design privacy-preserving protocols based on the use of temporary pseudonyms which could be traced back if needed.

8.1.2.2 Protection of Messages The messages exchanged in VANETs also need to be protected. The actual objectives, however, would depend on the types of messages. For example, safety messages are not expected to contain any confidential information. Hence, confidentiality would not be needed for such messages. Authentication, though, would be very much required since otherwise an adversary could easily create false messages. The authentication should also make accountability possible. This would make it possible to trace the creator of a message, which would be needed in case of accidents.

8.1.2.3 Key Management Vehicular networks are not expected to face the same resource constraints as faced by ad hoc and sensor networks. Therefore asymmetric key operations are very feasible. A problem would be related to privacy, however. Use of public keys can very easily lead to loss of privacy. This is because typically public keys can be used to infer the actual identity of the person or the vehicle. Hence the concept of anonymous keys has been proposed to address this [173]. An anonymous key pair is a public–private key pair that is authenticated by a certificate authority but does not contain information that allows an observer to discover the actual identity of the vehicle or its owner. The anonymity also needs to be conditional for liability purposes. Thus, this anonymity would not apply when incidents such as accidents occur. Hence the key distribution infrastructure has to be designed such that the information about these keys will only be available say with a court warrant.

The anonymity keys can be looked at as pseudonyms. Use of the same pseudonym all the time can also lead to information that should not be available to the adversary. Therefore, it would also be necessary to consider mechanisms to change the keys at various intervals. This also implies that the vehicle will have to store a large number of keys. Management of this large set is also a challenge.

8.1.2.4 Localization Another important problem here is related to secure determination of the location of a car using GPS or with the help of on-road infrastructure. Techniques for secure location determination as given in Chapter 7 will have to be adapted in order to achieve this. The authors in [175] proposed using either tamper-proof GPS or VM (Chapter 7) to make this possible.

Thus, we see that, while vehicular networks have many similar problems to sensor or ad hoc networks, there are also other problems that are more important in such networks. In addition, the solution strategies also need to be different given the features and constraints in such networks.

8.2 SUMMARY

The area of security in wireless ad hoc networks has been attracting much attention. While many problems have been addressed, there are many others that need attention. Even the problems that have been addressed might have to be revisited in the light of improvements in technology, both wireless and computational. The former makes possible the use of higher bandwidths while the latter makes possible the use of more complex cryptographic mechanisms. Thus, in summary, this is an area rife with open research problems that have to be addressed before we can start seeing the widespread use of such networks.

Acronyms

ACF	Authentication control function
ADVSIG	Advanced signature
AoA	Angle of arrival
AODV	Ad hoc on-demand distance vector
AP	Access point
APIT	Approximate point-in-triangulation
ARAN	Authenticated routing for ad hoc networks
BFS	Breadth-first-search
CA	Certificate authority
CDMA	Code division multiple access
CIM	Common information modeling
CoG	Center of gravity
COPS	Common open policy service
CSMA	Carrier sense multiple access
CVIS	Cooperative vehicle infrastructure system
DAD	Duplicate address detection
DES	Data encryption standard
DH	Diffie–hellman
DHCP	Dynamic host configuration protocol
DMTF	Distributed management task force
DoD	Department of Defense
DoS	Denial of service
DOT	Department of Transportation
DREAM	Distance routing effect algorithm for mobility
DSDV	Destination-sequenced distance vector routing
DSDV-SQ	Destination-sequenced distance vector routing—sequence number
DSR	Dynamic source routing
DSRC	Dedicated short-range communication
ERR	Error message
FDMA	Frequency division multiple access
GHT	Geographic hash tables
GPS	Global positioning system
GPSR	Greedy perimeter stateless routing
GST	Global system for telematics
HARPS	Hashed random preloaded subsets
HIDS	Host-based intrusion detection system

Security for Wireless Ad Hoc Networks, by Farooq Anjum and Petros Mouchtaris

HiRLoC	High-resolution range-independent localization
HMAC	Keyed-hashing for message authentication
HNA	Host and network association
ICMP	Internet control message protocol
IDS	Intrusion detection system
IETF	Internet Engineering Task Force
IKE	Internet key exchange
IPSec	IP security protocol
IPSP	IP security policy
IR	Infrared
KDC	Key distribution center
LDAP	Lightweight directory access protocol
LEAP	Localized encryption and authentication protocol
LSA	Link state advertisement
LSU	Link state update
MAC	Message authentication code
MANET	Mobile ad hoc network
MID	Multiple interface declaration
MMSE	Minimum mean squared estimate technique
MPR	Multipoint relay
NCW	Network centric warfare
NIC	Network interface card
NIDS	Network-based intrusion detection system
NOW	Network on wheels
OLSR	Optimized link-state routing protocol
PBNM	Policy-based network management
PDL	Policy description language
PDP	Policy decision point
PEP	Policy enforcement point
PIKE	Peer intermediaries for key establishment
PKI	Public key infrastructure
PMS	Policy management system
QoS	Quality of service
RAP	Resource allocation protocol
RBAC	Role-based access control
RDP	Route discovery message
REP	Reply packet
RERR	Route error
RF	Radio frequency
ROA	Region of acceptance
ROI	Region of intersection
ROPE	Robust position estimation
RREP	Route reply
RREQ	Route request
RSS	Received signal strength
RSSI	Received signal strength indicator
SA	Security association

SAODV	Secure ad hoc on-demand distance vector
SAR	Security aware ad hoc routing
SCTP	Stream control transmission protocol
SEAD	Secure efficient distance vector routing
SeRLoc	Secure range-independent localization
SLSP	Secure link state routing protocol
SNEP	Secure network encryption protocol
SNMP	Simple network management protocol
SRP	Secure routing protocol
TA	Trusted authority
TDMA	Time division multiple access
TDoA	Time difference of arrival
TESLA	Timed efficient stream loss-tolerant authentication
ToF	Time of flight
TTL	Time to live
TTP	Trusted third party
UAV	Unattended air vehicle
US	Ultrasound
UWB	Ultra wide band
VANET	Vehicular network
VII	Vehicular infrastructure integration
VIN	Vehicular identity number
VM	Verifiable multilateration
WLAN	Wireless local area network

References

[1] F. Stajano and R. Anderson, "The Resurrecting Duckling: Security Issues for Ad-hoc Wireless Networks," in *Proceedings of 7th International Workshop on Security Protocols*, Cambridge. Picture Notes in Computer Science, Vol. 1796, Springer, Berlin 1999, pp. 172–194.

[2] J. Douceur, "The Sybil Attack," in *Proceedings of IPTPS 2002*, March 2002, Cambridge, MA, pp. 251–260.

[3] A. Sabir, S. Murphy, and Y. Yang, "Generic Threats to Routing Protocols," draft-ietf-rpsec-routing-threats-07, October 2004.

[4] C.-Y. Tseng, P. Balasubramanyam, C. Ko, R. Limprasittiporn, J. Rowe, and K. Levitt, "A Specification-Based Intrusion Detection System For AODV," in *Workshop on Security in Ad Hoc and Sensor Networks (SASN)'03.*

[5] G. Vigna, S. Gwalani, K. Srinivasan, E. Belding-Royer, and R. Kemmerer, "An Intrusion Detection Tool for AODV-based Ad hoc Wireless Networks," in *Proceedings of the Annual Computer Security Applications Conference (ACSAC)*, Tucson, AZ, December 2004, pp. 16–27.

[6] P. Ning and K. Sun, "How to Misuse AODV: a Case Study of Insider Attacks against Mobile Ad-hoc Routing Protocols," in *Proceedings of the 2003 IEEE Workshop on Information Assurance*. United States Military Academy, West Point, NY, June 2003.

[7] A. J. Menezes, P. C. van Oorschot, and S. A. Vanstone, *Handbook of Applied Cryptography*, CRC Press, October 1996.

[8] J. Walker, "Unsafe at any Key Size; an Analysis of the WEP Encapsulation," IEEE P802.11, Wireless LANs, October 2000.

[9] X. Wang, Y. L. Yin, and H. Yu, "Finding Collisions in the Full SHA-1", *CRYPTO*, 2005.

[10] R. Canetti, D. Song, and D. Tygar, "Efficient and Secure Source Authentication for Multicast," *Proceedings of Network and Distributed System Security Symposium, NDSS 2001*, February 2001.

[11] A. Perrig, R. Szewczyk, V. Wen, D. Culler, and J. D. Tygar, "SPINS: Security Protocols for Sensor Networks," *Wireless Networks Journal (WINE)*, September 2002.

[12] D. Harkins and D. Carrel, "The Internet Key Exchange (IKE)," RFC 2409, November 1998; www.ietf.org/rfc/rfc2409.txt?number = 2409.

[13] W. Aiello, S. M. Bellovin, M. Blaze, R. Canettia, J. Ioannidis, A. D. Keromytis, and O. Reingold, "Efficient, DoS-Resistant, Secure Key Exchange for Internet Protocols," in *Proc. ACM Computer and Communications Security Conference*, Washington, DC, USA, 2000, pp. 48–58.

[14] W. Du, J. Deng, Y. S. Han, S. Chen, and P. K. Varshney, "A Key Management Scheme for Wireless Sensor Networks using Deployment Knowledge," in *INFOCOM*, 2004.

[15] W. Du, J. Deng, Y. Han, and P. Varshney, "A Pairwise Key Predistribution Scheme for Wireless Sensor Networks," in *Proceedings of the Tenth ACM Conference on Computer and Communications Security (CCS 2003)*, October 2003, pp. 42–51.

Security for Wireless Ad Hoc Networks, by Farooq Anjum and Petros Mouchtaris

[16] W. Du, J. Deng, Y. S. Han, P. Varshney, J. Katz, and A. Khalili, "A Pairwise Key Predistribution Scheme for Wireless Sensor Networks," *ACM Transactions on Information and System Security (TISSEC)*, 2005, pp. 228–258.

[17] L. Eschenauer and V. Gligor, "A Key-management Scheme for Distributed Sensor Networks." In *Proceedings of the 9th ACM Conference on Computer and Communications Security*, November 2002, pp. 41–47.

[18] D. Liu, P. Ning, and R. Li, "Establishing Pairwise Keys in Distributed Sensor Networks," *ACM Transactions on Information Systems Security*, **8**(1), 41–77 (2005).

[19] D. Balfanz, D. Smetters, P. Stewart, and H. Wong, "Talking to Strangers: Authentication in Ad Hoc Wireless Networks," in *Proceedings of 9th Annual Network and Distributed System Security Symposium*, San Diego, CA, 2002.

[20] H. Deng and D. P. Agrawal, "TIDS: threshold and identity based security scheme for wireless ad hoc networks," *Ad Hoc Networks*, **2**(3), 291–307 (2004).

[21] A. Khalili, J. Katz, and W. A. Arbaugh, "Toward Secure Key Distribution in Truly Ad-hoc Networks," in *Proceedings of IEEE Workshop on Security and Assurance in Ad-Hoc Networks*, 2003.

[22] N. Arora and R. K. Shyamasundar. UGSP: Secure Key Establishment Protocol for Ad-hoc Networks. In R. K. Ghosh and H. Mohanty, eds, *Distributed Computing and Internet Technology: First International Conference*, ICDCIT 2004, Bhubaneswar, India, December 22–24, 2004. Proceedings, Volume 3347 of Lecture Notes in Computer Science, Springer, Berlin, 2004, pp. 391–399.

[23] D. J. Malan, M. Welsh, and M. D. Smith, "A Public-key Infrastructure for Key Distribution in TinyOS Based on Elliptic Curve Cryptography," *IEEE Secon 2004*.

[24] L. Zhou and Z. Haas, "Securing Ad Hoc Networks," *IEEE Network*, **13**(6), 24–30 (1999).

[25] H. Luo, J. Kong, P. Zerfos, S. Lu, and L. Zhang, "URSA: Ubiquitous and Robust Access Control for Mobile Ad Hoc Networks," *IEEE/ACM Transactions on Networking*, December 2004, pp. 1049–1063.

[26] J. Kong, P. Zerfos, H. Luo, S. Lu, and L. Zhang, "Providing Robust and Ubiquitous Security Support for Mobile Ad Hoc Networks," in *Proceedings of Ninth International Conference on Network Protocols (ICNP)*, November 2001.

[27] S. Capkun, L. Buttyan, and J.-P. Hubaux, "Self-Organized Public-Key Management for Mobile Ad Hoc Networks," *IEEE Transactions on Mobile Computing*, **2**(1), 52–64 (2003).

[28] J.-P. Hubaux, L. Buttýan, and S. Capkun, "The Quest for Security in Mobile Ad hoc Networks," in *MobiHoc*. ACM, New York, 2001, pp. 146–155.

[29] G. O'Shea and M. Roe, "Child-proof Authentication for MIPv6 (CAM)," *ACM Computer Communications Review*, **31**(2), 4–8 (2001).

[30] G. Montenegro and C. Castelluccia, "Statistically Unique and Cryptographically Verifiable (SUCV) Identifiers and Addresses," in *Proceedings of NDSS*, 2002.

[31] M. Cagalj, S. Capkun, and J. P. Hubaux, "Key Agreement in Peer-to-Peer Wireless Networks," in *Proceedings of the IEEE (Special Issue on Cryptography and Security)*, February 2006, pp. 467–478.

[32] C. Ellison and S. Dohrmann, "Public-key Support for Group Collaboration," *ACM Transactions on Information Systems Security*, **6**(4), 547–565 (2003).

[33] C. Gehrmann, C. Mitchell, and K. Nyberg, "Manual Authentication for Wireless Devices," *RSA Cryptobytes*, **7**(1), 29–37 (2004).

[34] C. Gehrmann and K. Nyberg, "Enhancements to Bluetooth Baseband Security," in *Proceedings of 6th Nordic Workshop on Secure IT Systems*, Copenhagen, November 2001, pp. 39–53.

[35] J.-H. Hoepman, "The Ephemeral Pairing Problem," in *Intl Conference on Financial Cryptography*, Key West, FL, February 2004, Vol. 3110, pp. 212–226.

[36] A. Perrig and D. Song, "Hash Visualization: a New Technique to Improve Real-World Security," in *Proceedings of 1999 International Workshop on Cryptographic Techniques and E-commerce*, 1999, pp. 131–138.

[37] S. Basagni, K. Herrin, D. Bruschi, and E. Rosti, "Secure Pebblenet," in *Proceedings of the 2001 ACM Iternational Symposium on Mobile Ad Hoc Networking and Computing, MobiHoc 2001*, Long Beach, CA, 4–5 October 2001, pp. 156–163.

[38] S. Zhu, S. Setia, and S. Jajodia. "LEAP: Efficient Security Mechanisms for Large-Scale Distributed Sensor Networks," in *Proceedings of the 10th ACM Conference on Computer and Communications Security (CCS '03)*, Washington, DC, October 2003.

[39] H. Chan, A. Perrig, and D. Song, "Random Key Predistribution Schemes for Sensor Networks," in *IEEE Symposium on Security and Privacy*, May 2003, pp. 197–213.

[40] M. Ramkumar and N. Memon, "An Efficient Key Predistribution Scheme for Ad Hoc Network Security," *IEEE Journal of Selected Areas of Communication*, **23**(3), 611–621 (2005).

[41] D. Liu and P. Ning, "Establishing Pairwise Keys in Distributed Sensor Networks," in *Proceedings of the Tenth ACM Conference on Computer and Communications Security (CCS 2003)*, October 2003, pp. 52–61.

[42] C. Blundo, A. De Santis, A. Herzberg, S. Kutten, U. Vaccaro, and M. Yung, "Perfectly-secure Key Distribution for Dynamic Conferences," in E. F. Brickell, ed., *Advances in Cryptology—Crypto'*92. Lecture Notes in Computer Science Vol. 740. Springer, Berlin, 1992, pp. 471–480.

[43] R. Blom, "Non-Public Key Distribution," in *Advances in Cryptology: Proceedings of Crypto'82*, 1982, pp. 231–236.

[44] H. Chan and A. Perrig, "PIKE: Peer Intermediaries for Key Establishment in Sensor Networks," *IEEE Infocom*, 2005.

[45] D. Liu and P. Ning, "Location Based Pairwise Key Establishments for Static Sensor Networks," *SASN 2003*.

[46] D. Liu, P. Ning, and W. Du, "Group-based Key Pre-distribution in Wireless Sensor Networks," in *Proceedings of 2005 ACM Workshop on Wireless Security (WiSe 2005)*, September 2005.

[47] R. Perlman, *Interconnections: Bridges and Routers*. Addison-Wesley, Reading, MA, 1993.

[48] C. Perkins, E. Belding-Royer and S. Das, "Ad hoc On-Demand Distance Vector (AODV) Routing," IETF RFC, July 2003; //www.faqs.org/rfcs/rfc3561.html

[49] Manel Guerrero Zapata, "Draft-guerrero-manet-saodv-02.txt"; http://ietfreport.isoc.org/idref/draft-guerrero-manet-saodv/

[50] M. G. Zapata and N. Asokan, "Securing Ad hoc Routing Protocols," *WiSe*, September 2002.

[51] K. Sanzgiri, B. Dahill, B. N. Levine, C. Shields, and E. M. Belding-Royer, "A Secure Routing Protocol for Ad hoc Networks," in *International Conference on Network Protocols (ICNP)*, 2002.

[52] S. Yi, P. Naldurg, and R. Kravets, "A Security-Aware Routing Protocol for Wireless Ad Hoc Networks," in *6th World Multi-Conference on Systemics, Cybernetics and Informatics (SCI 2002)*, 2002.

[53] D. B. Johnson *et al.*, "The Dynamic Source Routing Protocol for Mobile Ad Hoc Networks (DSR)," IETF Draft, draft-ietf-manet-dsr-10.txt, July 2004.

[54] P. Papadimitratos and Z. J. Haas, "Secure Routing for Mobile Ad Hoc Networks", in *SCS Communication Networks and Distributed Systems (CNDS 2002)*, 27–31 January 2002.

[55] L. Buttyán and I. Vajda, "Towards Provable Security for Ad Hoc Routing Protocols," in *Proceedings of the 2nd ACM workshop on Security of Ad hoc and Sensor Networks*, 2004, pp. 94–105.

[56] Y.-C. Hu, A. Perrig, and Davic B. Johnson. "Ariadne: a Secure On-demand Routing Protocol for Ad Hoc Networks," in *Proceedings of the ACM Conference on Mobile Computing and Networking (Mobicom)*, 2002, pp. 12–23.

[57] C. E. Perkins and P. Bhagwat, "Highly Dynamic Destination-Sequenced Distance-vector Routing (DSDV) for Mobile Computers," *SIGCOMM*, UK, 1994.

[58] J. Broch, D. Maltz, D. Johnson, Y. Hu, and J. Jetcheva, "A Performance Comparison of Multi-hop Wireless Ad Hoc Network Routing Protocols," *Proceedings of the Fourth Annual ACM/IEEE International Conference on Mobile Computing and Networking (Mobicom 98)*, October 1998.

[59] Y.-C. Hu, D. B. Johnson, and A. Perrig. "SEAD: Secure Efficient Distance Vector Routing for Mobile Wireless Ad Hoc Networks," *Ad Hoc Networks*, **1**(1), 175–192 (2003).

[60] T. Wan, E. Kranakis, and P. Van Oorschot, "Securing the Destination Sequenced Distance Vector Routing Protocol (S-DSDV)," in *Proceedings of 6th International Conference on Information and Communications Security (ICICS'04)*, October 2004, Malaga. Lecture Notes in Computer Science, Vol. 3269, J. Lopez, S. Qing, and E. Okamoto, eds. Berlin, Springer, 2004.

[61] T. Clausen and P. Jacquet, eds, "Optimized Link State Routing Protocol (OLSR)," IETF RFC 3626, October 2003; //www.ietf.org/rfc/rfc3626.txt

[62] A. Halfslund, A. Tonnesen, R. B. Rotvik, J. Andersson, and O. Kure, "Secure Extension to the OLSR Protocol," *OLSR Interop and Workshop*, 2004.

[63] D. Raffo, T. Clausen, C. Adjih, and P. Muhlethaler, "An Advanced Signature System for OLSR," *SASN'04*, October 2004.

[64] C. Adjih, T. Clausen, P. Jacquet, A. Laouiti, P. Muhlethaler, and D. Raffo, "Securing the OLSR Protocol," *Proceedings of Med-Hoc-Net*, June 2003.

[65] P. Papadimitratos and Z. J. Haas, "Secure Link State Routing for Mobile Ad Hoc Networks," in *Proceedings of the 2003 Symposium on Applications and the Internet Workshops (SAINT'03 Workshops)*, 2003, p. 379.

[66] J. Kong and X. Hong, "ANODR: Anonymous On Demand Routing with Untraceable Routes for Mobile Ad-hoc Networks," *ACM MobiHoc*, June 2003.

[67] Y. Zhang, W. Liu, W. Lou, and Y. Fang, "MASK: an Anonymous Routing Protocol for Mobile Ad Hoc Networks," *IEEE Transaction on Wireless Communications*, in press.

[68] Y. Zhang, W. Liu, and W. Lou, "Anonymous Communications in Mobile Ad hoc Networks," *IEEE INFOCOM 2005*, Miami, FL, March 2005.

[69] D. Boneh and M. Franklin, "Identify-based Encryption from the Weil Pairing," in *Proc. CRYPTO 01*, Springer, Berlin, 2001.

[70] P. S. L. M. Barreto, H. Y. Kim, B. Bynn, and M. Scott, "Efficient Algorithms for Pairing-Based Cryptosystems," in *Proc. CRYPTO 02*, Springer Berlin, August 2002.

[71] Y. Hu, A. Perrig, and D. Johnson, "Packet Leashes: a Defense Against Wormhole Attacks in Wireless Ad Hoc Networks," in *Proceedings of INFOCOM*, San Francisco, CA, April 2003, pp. 1976–1986.

[72] L. Hu and D. Evans, "Using Directional Antennas to Prevent Wormhole Attacks," *Network and Distributed System Security Symposium*, San Diego, CA, 5–6 February 2004.

[73] R. Poovendran and L. Lazos, "A Graph Theoretic Framework for Preventing the Wormhole Attack in Wireless Ad Hoc Networks," *ACM Journal on Wireless Networks (WINET)*, in press.

[74] Y.-C. Hu, A. Perrig, and D. B. Johnson, "Rushing Attacks and Defense in Wireless Ad Hoc Network Routing Protocols," *Proceedings of the 2003 ACM Workshop on Wireless Security (WiSe 2003)*, ACM, San Diego, CA, September 2003, pp. 30–40.

[75] C. Karlof and D. Wagner, "Secure Routing in Wireless Sensor Networks: Attacks and Countermeasures," in *Proceedings of the 1st IEEE International Workshop on Sensor Network Protocols and Applications*, May 2002.

[76] J. Newsome, A. Perrig, E. Shi, and D. Song, "The Sybil Attack in Sensor Networks: Analysis and Defenses (PS, PDF, BIB)," *Third International Symposium on Information Processing in Sensor Networks (IPSN)*, 2004.

[77] Q. Zhang, P. Wang, D. S. Reeves, and P. Ning, "Defending Sybil Attacks in Sensor Networks," in *Proceedings of the International Workshop on Security in Distributed Computing Systems (SDCS-2005)*, June 2005, pp. 185–191.

[78] R. Stewart, Q. Xie, K. Morneault, C. Sharp, H. Schwarzbauer, T. Taylor, I. Rytina, M. Kalla, L. Zhang, and V. Paxson, "Stream Control Transmission Protocol," IETF RFC 2960, October 2000.

[79] E. Cole. *Hackers Beware*, Sams Publishing, Indianapolis, Indiana, 2001.

[80] D. Subhadrabandhu, S. Sarkar, and F. Anjum, "Efficacy of Misuse Detection in Ad Hoc Networks: Part I," *IEEE Journal for Selected Areas in Communications*, special issue on Security in Wireless Networks, **24**(2), 274–290 (2006).

[81] D. Subhadrabandhu, S. Sarkar, and F. Anjum, "Efficacy of Misuse Detection in Ad Hoc Networks: Part II," *IEEE Journal for Selected Areas in Communications*, special issue on Security in Wireless Networks, **24**(2), 290–305 (2006).

[82] D. Subhadrabandhu, S. Sarkar, and F. Anjum, "RIDA: Robust Intrusion Detection in Ad Hoc Networks," in *Proceedings of 2005 IFIP Networking Conference*, University of Waterloo, Ontario, 2–5 May 2005.

[83] D. Subhadrabandhu, S. Sarkar, and F. Anjum, "Efficacy of Misuse Detection in Adhoc Networks," in *Proceedings of the IEEE International Conference on Sensor and Ad hoc Communications and Networks (SECON)*, 2004, Santa Clara, CA, 4–7 October 2004.

[84] S. Marti, T. J. Giuli, K. Lai, and M. Baker, "Mitigating Routing Misbehavior in Mobile Ad Hoc Networks," in *Proceedings of the 6th International Conference on Mobile Computing and Networking*, Boston, MA, August 2000, pp. 255–265.

[85] D. Sterne, P. Balasubramanyam, D. Carman, B. Wilson, R. Talpade, C. Ko, R. Balupari, C. Tseng, T. Bowen, K. Levitt, and J. Rowe, "A General Cooperative Intrusion Detection Architecture for MANETs," in 3rd *IEEE International Workshop on Information Assurance*, March 2005.

[86] Y. Huang and W. Lee, "A Cooperative Intrusion Detection System for Ad Hoc Networks," in *Proceedings of the ACM Workshop on Security of Ad Hoc and Sensor Networks (SASN'03)*, Fairfax VA, October 2003.

[87] M. Chatterjee, S. K. Das, and D. Turgut, "WCA: a Weighted Clustering Algorithm for Mobile Ad hoc Networks," *Journal of Cluster Computing* (special issue on Mobile Ad hoc Networks), **5**(2), 193–204 (2002).

[88] S. Banerjee and S. Khuller, "A Clustering Scheme for Hierarchical Control in Multi-hop Wireless Networks," *IEEE Infocom*, 2001, pp. 1028–1037.

[89] P. Krishna, N. H. Vaidya, M. Chatterjee, and D.K. Pradhan, "A Cluster-based Approach for Routing in Dynamic Networks," *ACM SIGCOMM Computer Communications Review*, **27**(20), 49–64 (1997).

[90] R. Anderson and M. Kuhn, "Tamper Resistance—a Cautionary Note," in *Proceedings of the Second Usenix Workshop on Electronic Commerce*, 1996.

[91] P. Resnick, R. Zeckhauser, E. Friedman, and K. Kuwabara, "Reputation Systems," *Communications of the ACM*, **43**(12), 45–48 (2000).

[92] S. Buchegger and J. Boudec, "Performance Analysis of the CONFIDANT Protocol: Cooperation of Nodes—Fairness In Distributed Ad hoc NeTworks," in *Proceedings of IEEE/ACM Workshop on Mobile Ad Hoc Networking and Computing (MobiHOC)*, Lausanne, June 2002.

[93] P. Michiardi and R. Molva, "Core: a Collaborative Reputation mechanism to enforce node cooperation in Mobile Ad Hoc Networks," *Communication and Multimedia Security Conference*, 2002.

[94] P.-W. Yau and C. J. Mitchell, "Reputation Methods for Routing Security for Mobile Ad Hoc Networks," in *Proceedings of SympoTIC'03, Joint IST Workshop on Mobile Future and Symposium on Trends in Communications*, Bratislava, October 2003, IEEE Press, New York 2003, pp. 130–137.

[95] L. Buttyán and J.-P. Hubaux, "Enforcing Service Availability in Mobile Ad-Hoc WANs," *Proceedings of the IEEE/ACM Workshop on Mobile Ad Hoc Networking and Computing (MobiHOC)*, Boston, MA, August 2000.

[96] L. Buttyán and J.-P. Hubaux, "Stimulating Cooperation in Self-Organizing Mobile Ad Hoc Networks," Technical Report no. DSC/2001/046, Swiss Federal Institute of Technology, Lausanne, August 2001.

[97] L. Buttyán and J.-P. Hubaux, "Stimulating Cooperation in Self-organizing Mobile Ad Hoc Networks," *MONET Journal of Mobile Networks*, March 2002, pp. 579–592.

[98] A. Deligiannakis, Y. Kotidis, and N. Roussopoulos, "Compressing Historical Information in Sensor Networks," *ACM SIGMOD International Conference on Management of Data*, 2004.

[99] M. Little and C. Ko, "Detecting Coordinated Attacks in Tactical Wireless Networks using Cooperative Signature-based Detectors," *Milcom*, October 2005.

[100] C. Kruegel, T. Toth, and C. Kerer, "Decentralized Event Correlation for Intrusion Detection," in *Proceedings of Information Security and Cryptology (ICISC)*, Seoul, 6–7 December 2001, pp. 114–131.

[101] R. Rao and G. Kesidis, "Detecting Malicious Packet Dropping Using Statistically Regular Traffic Patterns in Multihop Wireless Networks that are not Bandwidth Limited," in *Proceedings of IEEE Globecom*, San Francisco, December 2003.

[102] F. Anjum and R. Talpade, "Packet-Drop Detection Algorithm for Ad hoc Networks," *Proceedings. of 60th IEEE Vehicular Technology Conference*, September 2004.

[103] V. D. Gligor, "Security of Emergent Properties in Ad-hoc Networks," in *Proceedings of International Workshop on Security Protocols*, April, 2004.

[104] Y. Zhang and W. Lee, "Intrusion Detection in Wireless Ad Hoc Networks," in *Proceedings of The Sixth International Conference on Mobile Computing and Networking (MobiCom 2000)*, Boston, MA, August 2000.

[105] Y. Zhang, W. Lee, and Y. Huang, "Intrusion Detection Techniques for Mobile Wireless Networks," in *Mobile Networks and Applications*, **9**(5), 545–556 (2003).

[106] M. Slomin and E. Lupu, "Security and Management Policy Specification," *IEEE Network*, **16**(2), 10–19 (2002).

[107] L. Kagal, T. Finin, and A. Joshi, "A Policy Language for a Pervasive Computing Environment," *IEEE 4th International Workshop on Policies for Distributed Systems and Networks, Policy 2003*, Lake Como, Italy, 2003.

[108] R. Ortalo, "A Flexible Method for Information System Security Policy Specification," in *Proceedings of 5th European Symposium Research in Computer Security*, Louvain-la-Neuve, 1998.

[109] J. Lobo, R. Bhatia, and S. Naqvi, "A Policy Description Language," in *Proceedings of AAAI*, Orlando, FL, July 1999.

[110] P. McDaniel and A Prakash, "Ismene: Provisioning and Policy Reconciliation in Secure Group Communication," Technical Report CSE-TR-438-00, Electrical Engineering and Computer Science, University of Michigan, December 2000.

[111] P. McDaniel, A. Prakash, J. Irrer, S. Mittal, and T. Thuang, "Flexibly Constructing Secure Groups in Antigone 2.0," *DISCEX 20001*, California, June 2001.

[112] *Information Survivability Conference and Exposition, DISCEX 2003*, Washington DC, April 2003.

[113] DARPA, *Information Survivability Conference and Exposition, DISCEX* 2001, Anaheim, CA, June 2001.

[114] R. Chadha, Y.-H. Cheng, J. Chiang, G. Levin, S.-W. Li, and A. Poylisher, "Policy-Based Mobile Ad Hoc Network Management," *Policy 2004*, Yorktown Heights, NY, June 2004.

[115] R. Chadha, Y.-H. Cheng, J. Chiang, G. Levin, S.-W. Li, and A. Poylisher, "Policy-Based Mobile Ad Hoc Network Management for DRAMA," *Milcom*, 2004.

[116] R. Chadha, Y.-H. Cheng, J. Chiang, G. Levin, S.-W. Li, A. Poylisher, L. LaVergn, and S. Newan, "Scalable Policy Management for Ad Hoc Networks," *Milcom*, 2005.

[117] S. Herzog, J. Boyle, R. Cohen, D. Durham, R. Rajan and A. Sastry, "COPS Usage for RSVP," IETF RFC 2749, January 2000.

[118] R. Braden, "Resource ReSerVation Protocol (RSVP)—Version 1 Functional Specification," *IETF RFC 2205*, September 1997.

[119] K. Chan, J. Seligson, D. Durham, S. Gai, K. McCloghrie, S. Herzog, F. Reichmeyer, R. Yavatkar, and A. Smith, "COPS Usage for Policy Provisioning (COPS-PR)," IETF RFC 3084, March 2001.

[120] M. Wahl, T. Howes, and S. Kille, "Lightweight Directory Access Protocol (v3)," IETF RFC 2251, December 1997.

[121] M. Blaze, J. Feigenbaum, and J. Lacy, "Decentralized Trust Management," *Proceedings of the 17th IEEE Symposium on Security and Privacy*, IEEE Computer Society, New York, 1996, pp. 164–173,

[122] M. Blaze, J. Feigenbaum, J. Ioannidis, and A. Keromytis, "The KeyNote Trust-Management System Version 2," IETF RFC 2704, September 1999.

[123] A. D. Keromytis, S. Ioannidis, M. Greenwald, and J. Smith, "The STRONGMAN Architecture," *DISCEX 20003*, Washington, DC, April 2003.

[124] J. Burns, A. Cheng, P. Gurung, S. Rajagopalan, P. Rao, D. Rosenbluth, A. V. Surendran, and D. M. Martin Jr., "Automatic Management of Network Security Policy," *DISCEX 2001*, California, June 2001.

[125] T. Markham and C. Payne, "Security at the Network Edge: A Distributed Firewall Architecture," *DISCEX 20001*, California, June 2001.

[126] Y. B. Ko, and N. Vaidya, "Location-aided Routing (LAR) in Mobile Ad hoc Networks," in *Proceedings of MobiCom*, 1998, pp. 66–75.

[127] S. Basagni, I. Chlamtac, V. Syrotiuk, and B. Woodward, "A Distance Routing Effect Algorithm for Mobility (DREAM)," in *Proceedings of MOBICOM*, Dallas, TX, October 1998, pp. 76–84.

[128] Karp, B. and Kung, H.T., "GPSR: Greedy Perimeter Stateless Routing for Wireless Networks," in *Proceedings of MobiCom*. 2000, pp. 243–254.

[129] J. C. Navas and T. Imielinski, "Geographic Addressing and Routing," in *MOBICOM*, Budapest, 26–30 September 1997.

[130] D. Johnson and D. Maltz, "Dynamic Source Routing in Ad Hoc Wireless Networks," *Mobile Computing*, 1996, pp. 153–181.

[131] V. Park and M. Corson, "A Highly Adaptive Distributed Routing Algorithm for Mobile Wireless Networks," *Proc. Infocom*, 1997.

[132] C. Perkins and E. Royer, "Ad-Hoc On-Demand Distance Vector Routing," in *Proceedings of Second IEEE Workshop Mobile Computing Systems and Applications*, February 1999, pp. 90–100.

[133] T. Camp, J. Boleng, B. Williams, L. Wilcox, and W. Navidi, "Performance Comparison of Two Location Based Routing Protocols for Ad Hoc Networks," *Proc. IEEE Infocom*, 2002.

[134] L. Lazos and R. Poovendran, "Energy-aware Secure Multicast Communication in Ad-hoc Networks Using Geographic Location Information," in *Proceedings of IEEE ICASSP*, Hong Kong, April 2003, Vol. 6, pp. 201–204.

[135] S. Ratnasamy, B. Karp, L. Yin, F. Yu, D. Estrin, R. Govindan, and S. Shenker, "GHT: a Geographic Hash Table for Data-Centric Storage," in *Proceedings of WSNA*, 2002.

[136] B. Przydatek, D. Song, and A. Perrig, "SIA: Secure Information Aggregation in Sensor Networks," *ACM SenSys*, 2003.

[137] Y.-C. Hu, A. Perrig, and D. Johnson, "Efficient Security Mechanisms for Routing Protocols," *Proceedings of the Tenth Annual Network and Distributed System Security Symposium (NDSS 2003).*

[138] D. E. Denning and P. F. MacDoran, "Location-based Authentication: Grounding Cyberspace for Better Security," in *Computer Fraud and Security*, Elsevier, Amsterdam, 1996, pp. 12–16; www.cosc.georgetown.edu/ ~ denning/infosec/Grounding.txt

[139] D. Son, A. Helmy, and B. Krishnamachari, "The Effect of Mobility-induced Location Errors on Geographic Routing in Mobile Ad Hoc and Sensor Networks: Analysis and Improvement Using Mobility Prediction," *IEEE Transactions on Mobile Computing*, **3**(3), 233–245 (2004).

[140] K. Seada, A. Helmy, and R. Govindan, "On the Effect of Localization Errors on Geographic Face Routing in Sensor Networks," in *IPSN'04*, 26–27 April 2004, Berkeley, CA, 2004.

[141] R. Volpe, T. Litwin and L. Matthies, "Mobile Robot Localization by Remote Viewing of a Colored Cylinder," in *Proceedings of IEEE/RSJ International Conference on Robots and Systems (IROS)*, 1995.

[142] L. Fang, W. Du, and P. Ning, "A Beacon-less Location Discovery Scheme for Wireless Sensor Networks," *Proceedings of the IEEE, INFOCOM'05*, 13–17 March 2005, Miami, FL.

[143] E. Kaplan, *Understanding GPS Principles and Applications*. Artech House, Norwood, MA, 1996.

[144] J. S. Warner and R. G. Johnston, "GPS Spoofing Countermeasures," *Homeland Security Journal*, December (2003).

[145] LORAN, www.navcen.uscg.mil/loran/Default.htm#Link

[146] M. G. Kuhn, "An Asymmetric Security Mechanism for Navigation Signals," in *Proceedings of the Information Hiding Workshop*, 2004.

[147] A. Savvides, C. Han, and M. Srivastava, "Dynamic Fine-grained Localization in Ad-hoc Networks of Sensors," in *Proceedings of ACM MOBICOM*, Rome, July 2001, pp. 166–179.

[148] M. Maroti, B. Kusy, G. Balogh, P. Volgyesi, A. Nadas, K. Molnar, S. Dora, and A. Ledeczi (Vanderbilt University), "Radio Interferometric Geolocation," *ACM Sensys 2005*.

[149] D. Niculescu and B. Nath, "Ad Hoc Positioning System (APS) using AoA," *INFOCOM 2003*, San Francisco, CA.

[150] D. Niculescu and B. Nath, "DV Based Positioning in Ad Hoc Networks," in *Journal of Telecommunication Systems*, **22**(1–4), 267–280 (2003).

[151] R. Nagpal, H. Shrobe, and J. Bachrach, "Organizing a Global Coordinate System from Local Information on an Ad Hoc Sensor Network," in *Proceedings of IPSN*, Vol. 2634, 2003, pp. 333–348.

[152] N. Bulusu, J. Heidemann, and D. Estrin, "Gps-less Low Cost Outdoor Localization for Very Small Devices," *IEEE Personal Communications Magazine* **7**(5), 28–34 (2000).

[153] N. Bulusu, V. Bychkovskiy, D. Estrin and J. Heidemann, "Scalable, Ad Hoc Deployable RF-based Localization," in *Proceedings of the Grace Hopper Celebration of Women in Computing Conference 2002*, Vancouver, October 2002.

[154] T. He, C. Huang, B. Blum, J. Stankovic, and T. Abdelzaher, "Range-free Localization Schemes in Large Scale Sensor Network, in *Proceedings of ACM MOBICOM*, 2003, pp. 81–95.

[155] S. Brands and D. Chaum, "Distance-bounding Protocols," in *Workshop on the Theory and Application of Cryptographic Techniques on Advances in Cryptology*. Springer, New York, 1994, pp. 344–359.

[156] S. Capkun and J. Hubaux, "Secure Positioning of Wireless Devices with Application to Sensor Networks," in *Proceedings of Infocom*, 2005.

[157] R. J. Fontana, E. Richley, and J. Barney, "Commercialization of an Ultra Wideband Precision Asset Location System," in *Proceedings of IEEE Conference on Ultra Wideband Systems and Technologies*, November 2003.

[158] N. Sastry, U. Shankar, and D. Wagner, "Secure Verification of Location Claims," in *Proceedings of ACM WISE*, 2002, pp. 1–10.

[159] B. Waters and E. Felten, "Proving the Location of Tamper Resistent Devices," //www.cs.princeton.edu/~bwaters/research/location_proving.ps

[160] S. Capkun and J. Hubaux, "Secure Positioning in Wireless Networks," *IEEE JSAC*, special issue on security in ad-hoc networks, **24**(2) 221–232 (2006).

[161] S. Capkun, M. Srivastava, and M. Cagalj, "Securing Localization With Hidden and Mobile Base Stations," NESL-UCLA Technical Report, March 2005.

[162] D. Shaw and W. Kinsner, "Multifractal Modeling of Radio Transmitter Transients for Clasification," in *Proceedings of the IEEE Conference on Communications, Power and Computing*, May 1997, pp. 306–312.

[163] L. Lazos and R. Poovendran, "SeRLoc: Robust Localization for Wireless Sensor Networks," ACM Sensor Networks, 2005.

[164] L. Lazos and R. Poovendran, "HiRLoc: Hi-resolution Robust Localization for Wireless Sensor Networks," in *IEEE JSAC*, special issue on wireless security, **24**(2), 233–246 (2006).

[165] D. Niculescu and B. Nath, "Ad-hoc Positioning Systems (APS)," in *Proceedings of IEEE GLOBECOM*, Vol. 5, 2001, pp. 2926–2931.

[166] F. Anjum, S. Pandey, and P. Agrawal, "Secure Localization in Sensor Networks using Transmission Range Variation," *1st International Workshop on Resource Provisioning and Management in Sensor Networks (RPMSN05), IEEE Conference on Mobile Ad Hoc and Sensor Systems*, November 2005.

[167] Wireless Integrated Network Systems (WINS), http://wins.rsc.rockwell.com/

[168] L. Lazos, S. Capkun, and R. Poovendran, "ROPE: Robust Position Estimation in Wireless Sensor Networks," *The Fourth International Conference on Information Processing in Sensor Networks (IPSN '05)*, April 2005.

[169] D. Liu, P. Ning, and W. Du, "Detecting Malicious Beacon Nodes for Secure Location Discovery in Wireless Sensor Networks," in *Proceedings of the The 25th International Conference on Distributed Computing Systems (ICDCS '05)*, June 2005, pp. 609–619.

[170] D. Liu, P. Ning, and W. Du, "Attack-resistant Location Estimation in Sensor Networks," in *Proceedings of The Fourth International Symposium on Information Processing in Sensor Networks (IPSN '05)*, April 2005, pp. 99–106.

[171] Z. Li, Y. Zhang, W. Trappe, and B. Nath, "Robust Statistical Methods for Securing Wireless Localization in Sensor Networks," in *International Symposium on Information Processing in Sensor Networks (IPSN)*, 2005, pp. 91–98.

[172] W. Jones, "Building Safer Cars," *IEEE Spectrum*, **39**(1), 82–85 (2002).

[173] B. Parno and A. Perrig, "Challenges in Securing Vehicular Networks," in *Workshop on Hot Topics in Networks (HotNets-IV)*, 2006.

[174] P. Samuel, "Of Sticker Tags and 5.9 GHz," *ITS International*, 2004.

[175] J. P. Hubaux, S. Capkun, and J. Luo, "The Security and Privacy of Smart Vehicles," *IEEE Security and Privacy*, May/June 2004.

[176] F. Anjum, D. Subhadrabandhu, and S. Sarkar, "Intrusion Detection for Wireless Adhoc Networks," *Proceedings of Vehicular Technology Conference, Wireless Security Symposium*, Orlando, Florida, October 2003.

[177] N. Asokan and P. Ginzboorg, "Key Agreement in Ad Hoc Networks," *Computer Communications*, **23**, 1627–1637, (2000).

[178] S. Basagni, "Distributed Clustering in Ad Hoc Networks," *Proceedings of the 1999 International Symposium on Parallel Architectures, Algorithms and Networks (I-SPAN'99)*, Freemantle, Australia, 1999.

[179] S. Capkun, L. Buttyan, and J.-P. Hubaux, "SECTOR: Secure Tracking of Node Encounters in Multi-hop Wireless Networks," in *Proceedings of SASN*, Washington, DC, October 2003.

[180] S. Capkun, M. Hamdi, and J. Hubaux, "Gps-free Positioning in Mobile Ad-hoc Networks," in *Proceedings of HICCSS*, 2001, pp. 3481–3490.

[181] S. Capkun and J.-P. Hubaux, "BISS: Building Secure Routing out of an Incomplete Set of Secure Associations," in *Procedings of 2nd ACM Wireless Security (WiSe'03)*, San Diego, CA, September 2003, pp. 21–29.

[182] S. Capkun, J. P. Hubaux and L. Buttyán, "Mobility Helps Security in Ad Hoc Networks," in *Proceedings of the 4th ACM Symposium on Mobile Ad Hoc Networking and Computing (MobiHOC 2003).*

[183] R. Chadha, G. Lapiotis, and S. Wright, "Policy-Based Networking," *IEEE Network*, special issue, **16**(2), (2002).

[184] M. Corner and B. Noble, "Protecting File Systems with Transient Authentication," *Wireless Networks*, **11**(1–2), 7–19 (2005).

[185] P. T. Dinsmore, D. M. Balenson, M. Heyman, P. S. Kruus, C. D. Scace, and A. T. Sherman, "Policy-based Security Management for Large Dynamic Groups: an Overview of the FCCM Project," *DISCEX 2000*, South Carolina, January 2000.

[186] H. Fuessler, M. Mauve, H. Hartenstein, M. Kaesemann, and D. Vollmer, "A Comparison of Routing Strategies for Vehicular Ad Hoc Networks," Department of Computer Science, University of Mannheim, Mannheim, Technical Report no. TR-3-2002, 2002.

[187] U. C. Guard, "Navstar GPS User Equipment Introduction (Public Release Version)," Technical Report September 1996.

[188] A. Harter, A. Hopper, P. Steggles, A. Ward, and P. Webster, "The Anatomy of a Context-aware Application," in *Proceedings of MOBICOM '99*, Seattle, WA, 1999.

[189] T. Hardjono and B. Weis, "The Multicast Group Security Architecture," IETF RFC 3740, March 2004.

[190] B. Hofmann-Wellenhof, H. Lichtenegger, and J. Collins, *Global Positioning System: Theory and Practice*. Springer, Berlin, 1997.

[191] T. Leighton and S. Micali, "Secret-key Agreement without Public-key Cryptography," in *Advances in Cryptology—Crypto '93*, 1993, pp. 456–479.

[192] C. R. Lin and M. Gerla, "Adaptive Clustering for Mobile Networks," *IEEE Journal on Selected Areas in Communications*, **15**, 1265–1275 (1997).

[193] P. McDaniel, H. Harney, P. Dinsmore, and A. Prakash, "Multicast Security Policy," IRTF Internet draft, draft-irtf-smug-mcast-policy-01.txt, November 2000.

[194] P. McDaniel and A. Prakash, "Antigone: Implementing Policy in Secure Group Communication," Technical Report CSE-TR-426-00, Electrical Engineering and Computer Science, University of Michigan, May 2000.

[195] A. B. McDonald and T. Znati, "A Mobility Based Framework for Adaptive Clustering in Wireless Ad-Hoc Networks," *IEEE Journal on Selected Areas in Communications*, 17, 1466–1487 (1999).

[196] D. Niculescu and B. Nath, "Ad Hoc Positioning System (APS) using AoA," *INFOCOM' 03*, Vol. 3, San Francisco, CA, 2003, pp. 1734–1743.

[197] N. Priyantha, H. Balakrishnan, E. Demaine, and S. Teller, "Anchor-free Distributed Localization in Sensor Networks," in *Proceedings of ACM SenSys*, 2003, pp. 340–341.

[198] N. B. Priyantha, A. Chakraborty, and H. Balakrishnan, "The Cricket Location Support System," in *Proceedings of MOBICOM'00*, New York, August 2000.

[199] A. Savvides, H. Park, and M. Srivastava, "The Bits and Flops of the N-Hop Multilateration Primitive for Node Localization Problems," in *First ACM International Workshop on Wireless Sensor Networks and Application*, Atlanta, GA, September 2002.

[200] C. Savarese, J. Rabay, and K. Langendoen, "Robust Positioning Algorithms for Distributed Ad-Hoc Wireless Sensor Networks," *USENIX Technical Annual Conference*, Monterey, CA, June 2002.

[201] A. Shamir, "How to Share a Secret," *Communications of the ACM*, **22**(11), 612–612 (1979).

[202] W. Stallings, *Cryptography and Network Security: Principles and Practice*, Prentice Hall, Engelwood Cliffs, NJ, 1999.

[203] D. Subhadrabandhu, F. Anjum, S. Sarkar, and R. Shetty, "On Optimal Placement of Intrusion Detection Modules in Wireless Sensor Networks," Invited Paper, in *Proceedings of Broadnets*, San Jose, CA, 25–29 October 2004.

[204] R. Szewczyk, E. Osterweil, J. Polastre, M. Hamilton, A. Mainwaring, and D. Estrin, "Habitat Monitoring with Sensor Networks," *Communications of ACM*, **47**, 34–40 (2004).

[205] A. Uszok, J. Bradshaw, R. Jeffers, N. Suri, P. Hayes, M. Breedy, L. Bunch, M. Johnson, S. Kulkarni, and J. Lott, "KAoS Policy and Domain Services Towards a Description-Logic Approach to Policy Representation, Deconfliction, and Enforcement," in *Policy 2003, Workshop on Policies for Distributed Systems and Networks*.

[206] D. Verma, "Simplifying Network Administration Using Policy-Based Management," *IEEE Network*, **16**(2), 20–26 (2002).

[207] R. Yavatkar, D. Pendarakis, and R. Guerin, "A Framework for Policy based Admission Control," IETF RFC 2753, January 2000.

[208] Y. Zhang, W. Liu, W. Lou, and Y. Fang "Location-based Compromise-tolerant Security Mechanisms for Wireless Sensor Networks," *IEEE JSAC*, special issue on Security in Ad-Hoc Networks, **24**(2), 247–260 (2006).

[209] J. Jason, L. Rafalow, and E. Vyncke, "IPsec Configuration Policy Information Model," IETF RFC 3585, August 2003.

[210] J. Hightower and G. Borriella, "A Survey and Taxonomy of Location Systems for Ubiquitous Computing," IEEE Computer, **34**(8), 57–66 (2001).

INDEX

Security for Wireless Ad Hoc Networks, by Farooq Anjum and Petros Mouchtaris